T0282674

CAMBRIDGE MONOGRAPHS ON MATHEMATICAL PHYSICS

General editors: P.V. Landshoff, D.R. Nelson, D.W. Sciama, S. Weinberg

STATISTICAL FIELD THEORY

Volume 2: Strong coupling, Monte Carlo methods, conformal field theory, and random systems

STATISTICAL FIELD THEORY

Volume 2
Strong coupling, Monte
Carlo methods, conformal
field theory, and random systems

CLAUDE ITZYKSON

JEAN-MICHEL DROUFFE

Commissariat à l'Energie Atomique
Centre d'Etudes Nucléaires de Saclay
Service de Physique Théorique

CAMBRIDGE
UNIVERSITY PRESS

CAMBRIDGE UNIVERSITY PRESS
Cambridge, New York, Melbourne, Madrid, Cape Town, Singapore,
São Paulo, Delhi, Dubai, Tokyo, Mexico City

Cambridge University Press
The Edinburgh Building, Cambridge CB2 8RU, UK

Published in the United States of America by
Cambridge University Press, New York

www.cambridge.org
Information on this title: www.cambridge.org/9780521408066

First published 1989
First paperback edition 1991
Reprinted 1992,1995

A catalogue record for this publication is available from the British Library

Library of Congress Cataloguing in Publication Data

ISBN 978-0-521-37012-7 Hardback
ISBN 978-0-521-40806-6 Paperback

Contents

Contents of Volume 1   *ix*

Preface   *xi*

7   Diagrammatic methods   405

7.1   General Techniques   405

     7.1.1   Definitions and notations   405

     7.1.2   Connected graphs and cumulants   411

     7.1.3   Irreducibility and Legendre transformation   415

7.2   Series expansions   421

     7.2.1   High temperature expansion   421

     7.2.2   The role of symmetries   427

     7.2.3   Low temperature expansion – discrete case   430

     7.2.4   Low temperature expansion – continuous case   434

     7.2.5   Strong field expansions   437

     7.2.6   Fermionic fields   438

7.3   Enumeration of graphs   439

     7.3.1   Configuration numbers with exclusion constraint   439

     7.3.2   Multiply connected graphs   442

7.4   Results and analysis   446

     7.4.1   Series analysis   446

     7.4.2   An example: the Ising series on a body centered cubic lattice   450

     Notes   454

8   Numerical simulations   456

8.1   Algorithms   456

     8.1.1   Generalities   456

     8.1.2   The classical algorithms   460

     8.1.3   Microcanonical simulations   465

     8.1.4   Practical considerations   466

8.2	Extraction of results in a simulation	475
	8.2.1 Determination of transitions	475
	8.2.2 Finite size effects	478
	8.2.3 Monte Carlo renormalization group	481
	8.2.4 Dynamics and the Langevin equation	486
8.3	Simulating fermions	489
	8.3.1 The quenched approximation	491
	8.3.2 Dynamical fermions	493
	8.3.3 Hadron mass calculation in lattice gauge theory	494
	Notes	499
9	**Conformal invariance**	501
9.1	Energy-momentum tensor – Virasoro algebra	502
	9.1.1 Conformal invariance	502
	9.1.2 Energy–momentum tensor	507
	9.1.3 Two-dimensional conformal transformations	510
	9.1.4 Central charge	515
	9.1.5 Virasoro algebra	521
	9.1.6 The Kac determinant	532
	9.1.7 Unitary and minimal representations	543
	9.1.8 Characters of the Virasoro algebra	547
9.2	Examples	549
	9.2.1 Gaussian model	550
	9.2.2 Ising model	554
	9.2.3 Three state Potts model	557
9.3	Finite size effects and modular invariance	564
	9.3.1 Partition functions on a torus	564
	9.3.2 Kronecker's limit formula	567
	9.3.3 Ising model	573
	9.3.4 The A–D–E classification of minimal models	579
	9.3.5 Frustrations and discrete symmetries	588
	9.3.6 Nonminimal models	592
	9.3.7 Correlations in a half plane	599
	9.3.8 The vicinity of the critical point	603
9.A	Jacobian θ-series and products	609
9.B	Superconformal algebra	613
9.C	Current algebra	619
	9.C.1 Simple Lie algebras	619

9.C.2 The Wess–Zumino–Witten model   626
9.C.3 Representations and characters of Kac–
        Moody algebras   639
Notes   642

10 Disordered systems and fermionic methods   646
10.1 One-dimensional models   647
      10.1.1 Gaussian random potential   647
      10.1.2 Fokker–Planck equation   649
      10.1.3 The replica trick   658
      10.1.4 Random one-dimensional lattice   664
10.2 Two-dimensional electron gas in a strong field 675
      10.2.1 Landau levels – Quantum Hall effect   675
      10.2.2 One particle spectrum in the presence of
              impurities   679
10.3 Random matrices   690
      10.3.1 Semicircle law   690
      10.3.2 The fermionic method   694
      10.3.3 Level spacings   696
10.4 The planar approximation   703
      10.4.1 Combinatorics   704
      10.4.2 The planar approximation in quantum
              mechanics   712
10.5 Spin systems with random interactions   715
      10.5.1 Random external field and dimensional
              transmutation   716
      10.5.2 The two-dimensional Ising model with
              random bonds   719
10.A The Hall conductance as a topological invariant 729
Notes   735

11 Random geometry   738
11.1 Random lattices   739
      11.1.1 Poissonian lattices and cell statistics 739
      11.1.2 Field equations   755
      11.1.3 The spectrum of the Laplacian   760
11.2 Random surfaces   764
      11.2.1 Piecewise linear surfaces   765
      11.2.2 The conformal anomaly and the Liouville
              action   773

11.2.3 Sums over smooth surfaces   780
11.2.4 Discretized models   799
Notes   809

Index

Contents of Volume 1

1   From Brownian motion to Euclidean fields   1
2   Grassmannian integrals and the two-dimensional
        Ising model   48
3   Spontaneous symmetry breaking, mean field   107
4   Scaling transformations and the XY-model   162
5   Continuous field theory and the renormalization
        group   233
6   Lattice gauge fields   328

Preface

Some ten years ago, when completing with J.-B. Zuber a previous
text on *Quantum Field Theory*, the senior author was painfully
aware that little mention was made that methods in statistical
physics and Euclidean field theory were coming closer and closer,
with common tools based on the use of path integrals and the
renormalization group giving insights on global structures. It was
partly to fill this gap that the present book was undertaken. Alas,
over the five years that it took to come to life, both subjects have
undergone a new evolution. Disordered media, growth patterns,
complex dynamical systems or spin glasses are among the new
important topics in statistical mechanics, while superstring theory
has turned to the study of extended systems, Kaluza–Klein
theories in higher dimensions, anticommuting coordinates ... in
an attempt to formulate a unified model including all known
interactions. New and sophisticated techniques have invaded
statistical physics, ranging from algebraic methods in integrable
systems to fractal sets or random surfaces. Powerful computers
or special devices provide "experimental" means for a new brand
of theoretical physicists. In quantum field theory, applications of
differential topology, geometry, Riemannian manifolds, operator
theory ... require a deeper background in mathematics and a
knowledge of some of its most recent developments. As a result,
when surveying what has been included in the present volume
in an attempt to uncover the basic unity of these subjects, the
authors have the same unsatisfactory feeling of not being able to
bring the reader really up to date. It is presumably the fate of such
endeavours to always come short of accomplishing their purpose.

 With these shortcomings fully admitted, we have tried to
present to the reader an overview of the main themes which justify
the title "Statistical field theory." This interpretation of Euclidean
field theory offers a new language, effective computing means, as

well as a natural and consistent short-distance cutoff. In other
words, it allows one to give a global meaning to path integrals, to
discover possible anomalies arising from integration measures, or
to understand in simple terms systems with redundant variables
such as gauge models. The theory of continuous phase transitions
provides a bridge between probabilistic mechanics and continuous
field theory, using the renormalization group to filter out relevant
operators and interactions. Many authors contributed to these
views, culminating in the work of K. Wilson and his collaborators
and followers, which promoted the renormalization group as a
universal tool to analyse the large-distance behaviour. It still
retains its value, while new developments take place, particularly
with conformal, or local scale invariance coming to prominence in
the study of two-dimensional systems.

The content of this book is naturally divided into two parts. The
first six chapters describe in succession Brownian motion, its anticommutative counterpart in the guise of Onsager's solution to the
two-dimensional Ising model, the mean-field or Landau approximation, scaling ideas exemplified by the Kosterlitz–Thouless theory for the XY-transition, the continuous renormalization group
applied to the standard φ^4 theory, the simplest typical case, and
lattice gauge theory as an attempt to understand quark confinement in chromodynamics.

The next five chapters (in volume 2) cover more diverse subjects.
We give an introduction to strong coupling expansions and various
means of analyzing them. We then briefly introduce Monte
Carlo simulations with an emphasis on the applications to gauge
theories. Next we turn to the significant advances in two-
dimensional conformal field theory, with a lengthy presentation
of the methods as well as early results. A chapter on simple
disordered systems includes sample applications of fermionic
techniques with no pretence at completeness. The final chapter is
devoted to random geometry and an introduction to the Polyakov
model of random surfaces which illustrates the relations between
string theory and statistical physics.

At the price of being perhaps a bit repetitive, we have tried in
the first part to introduce the subject in an elementary fashion.
It is, however, assumed that the reader has some familiarity with
thermodynamics as well as with quantum field theory. We often
switch from one to the other interpretation, assuming that it will

not be disturbing once it is realized that the exponential of the action plays the role of the Boltzmann–Gibbs statistical weight. The last chapters cover subjects still in fast evolution.

Many important subjects could unfortunately not be covered. In random order they include dynamical critical phenomena, renormalization of σ-models or non-Abelian gauge fields except for a mention of lowest order results, topological aspects, classical solutions, instantons, monopoles, anomalies (except for the conformal one). Integrable systems are missing apart from the two-dimensional Ising model. Quantum gravity *à la Regge* is only mentioned. The list could, of course, be made much longer. Our involvement in some of the topics has certainly produced obvious biases and overemphases in certain instances. We have tried, as much as possible, to correct for these defects as well as for the numerous omissions by including at the end of each chapter a section entitled "Notes." Here we quote our sources, original articles, reviews, books and complementary material. These notes are purposely scattered through the volume, as we are sure that our quotations are very incomplete. A fair bibliography in such a large domain is beyond human capacities. Should any one feel that his or her work has not been reported or not properly mentioned, he or she is certainly right and we present our most sincere apologies. On the other hand we did not hesitate to use and sometimes follow very closely some articles or reviews which served our purpose. For instance chapter 5 is built around the definitive contributions of E. Brézin, J.-C. Le Guillou, J. Zinn-Justin and G. Parisi. Except for some further elaboration by the authors themselves, it was futile to try to improve on their work. Further examples are mentioned in the notes. It is the very nature of a survey such as this one to be inspired largely by other people's works. We hope that we did not distort or caricature them.

A book might give the illusion, especially to students, that some knowledge has become definitive and that the authors understand every part of it. This is a completely false view. No one can really fully master even his own subject, and this is luckily a source of progress. It is in the process of learning, of objecting, of finding misprints and errors, in rediscovering for oneself, that one gets the real benefits. It is very likely that, in spite of our care, many errors have crept in here and there. We welcome gladly comments and criticisms.

It was very hard to keep uniform notation throughout the text, even sometimes in the same chapter. This is a standard difficulty, especially when traditional notation in a given domain comes into conflict with those used in another one, and a compromise is necessary. We hope that this will not be a source of confusion for the reader.

We have added appendices which generally gather material in very concise form. They should be supplemented by further reading. For instance appendix C of chapter 9 is obviously insufficient to describe finite and infinite Lie algebras and their representations. This appendix is, rather, meant to induce the interested reader to study the subject further. This is also the nature of several sections where the degree of mathematical sophistication seems to increase beyond the standard background, reflecting recent trends. It was felt difficult to omit these developments but also impossible to give a proper complete introduction.

Included in small type here and there are comments, exercises and short complements ... It was felt inappropriate to develop a scholarly set of problems. In this respect the whole text can be read as a problem book.

One of the "heroes" of the whole subject of statistical physics, in one guise or another, is still to this day our old friend the Ising model. We keep a few bottles of good old French wine for the lucky person who solves it in three dimensions. It would seem appropriate to create in the theoretical physics community a prize for its solution, analogous to the one founded at the beginning of the century for the proof of Fermat's theorem. Both subjects have a similar flavour, being elementary to formulate. While it is to be presumed that the answer itself is to a large extent inessential, they motivated creative efforts (and still do) which go largely beyond the goal of solving the problem itself.

Among the many books which either overlap or amply complement the present one, the foremost are of course those in the series edited by C. Domb and M.S. Green and now J. Lebowitz, entitled *Phase transitions and critical phenomena* and published through the years by Academic Press (New York). We freely refer to this series in the notes. Let us also quote here a few others, again without pretence at exhaustivity. On the statistical side, K. Huang, *Statistical mechanics*, Wiley, New York (1963); H.E. Stanley, *In-*

troduction to phase transitions and critical phenomena, Oxford University Press (1971); S.K. Ma, *Modern theory of critical phenomena*, Benjamin, New York (1976) and *Statistical mechanics*, World Scientific, Singapore (1985); D.J. Amit, *Field theory, the renormalization group and critical phenomena*, 2nd edition, World Scientific, Singapore (1984).

Books on the path integral approach to field theory are by now numerous. Among them, the classical one is R.P. Feynman and A.R. Hibbs, *Quantum mechanics and path integrals*, McGraw Hill, New York (1965). Further aspects are covered in C. Itzykson and J.-B. Zuber, *Quantum field theory*, McGraw Hill, New York (1980); P. Ramond, *Field theory, a modern primer*, Benjamin/Cummings, Reading, Mass. (1981); J. Glimm and A. Jaffe, *Quantum physics*, Springer, New York (1981). To fill some gaps on other developments in field theory, see S. Coleman, *Aspects of symmetries*, Cambridge University Press (1985); S. Treiman, R. Jackiw, B. Zumino, E. Witten *Current algebra and anomalies*, World Scientific, Singapore (1985), and to learn about integrable systems, R. Baxter *Exactly solved models in statistical mechanics*, Academic Press, New York (1982), and M. Gaudin *La fonction d'onde de Bethe*, Masson, Paris (1983). Of course, many more books are mentioned in the notes. We are also aware that several important texts are either in preparation or will appear in the near future.

Our knowledge of English remains to this day very primitive and we apologize for our cumbersome use of a foreign language. This lack of fluency has prevented us of any attempt at humour which would have been sometimes more than welcome.

We would have never undertaken writing, were it not for the teaching opportunities that we were given by several universities and schools. One of the authors (C.I.) is grateful to his colleagues from the "Troisième cycle de Suisse Romande" in Lausanne and Geneva, from the "Département de Physique de l'Université de Louvain La Neuve" and from the "Troisième cycle de physique théorique" in Marseille for giving him the possibility to teach what became parts of this text, as well as to the staff of these institutions for providing secretarial help in preparing a French unpublished manuscript. The other author (J.M.D.) acknowledges similar opportunities afforded by the "Troisième cycle de physique théorique" in Paris.

The final and certainly most pleasant duty is, of course, to thank all those, friends, colleagues, collaborators, students and secretaries who have helped us through the years. A complete list should include all the members of the Saclay "Service de physique théorique", together with its numerous visitors and the members of the many departments, institutions and meetings which offered us generous hospitality and stimulation.

Particular thanks go to our very long time friends and colleagues R. Balian, M. Bander, M. Bauer, D. Bessis, E. Brézin, A. Cappelli, A. Coste, F. David, J. des Cloizeaux, C. De Dominicis, E. Gardner, M. Gaudin, B. Derrida, J.-M. Luck, A. Morel, P. Moussa, H. Orland, G. Parisi, Y. Pomeau, R. Lacaze, H. Saleur, R. Stora, J. Zinn-Justin, and J.-B. Zuber for friendly collaborations, endless discussions and generous advice. The final form of the manuscript owes a great deal to Dany Bunel. Let her receive here our warmest thanks for her tireless help. We are also very grateful to M. Porneuf and to the documentation staff, M. Féron, J. Delouvrier and F. Chétivaux.

Last but not least, we thank the Commissariat à l'Energie Atomique, the Institut de Recherche Fondamentale and the Service de Physique Théorique for their support.

Saclay, 1988

7

DIAGRAMMATIC METHODS

This chapter is devoted to technicalities related to various expansions already encountered in volume 1, mostly those that derive from the original lattice formulation of the models, be it high or low temperature, strong coupling expansions and to some extent those arising in the guise of Feynman diagrams in the continuous framework. We shall not try to be exhaustive, but rather illustrative, relying on the reader's interest to investigate in greater depth some aspects inadequately treated. Nor shall we try to explore with great sophistication the vast domain of graph theory. There are, however, a number of common features, mostly of topological nature, which we would like to present as examples of the diversity of what looks at first sight like straightforward procedures.

7.1 General Techniques

7.1.1 Definitions and notations

Let a *labelled graph* $\mathcal{G}$ be a collection of v elements from a set of indices, and l pairs of such elements with possible duplications (i.e. multiple links). We shall also interchangeably use the word *diagram* instead of graph. This abstract object is represented by v points (vertices) and l links. Each vertex is labelled by its index.

The problem under consideration will define a set of *admissible graphs*, with a corresponding *weight* $\omega(\mathcal{G})$ (a real or complex number) according to a well-defined set of rules. We wish to find the sum of weights over all admissible graphs.

Possible constraints on the graphs may be

i) the *exclusion* constraint, preventing two vertices from carrying the same index

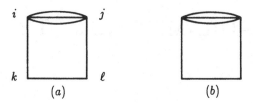

Fig. 1 (*a*) a labelled graph, (*b*) the corresponding free graph.

ii) *simplicity* when two vertices are joined by at most one link
(the graph in figure 1(*a*) is not simple).

Take for instance the straightforward high temperature expansion of the Ising partition function

$$
\begin{aligned}
Z &= 2^{-N} \sum_{\{\sigma_i=\pm1\}} \exp\left(\beta\sum_{\langle ij\rangle}\sigma_i\sigma_j\right) \\
&= 2^{-N} \sum_{\{\sigma_i=\pm1\}} \sum_{\langle ij\rangle}\left(1+\sum_{n_{ij}=1}^{\infty}\frac{\beta^{n_{ij}}}{n_{ij}!}\sigma_i\sigma_j\right)
\end{aligned}
\tag{1}
$$

Expanding the products, keeping terms with a finite power of
β, and averaging over $\sigma_i = \pm1$, leads to a straightforward high
temperature series encountered in volume 1. The successive
contributions are associated with graphs defined as follows. A
graph has n_{ij} lines joining vertices i and j. Isolated points are
not represented as vertices. Since only even powers of σ_i have
a nonvanishing unit average, admissible graphs have to obey the
following three rules

  i) a line can only join vertices indexed by neighbouring sites, and
     we may think of the graph as drawn on the lattice,
 ii) an even number of links are incident on a vertex,
iii) two vertices have distinct labels (the exclusion constraint).

Given an admissible graph, its weight is obtained by associating
a factor β to each line, and dividing by the product $\prod_{(i,j)} n_{ij}!$ i.e.
the order of the symmetry group of the graph under permutation
of equivalent links.
We can also write

$$
Z = (\cosh\beta)^{Nd}\frac{1}{2^N}\sum_{\{\sigma_i=\pm1\}}\prod_{\langle ij\rangle}(1+\sigma_i\sigma_j\tanh\beta)
\tag{2}
$$

which leads for $Z/(\cosh \beta d)^N$ to a different expansion. Admissible graphs are simple with a factor $\tanh \beta$ for each link. Both series are useful in applications.

Two graphs are *isomorphic* when a one-to-one correspondence can be set among vertices and links preserving the incidence relations. The difference lies therefore in the labels of the vertices. Isomorphism leads to equivalence classes called *free graphs* and denoted G. In a pictorial representation, vertices do not carry indices anymore (fig. 1(b)). Conventionally, the corresponding weight $\omega(G)$ will be the average over the equivalent labelled graphs. Call number of configurations $n(G)$ the cardinal of the equivalence class, then

$$\sum_{\mathcal{G}_i \in G} \omega(\mathcal{G}_i) = n(G)\omega(G) \tag{3}$$

This definition is useful whenever the weight of a graph is independent of the labelling of its vertices. In any case, it allows one to disentangle the part $\omega(G)$ that is specific to the model, from the geometry of the lattice, which yields $n(G)$. The following two sections will treat these problems separately.

The above definitions can be extended in various ways.

i) Vertices may be of several types.

ii) Links may have to be oriented.

iii) A generalization may be envisioned, where instead of dealing with 0 and 1 dimensional simplices (vertices and links), one may be required to consider higher dimensional elements (two dimensional plaquettes in the gauge case).

iv) Indices may be compound ones, and links may have to carry indices at their extremities.

This list is of course just indicative of possible extensions.

In some applications, the computation of correlations for instance, a subset of vertices carries fixed indices. Equivalence classes of such graphs will be called *rooted graphs*.

Two vertices x and y on G are *linked* if they can be joined by a path along links of the graph $xz_1, z_1z_2,..., z_ny$. This provides again an equivalence relation on vertices, and the corresponding classes

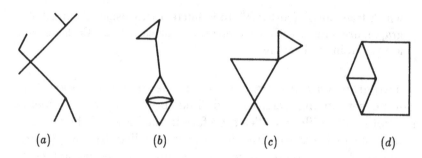

(a) (b) (c) (d)

Fig. 2 (a) a tree (b) a graph with four loops (c) a graph with two articulation points (d) a multiply connected graph.

define the connected disjoint parts of the graph. A *connected* graph has a unique connected part.

A *cycle* is a closed path of n links, and n vertices, all distinct, starting and ending at the same vertex. A connected graph without cycles is a *tree* (figure 2(a)). The *number of loops* in a connected graph is the minimum number of links which, when removed, leave a tree (figure 2(b)).

An *articulation point* (figure 2(c)) is such that its omission, together with incident links, increases the number of connected parts. It is therefore a vertex which appears on any path linking certain pairs of vertices. In particular, on a tree, all vertices but the external ones (joined to the graph by only one link) are articulation points. A connected graph without articulation points is a *multiply connected* graph: any two vertices belong to a cycle and can therefore be linked by at least two totally distinct paths.

In terms of the following notation

v_k, number of vertices with k incident links

$v = \sum_k v_k$, total number of vertices

l, number of links

b, number of loops

c, number of connected parts

we have the relation

$$2l = \sum_k k v_k, \tag{4}$$

expressing that each link joins two vertices, thus twice the number of links is equal to the sum over vertices weighted by the number of incident links. On the other hand, Euler's relation

$$v + b = c + l \qquad (5)$$

follows by successively suppressing the links until only v isolated vertices are left. At each stage, the number of loops decreases or the number of connected parts increases by unity.

i) Compute up to fourth order the Ising partition function on a d-dimensional hypercubic lattice.

Admissible graphs with at most four links are shown on figure 3. The corresponding numbers of configurations on a finite periodic lattice of N sites are respectively Nd, Nd, $Nd(2d-1)$, $\frac{1}{2}Nd(d-1)$ and $\frac{1}{2}Nd(Nd-4d+1)$. Their weights, taking into account the symmetry factor, are respectively $\frac{1}{2}\beta^2$, $\frac{1}{24}\beta^4$, $\frac{1}{4}\beta^4$, β^4, $\frac{1}{4}\beta^4$. Summing these contributions yields

$$Z = 1 + \tfrac{1}{2}Nd\beta^2 + \left[\tfrac{1}{12}Nd(6d-7) + \tfrac{1}{8}N^2d^2\right]\beta^4 + \mathcal{O}(\beta^6) \qquad (6)$$

To this order, one verifies the extensive character of the free energy. Indeed

$$-\frac{\beta\mathcal{F}}{N} = \frac{F}{N} = \frac{\ln Z}{N} = \tfrac{1}{2}d\beta^2 + \tfrac{1}{12}d(6d-7)\beta^4 + \mathcal{O}(\beta^6) \qquad (7)$$

is N-independent. The notation $\mathcal{F}$ refers to the traditional thermodynamic definition.

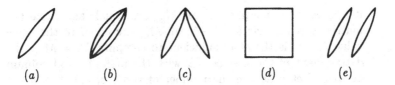

$\quad$ (a) $\qquad\qquad$ (b) $\qquad\qquad$ (c) $\qquad\qquad$ (d) $\qquad\qquad$ (e)

Fig. 3   Graphs for the Ising model up to order β^4.

This example illustrates the convenience of using $\tanh\beta$ rather than β as a small parameter in a high temperature expansion. Indeed, the corresponding graphs have now to satisfy the constraint of simplicity. Only the graph of figure 3(d) leads to a nonvanishing contribution in an expansion up to order $\tanh\beta^4$, and we find

$$Z = (\cosh\beta)^{Nd}[1 + \tfrac{1}{2}Nd(d-1)\tanh^4\beta + \mathcal{O}(\tanh^6\beta)] \qquad (8)$$

Equations (6) and (8) are easily shown to agree.

This simplification is useful in avoiding the proliferation of graphs, when investigating the higher order contributions. Other techniques will be discussed below.

ii) *Kirchoff's theorem.* The previous definitions allow us to recall a theorem due to Kirchoff giving in closed form the number of distinct trees that can be drawn on a connected graph, which uses the same set of vertices, the so called *spanning trees*. To a connected graph G we associate the incidence matrix which is a topological equivalent of (minus) the Laplacian. Along the main diagonal, $(-\Delta)_{ii}$ is equal to the number of links incident on vertex i, while $(-\Delta)_{ij}$ for distinct i and j is minus the number of links joining the vertices i and j. Elements in each line or column of $-\Delta$ add to zero, and therefore $\det(-\Delta)$ is zero, corresponding to the existence of a unique zero mode, a constant, since the graph is connected. The claim is that the determinant of any principal minor ($(-1)^{i+j}$ times the determinant obtained by deleting the ith line and jth column) is equal to the number of spanning trees. From the fact that the graph is connected, there exists a unique vector (the zero mode) with equal components corresponding to the zero eigenvalue of $-\Delta$ up to an overall factor. Let M_{ij} be the principal minor of the element ij (including the sign). From

$$\sum_j (-\Delta)_{ij} M_{kj} = \delta_{ik} \det(-\Delta) = 0$$

it follows that, for fixed k, all M_{kj} are equal, and since the matrix is symmetric as is $-\Delta$, all M_{kj} are equal to the same value M. It is therefore sufficient to compute $M = M_{11}$, the determinant of the matrix $-\Delta$ with the first line and column deleted. Let v denote the number of vertices, and $\ell \geq v - 1$ the number of links. Define a $l \times v$ matrix $L_{\alpha i}$ where α labels links, i vertices, after giving to each link an arbitrary orientation, through

$$L_{\alpha i} = \begin{cases} +1 & \text{if the link } \alpha, \text{ incident on } i \text{ is oriented off } i \\ -1 & \text{if the link } \alpha, \text{ incident on } i \text{ is oriented towards } i \\ 0 & \text{if the link } \alpha \text{ is not incident on } i \end{cases}$$

Then $(-\Delta) = L^{\mathrm{T}} L$. Call L' the matrix L with its first column deleted, in such a way that

$$M = \det L'^{\mathrm{T}} L' = \sum_{\{\alpha_2 < \cdots < \alpha_v\}} \det L'^{\mathrm{T}}_{\alpha_2 \ldots \alpha_v} L'_{\alpha_2 \ldots \alpha_v}$$

with the sum running over all $(v-1) \times (v-1)$ matrices $L'_{\alpha_2 \ldots \alpha_v}$ obtained by selecting $v-1$ rows in L' labelled $\alpha_2 < \alpha_3 < \cdots < \alpha_v$. The above equality is a classical identity in the theory of determinants. Each term in the sum is of the form $(\det L'_{\alpha_2 \ldots \alpha_v})^2$, and is zero unless the map $i \rightarrow \alpha_i$ assigns to each vertex $i = 2, \ldots, v$ an incident link, in which case the matrix $L'_{\alpha_2 \ldots \alpha_v}$ differs from a permutation matrix only by the fact that its entries are ± 1 instead of $+1$. This has two consequences. The first is that $(\det L'_{\alpha_2 \ldots \alpha_v})^2$ is equal to unity, and the second that it is in one-to-one correspondence with a spanning tree. This proves Kirchoff's theorem. It gives a topological meaning to the Laplacian which proves useful in percolation and polymer problems. We shall encounter an application in chapter 11.

7.1.2   Connected graphs and cumulants

The fundamental property of exponentiation relies on the following conditions.

i) The empty graph (no vertex, no link) is admissible, with a weight equal to 1. It has no connected parts $(c = 0)$ and is therefore not connected $(c \neq 1)$.

ii) Every union of admissible graphs is admissible.

iii) The weight of a graph factors into contributions from its connected parts.

If these hypotheses are valid, the sum over all admissible graphs is equal to the exponential of the sum over connected graphs. The exclusion constraint is not compatible with condition (ii). The reader will check on the above example of the β-expansion up to β^4 for the Ising model, that the preceding property is wrong, namely the free energy is not directly given in terms of contributions from connected graphs. It is our present task to modify the rules in order to find a direct expansion for the free energy.

The proof is quite simple. An arbitrary graph is built by drawing successively and independently its c connected parts $\mathcal{G}_1, \ldots \mathcal{G}_c$. The order being immaterial, each disconnected graph is therefore obtained $c!$ times instead of one. Using factorization of weights and summing on c, one finds

$$\sum_{\mathcal{G}} \omega(\mathcal{G}) = \sum_{c=0}^{\infty} \frac{1}{c!} \sum_{\substack{\mathcal{G}_1, \ldots \mathcal{G}_c \\ connected}} \omega(\mathcal{G}_1) \cdots \omega(\mathcal{G}_c) \tag{9}$$

On the right-hand side, one recognizes the expansion of an exponential, namely

$$\sum_{\mathcal{G}} \omega(\mathcal{G}) = \exp\left(\sum_{\mathcal{G} \text{ connected}} \omega(\mathcal{G})\right) \qquad (10)$$

The above exponential property is crucial in the computation of extensive quantities as well as in the study of various asymptotic behaviours, correlation lengths, or boundary effects. Unfortunately the exclusion rule forbids its direct application. We will now modify these rules, using the *cumulant* method, to restore exponentiation.

Assume that another set of rules can be found avoiding the exclusion constraint. To distinguish them, we represent vertices of the new graphs by open instead of full circles. A contribution of a new graph represents part of the former ones, obtained by identifying vertices with the same labels. If one requires that the new expansion reproduces the previous results, one obtains a set of equations, written symbolically as

$$\frac{1}{1!}\circ \; + \frac{1}{2!}\circ\circ + \frac{1}{3!}\circ\circ\circ + \cdots = \bullet$$

$$(1+\bullet)(\circ\!\!-\!\!) = \bullet\!\!-\!\!$$

$$(1+\bullet)(\,-\!\circ\!-\!\! + -\!\circ\circ\!-\,) = -\!\bullet\!-$$

$$(1+\bullet)\left(\; \curlyvee + \curlyvee\!\!' + \curlyvee\!\!' + \curlyvee\!\!' + \curlyvee\!\!' \;\right) = \curlyvee$$

$$(1+\bullet)\Big(\; \rtimes + \sum_{4\ \text{terms}} \rtimes\!' + \sum_{3\ \text{terms}} \rtimes\!'$$

$$+ \sum_{6\ \text{terms}} \rtimes\!' + \rtimes\!'\Big) = \rtimes \qquad (11)$$

$$\cdots$$

The factor $(1 + \bullet)$ follows from the possibility of identifying as many isolated vertices as one may wish. In this relation, all vertices carry the same index, which has been omitted for clarity.

To solve these equations, one must of course define the associated weights. Take the Ising case as a typical example. In the standard expansion which enforces the exclusion constraint, we have

i) a factor z_k for a vertex with k incident links

ii) a factor β for each link

iii) the product of these factors is divided by the order of the symmetry group of the graph.

The expansion without the exclusion constraint follows from similar rules, except for the fact that the factors z_k are to be replaced by cumulants u_k. The equations (11) read explicitly

$$
\begin{aligned}
e^{u_0} - 1 &= z_0 - 1 \\
u_1 &= z_1/z_0 \\
u_2 + u_1^2 &= z_2/z_0 \\
u_3 + 3u_2 u_1 + u_1^3 &= z_3/z_0 \\
u_4 + 4u_3 u_1 + 3u_2^2 + 6u_2 u_1^2 + u_1^4 &= z_4/z_0 \\
&\cdots
\end{aligned}
\tag{12}
$$

The cumulants u_k are obtained by inverting this system as

$$
\begin{aligned}
u_0 &= \ln z_0 \\
u_1 &= \frac{z_1}{z_0} \\
u_2 &= \frac{z_2}{z_0} - \left(\frac{z_1}{z_0}\right)^2 \\
u_3 &= \frac{z_3}{z_0} - 3\frac{z_2}{z_0}\frac{z_1}{z_0} + 2\left(\frac{z_1}{z_0}\right)^3 \\
u_4 &= \frac{z_4}{z_0} - 4\frac{z_3}{z_0}\frac{z_1}{z_0} - 3\left(\frac{z_2}{z_0}\right)^2 + 12\frac{z_2}{z_0}\left(\frac{z_1}{z_0}\right)^2 - 6\left(\frac{z_1}{z_0}\right)^4 \\
&\cdots
\end{aligned}
\tag{13}
$$

To obtain a compact form, write $z(h)$ and $u(h)$ for the generating functions

$$
z(h) = \sum_{k=0}^{\infty} \frac{z_k}{k!} h^k
\tag{14}
$$

$$
u(h) = \sum_{k=0}^{\infty} \frac{u_k}{k!} h^k
\tag{15}
$$

The above relations are then simply

$$
u(h) = \ln z(h)
\tag{16}
$$

i) Justify in more detail the cumulant procedure and check equation (16) up to order four against (13).

ii) Rederive the high temperature expansion in the Ising case using cumulants. The initial rules assume $z_{2k} = 1$, $z_{2k+1} = 0$, thus

$$z(h) = \cosh h \qquad (17)$$
$$u(h) = \ln \cosh h = \tfrac{1}{2}h^2 - \tfrac{1}{12}h^4 + \cdots \qquad (18)$$

The connected graphs are the first four on figure 3 up to fourth order. Since the exclusion rule no longer applies, the configuration numbers are modified to Nd, Nd, $\tfrac{1}{2}N(2d)^2$, $\tfrac{3}{4}Nd(2d-1)$ respectively. These numbers are all proportional to the number of sites N, because of translational invariance, and this insures that the free energy is extensive. Since $u_2 = 1$, $u_4 = -2$, the new weights are $\tfrac{1}{2}\beta^2$, $\tfrac{1}{6}\beta^4$, $-\tfrac{1}{2}\beta^4$, β^4. The weights are no longer positive, reflecting a similar loss of positivity of the cumulants. Summation over connected graphs yields immediately formula (7) and confirms the general property. The same method applies to the expansion in $\tanh\beta$. One has to introduce factors z_k which depend on the relative direction of incident links, and the graphs obtained in this way are no longer simple, which limits the interest of the method. The choice between simplicity and connectivity depends on the problem at hand. For gauge theories, we shall see that simplicity is more advantageous.

For rooted graphs, the distribution of fixed indices over connected parts leads to equations generalizing (11). Let $Z \langle i_1 \cdots i_k \rangle$ (Z is the partition function) denote the sum over all graphs (connected or disconnected) with roots $i_1, \ldots, i_k$, and $\langle i_1 \cdots i_k \rangle_c$ the sum over connected graphs using cumulants, one has in general

$$< i_1 > = < i_1 >_c$$
$$< i_1 i_2 > = < i_1 i_2 >_c + < i_1 >_c < i_2 >_c$$
$$< i_1 i_2 i_3 > = < i_1 i_2 i_3 >_c + < i_1 i_2 >_c < i_3 >_c + < i_1 i_3 >_c < i_2 >_c$$
$$+ < i_2 i_3 >_c < i_1 >_c + < i_1 >_c < i_2 >_c < i_3 >_c$$
$$\cdots$$

$$(19)$$

Disconnected graphs factorize partly into connected parts involving no roots, the sum of which yields the factor Z, and partly into connected rooted graphs which will realize all possible partitions of the set of indices $i_1, \ldots, i_k$. These properties become more obvious when one introduces a source j. Rooted graphs appear in the expansion of correlation functions, corresponding to

$$< \varphi_{i_1} \cdots \varphi_{i_k} >= \frac{1}{Z} \frac{\partial^k Z}{\partial j_{i_1} \cdots \partial j_{i_k}} \qquad (20)$$

or their connected counterparts

$$< \varphi_{i_1} \cdots \varphi_{i_k} >_c = \frac{\partial^k \ln Z}{\partial j_{i_1} \cdots \partial j_{i_k}} \qquad (21)$$

The relations (19) follow easily as a consequence of (20) and (21).

7.1.3   *Irreducibility and Legendre transformation*

To decrease further the number of graphs, we now decompose the connected graphs into irreducible parts. This notion is flexible and depends to some degree on the context. It can for instance be extended to multiply connected graphs (see exercise below). A compromise has to be found in order to obtain manageable calculations. Here we shall introduce irreducibility with respect to links.

A connected graph is *reducible* if it contains at least one link, the omission of which would disconnect it. This is shown on figure 4(a), where we have indicated the links with respect to which the graph can be reduced. We observe that a tree graph can be reduced with respect to each of its links. Having suppressed all these links, the graph is replaced by a collection of irreducible parts, which range from the simple vertex to "cactus" parts built from multiply connected parts, connected at articulation points. If these blocks are represented as hatched blobs, any graph takes the form of a tree with vertices replaced by blobs. Figure 4(b) illustrates symbolically this decomposition. The left-hand side is the sum over connected graphs represented by a white blob.

Assuming the conditions listed in the previous paragraph to be valid, we will show that it is possible to give diagrammatic rules for these generalized tree graphs, and then present a method of resummation.

Let $C(\{h_i\})$ be the sum over the weights of labelled connected graphs, where the vertex labelled i, with k incident links, is assigned a factor $\partial^k u(h_i)/\partial h_i^k$. Later on we shall let the h_i's go to zero to recover the initial graphs. The derivative operation

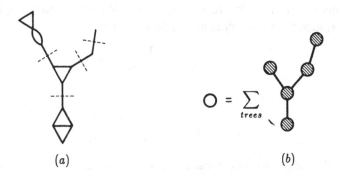

(a)   (b)

Fig. 4   (a) a reducible graph and its decomposition into irreducible parts (b) general type of connected graphs.

$\partial/\partial h_i$ acting on C amounts to choosing successively in all graphs the vertices with index i, and replacing the factor $\partial^k u(h_i)/\partial h_i^k$ by $\partial^{k+1}u(h_i)/\partial h_i^{k+1}$. Diagrammatically this means adding an external link emanating from vertex i. Symbolically

$$m_i = \frac{\partial C(\{h_i\})}{\partial h_i} = \frac{\partial}{\partial h_i} \bigcirc = \quad \quad (22)$$

Similarly we introduce a function (or rather a functional) $I(h_i)$ corresponding to the hatched blob, i.e. the sum over irreducible graphs with similar weights $\partial^k u(h_i)/\partial h_i^k$ for the labelled vertex i. The interpretation of the derivative $\partial/\partial h_i$ as in (22) remains valid when acting on I. In the case of a uniform field $h_i = h$, the contribution of hatched blobs in the trees of figure 4(b) is given by the generating function $I(h) = I(\{h_i\})|_{h_i=h}$. These graphs are unlabelled. It is necessary to include in the diagrammatic rules a convention of dividing out by a symmetry factor to recover once, and only once, each labelled connected graph of the original expansion.

In order to find a resummation procedure for trees, we shall first generate all of them with an incorrect weight, then amend the result using the topological relations discussed in section 1.1. Let us first consider the sum over all graphs, with one vertex singled out. This contribution, denoted $C_\bullet$, is obtained by weighting each tree by its number of vertices. We build the tree starting from the specified vertex, from which $k = 0, 1, 2,...$ links can emerge. The generalized vertex, represented by the hatched blob, contains generally several simple vertices. It is therefore necessary to "dress" each of them individually. The operation of adding k_i lines incident on the vertex i is represented by the derivative $\partial^{k_i}/\partial h_i^{k_i}$. Each of these lines carries a factor β and has to be joined to a neighbour of the site i. In other words, the dressed vertex is obtained from the bare one by the operator

$$\prod_i \sum_{k_i=0}^{\infty} \frac{\beta^{k_i}}{k_i!} \left(\sum_{j(i)} m_j\right)^{k_i} \frac{\partial^{k_i}}{\partial h_i^{k_i}}$$

which amounts to the replacement $h_i \rightarrow h_i + \beta \sum_{j(i)} m_j$. Assuming now a uniform situation with all $h_i \equiv h$, $m_i \equiv m$, this means that the generalized vertex is (recall that q is the coordination number).

$$C_{\bullet} = \oslash + \oslash + \oslash + \oslash + \cdots \tag{23}$$

or, more explicitly

$$C_{\bullet} = \sum_{k=0}^{\infty} \frac{d^k I(h)}{dh^k} (\beta m)^k = I(h + q\beta m). \tag{24}$$

One can perform a similar calculation if instead one singles out one link. Each graph is then weighted by its number of links. The corresponding quantity is C_l with

$$C_l = \text{O---O} = \sum_{links} \beta m^2 = N \frac{q}{2} m^2 \tag{25}$$

Equation (5) reads in the present case (a tree)

$$v - l = 1 \tag{26}$$

so that each graph is counted exactly once in the difference $C_{\bullet} - C_l$. The final result is therefore

$$C(h) = I(h + q\beta m) - \sum_{links} \beta m^2 \tag{27}$$

where the two terms are proportional to the total number of sites N and m is given by the uniform solution of

$$m_i = \frac{\partial C(\{h\})}{\partial h_i} \tag{28}$$

or equivalently, according to equation (27)

$$m = \frac{1}{N} \frac{\partial I(h + q\beta m)}{\partial h} \tag{29}$$

Equations (27) and (29) state that the functions C and I are Legendre transforms of each other. Equation (29) implies that in (27) the right-hand side is stationary with respect to variations in m for fixed h. This leads to the extremization procedure

$$C(h) = \underset{m}{\text{Extr}} \left(I(h + q\beta m) - \sum_{links} \beta m^2 \right) \tag{30}$$

If we wish, we can now set the external field equal to zero. This Legendre transformation enables one to extend the mean field procedure to all orders (in principle). Written in the above form of an extremum principle and properly interpreted, it exhibits the convex character of the effective action $C(h)$.

i) It would seem at first sight that in the example of the Ising model in the absence of an external field, the above technique is useless, since all finite graphs are irreducible, given the fact that an even number of links are incident on each site. The magnetization per site

$$\langle \sigma \rangle = \frac{1}{N} \frac{\partial \ln Z}{\partial h} \bigg|_{h=0} = m$$

vanishes indeed to all finite orders. Of course, as we know, this result is wrong at sufficiently low temperature (for $d \geq 2$). It is interesting therefore to note that formula (30), which amounts to a resummation over an infinite number of graphs, is capable of reproducing the phenomenon of spontaneous symmetry breaking by replacing the external field $h = 0$ by an effective mean field $q\beta m$. To see this, we observe that the series expressing I can be reorganized using other parameters, $1/d$ for instance. It is then possible to approximate $I(h)$ at any temperature by the contribution from its leading order corresponding to the lowest order graph, i.e. a vertex with a weight $\sum_i u(h_i) = \sum_i \ln \cosh h_i$. Formula (30) becomes

$$\frac{\ln Z}{N} = \underset{m}{\text{Extr}} \left(\ln \cosh 2d\beta m - \beta d m^2 \right) \tag{31}$$

when all m_i' s are equal, with a unique maximum at $m = 0$ for small enough β. As β increases, this maximum becomes a local minimum, and two new maxima develop symmetrically at a value corresponding to $\pm m_{sp}$, where m_{sp} is the spontaneous magnetization. Of course, the diagrammatic method is not adapted to provide a choice between extrema; the latter results from a global study, as discussed in volume 1. The value of β where the second derivative in m of the r.h.s. in (30) vanishes yields the critical temperature $\beta_c d = 1$ for a second order transition, and we recover the classical mean field expressions. The interest of the method is that it enables one to study the corrections systematically.

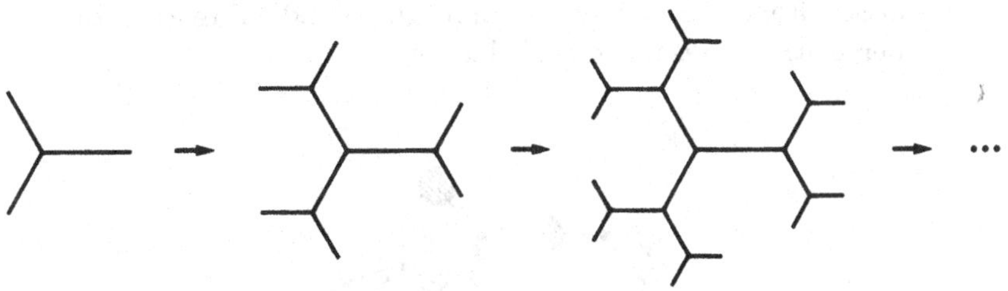

Fig. 5   Recursive construction of a Bethe lattice.

ii) Number of trees on a Bethe lattice. A Bethe lattice is an infinite tree where each site has q neighbours constructed recursively as shown on figure 5. To be precise, a recursively constructed tree – a Cayley tree – has the unfortunate property that in the "thermodynamic" limit, the number of boundary sites is comparable to the total number of sites. As a result, surface effects cannot be ignored in general, unless one considers only those quantities pertaining to sites deep "inside" the tree (meaning that the infinite volume limit has to be taken prior to any computation). This is what is meant by a Bethe lattice. Many problems admit of an exact solution on such a lattice, devoid of loops, where one can set up recursion relations. Let us count here the number N_k of trees with k branches, per site. We shall use a generating function $G(t) = \sum_{k=1}^{\infty} N_k t^k$. These trees have of course at most q links incident on a site. The equation corresponding to (27) which expresses this fact is

$$
\begin{aligned}
G(t) &= \sum_{k=1}^{q} \binom{q}{k} (mt)^k - \tfrac{1}{2} q t m^2 \\
&= (1 + mt)^q - 1 - \tfrac{1}{2} q t m^2
\end{aligned}
\tag{32}
$$

The quantity m represents the contribution of trees growing out a given link

$$m = \sum_{k=0}^{q-1} \binom{q-1}{k} (mt)^k = (1 + mt)^{q-1} \qquad (33)$$

The solution

$$G(t) = \mathop{\mathrm{Extr}}_{m} \left\{ (1 + tm)^q - 1 - \tfrac{1}{2} q t m^2 \right\} \qquad (34)$$

becomes singular at $t_c = (q-2)^{q-2}/(q-1)^{q-1}$. The singularity arises from the weight of large trees and is related to a percolation threshold on this lattice.

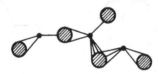

Fig. 6   Decomposition of a graph into multiply connected parts.

iii) Reduction to multiply connected graphs. To implement such a reduction, the number of variables on which to perform a Legendre transform becomes infinite. We want to resum "cactus" graphs (figure 6) with labelled articulation points. The relevant topological relation is now

$$1 = n_b + n_s - n_m \qquad (35)$$

where n_b is the number of blobs, n_s the number of articulation points and n_m the sum over articulation points weighted by the number of incident blobs. Let $B(\{u_i^{(n)}\})$ denote the contribution from multiply connected graphs, such that, to each vertex i with n incident links, is assigned a factor $u_i^{(n)}$ instead of the $\partial^n u(h_i)/\partial h_i^n$ assigned previously. To go from a blob to a cactus, we want to replace each vertex by a "dressed" one

$$u_i^{(n)} = \left[\exp\left(\sum_{k=1}^{\infty} g_i^{(k)} \frac{\partial^k}{\partial h_i^k} \right) \right] \frac{\partial^n u(h_i)}{\partial h_i^n} \qquad (36)$$

where $g_i^{(k)}$ is the contribution from all connected graphs such that k links are incident on vertex i

$$g_i^{(k)} = \frac{\partial B(\{u_j^{(n)}\})}{\partial u_i^{(k)}} \tag{37}$$

As above, we use relation (35) to obtain the sum C over all connected graphs in the form

$$C = B(\{u_i^{(n)}\}) + \sum_i \left[\exp\left(\sum_{k=1}^{\infty} g_i^{(k)} \frac{\partial^k}{\partial h_i^k} \right) \right] u(h_i) - \sum_i \sum_{k=1}^{\infty} u_i^{(k)} g_i^{(k)} \tag{38}$$

Again equations (36) and (37) yield conditions under which the right-hand side, considered as a functional of the independent variables $u_i^{(n)}$ and $g_i^{(n)}$, is extremal.

Although these reduction techniques look very attractive in principle, the corresponding expressions are so cumbersome in lattice models that their practical use is very limited. However, in the continuous field theoretic case, analogous irreducible kernels are necessary in several instances. They occur in the Schwinger–Dyson integral equations, in the analysis of bound states using the Bethe–Salpeter equation, as well as in the formal renormalization theory.

7.2 Series expansions

7.2.1 *High temperature expansion*

Consider a general model with an action (the Boltzmann weight is $\exp S$) of the form

$$S(J) = \beta \sum_{k \geq 2} \frac{1}{k!} V_{i_1 \dots i_k}^{\alpha_1 \dots \alpha_k} \varphi_{i_1}^{\alpha_1} \cdots \varphi_{i_k}^{\alpha_k} + \sum_{i,\alpha} J_i^{\alpha} \varphi_i^{\alpha} \tag{39}$$

The fields φ_i with components φ_i^{α} can be either bosonic or fermionic (in the latter case for instance the source J would also include Grassmannian variables). The *a priori* measure is denoted $\prod_i D\varphi_i$. We want to evaluate the generating function of correlations

$$Z(J) = \int \prod D\varphi_i \, \exp S(J) \tag{40}$$

In the high temperature expansion the integrand is expanded in a series in the variable β. Each term is then integrated over the fields φ_i. With J restricted to its value at a given site, we have to evaluate the local integrals

$$z^{\alpha_1 \cdots \alpha_n}(J) = z^{\alpha_1 \cdots \alpha_n}(J) = \int \varphi^{\alpha_1} \cdots \varphi^{\alpha_n} \ \exp J \cdot \varphi \ \mathrm{d}\varphi$$

$$= \frac{\partial^n}{\partial J^{\alpha_1} \cdots \partial J^{\alpha_n}} \int e^{J \cdot \varphi} \, \mathrm{d}\varphi = \frac{\partial^n z(J)}{\partial J^{\alpha_1} \cdots \partial J^{\alpha_n}} \qquad (41a)$$

Thus we require the value of the function

$$z(J) = \int D\varphi \ \exp J \cdot \varphi \qquad (41b)$$

associated to a site. Once this is known, let us list the rules to compute the partition function Z as an expansion in β. We consider all diagrams involving the following ingredients.

• Vertices corresponding to sites, with $n \geq 1$ branches, indexed by the site i and representing a product of fields $\varphi_i^{\alpha_1} \cdots \varphi_n^{\alpha_n}$ with a contribution to the weight

$$z^{\alpha_1 \cdots \alpha_n}(J_i)/z(J_i)$$

• Vertices corresponding to interactions with $k \geq 2$ branches, without an index and contributing a factor

$$\beta V_{i_1 \cdots i_k}^{\alpha_1 \cdots \alpha_k}$$

where $i_1 \ldots i_k$ denote the site indices of those previous vertices to which this vertex is linked.

• Links joining an interaction vertex to a site vertex, with a label α, the same index α which appears at both ends. Summation over those α's will have to be performed.

One has to sum over the contributions of all possible diagrams, connected or not, with an exclusion constraint on sites vertices. Applying the methods of the previous section, we give below some examples.

In the particular case of a homogeneous (i.e. translationally invariant) quadratic action, interaction vertices with two branches can be ignored, and one finds the typical graphs of spin models, in particular those of the Ising model used as an illustration in the previous section.

i) Establish an expansion valid in large dimension for the $O(n)$-spin model using irreducibility.

The fields φ_i are unit n-component vectors, and the corresponding action reads $\beta \sum_{(ij)} \varphi_i \cdot \varphi_j$. The generating function of cumulants is therefore

$$z(h) \equiv \exp u(h) = \int d\varphi \ \exp h \cdot \varphi = \Gamma(n/2)(n/2)^{1-n/2} I_{n/2-1}(h) \tag{42}$$

Here $I_k(h)$ is the modified Bessel function, not to be confused with the generating functional of irreducible graphs $I(\{h\})$. In equation (42), the measure over φ is normalized to unity. Setting $H = h + 2d\beta m$ in relation (30), we find

$$\frac{\ln Z}{N} = I(H) - \frac{H^2}{4\beta d} \quad , \quad H = 2\beta d \frac{dI}{dH} \tag{43}$$

As was stated before, the solution $H = 0$ ceases to be the physical one when it corresponds to a minimum of the right-hand side of (43). The location of the transition is given by requiring the coefficient of the second derivative in H at $H = 0$ to vanish. To lowest order $I(H) = u(H)$, thus, in the above units, $\beta_c = n/2d$ to this order, which vanishes like $1/d$ for large dimension. On figure 7, we show all diagrams up to order 6, as well as their second derivative in H at the origin. Since β_c behaves like d^{-1}, a diagram of order k gives a contribution behaving as $d^{-k+[k/2]}$. This is due to the fact that irreducibility forces the diagram to involve only cycles, and a cycle of length $2k$ can only use k distinct directions on a lattice. The configuration number is a polynomial in d of degree k. If the series for I is ordered in powers of βd and d^{-1}, only a finite number of diagrams will contribute to a given order in d^{-1}. Thus one derives an equation giving $\beta_c d$ as a series in d^{-1}, of the form

$$\frac{n}{2\beta_c d} = 1 - \frac{1}{2d} - \frac{2 - 2/(n+2)}{4d^2} - \frac{7 - 8/(n+2)}{8d^3} + \cdots \tag{44}$$

to be compared with the large n limit (with β_c/n rescaled as β_c) given in (3.104).

ii) It is interesting to rederive in this framework the large n limit. From the above discussion, one should keep β/n finite as $n \to \infty$. Inspecting figure 7, we see that the only contributions in the limit $n \to \infty$ arise from diagrams involving a single loop. This suggests the use of the formalism introduced in exercise

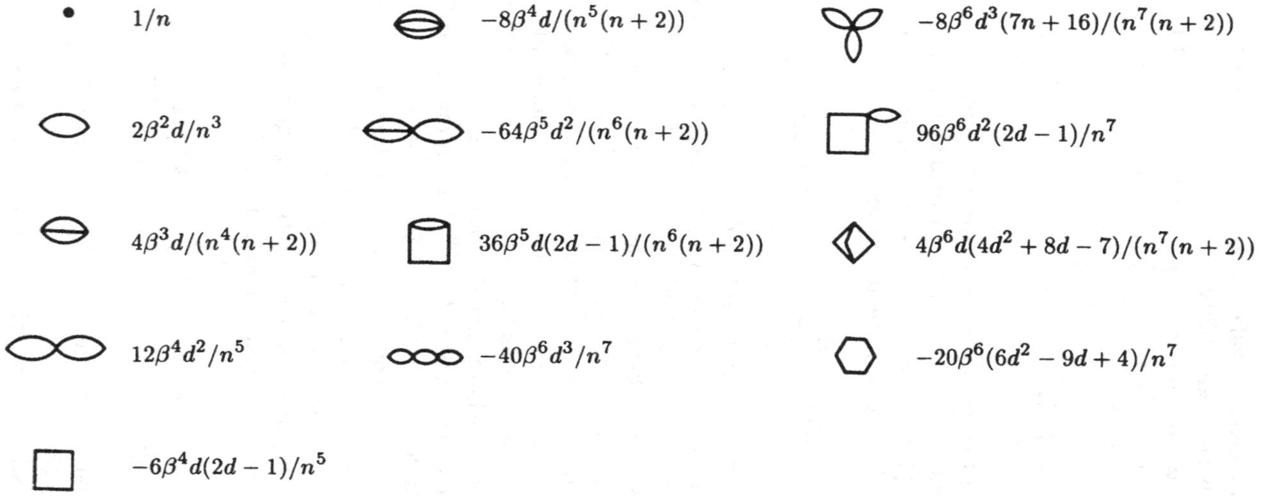

$$1/n$$

$$2\beta^2 d/n^3$$

$$4\beta^3 d/(n^4(n+2))$$

$$12\beta^4 d^2/n^5$$

$$-6\beta^4 d(2d-1)/n^5$$

$$-8\beta^4 d/(n^5(n+2))$$

$$-64\beta^5 d^2/(n^6(n+2))$$

$$36\beta^5 d(2d-1)/(n^6(n+2))$$

$$-40\beta^6 d^3/n^7$$

$$-8\beta^6 d^3(7n+16)/(n^7(n+2))$$

$$96\beta^6 d^2(2d-1)/n^7$$

$$4\beta^6 d(4d^2+8d-7)/(n^7(n+2))$$

$$-20\beta^6(6d^2-9d+4)/n^7$$

Fig. 7 Diagrams for the computation of the irreducible kernel $I(H)$ up to sixth order, and their contribution to $d^2 I(H)/dH^2\big|_{H=0}$.

(iii) of section 1.3. The function $u(h)$ behaves as

$$u(h) \sim \tfrac{1}{2}n \left\{ \sqrt{1 + 4h^2/n^2} - 1 - \ln \left(\tfrac{1}{2} + \tfrac{1}{2}\sqrt{1 + 4h^2/n^2} \right) \right\} \quad (45)$$

showing that the functions $g^{(3)}$, $g^{(4)}$... do not appear in equation (36); only $\mathbf{g}^{(1)}$ (a vector) and $\bar{g}^{(2)}$ (a matrix) play a role. The right-hand side of equation (36) reads

$$\exp \left(\frac{\partial}{\partial h_\alpha} \bar{g}^{(2)}_{\alpha\beta} \frac{\partial}{\partial h_\beta} \right) u^{(n)} \left(\left| \mathbf{h} + \mathbf{g}^{(1)} \right| \right)$$

$$\simeq u^{(n)} \left(\sqrt{\left| \mathbf{h} + \mathbf{g}^{(1)} \right| + 2 \operatorname{Tr} \bar{g}^{(2)}} \right) \quad (46)$$

As for the quantity B, we have only to keep the one link diagram, as well as those involving a single loop. It is shown in section 3 how to compute the number of loops with p links drawn on a lattice. This leads to the expression

$$B(\mathbf{u}^{(1)}, \bar{u}^{(2)}) = \beta d \mathbf{u}^{(1)^2} + \tfrac{1}{2} \operatorname{Tr} \left\{ \int_0^\infty \frac{ds}{s} \left[I_0 (2\beta\bar{u}^{(2)})^d - 1 \right] e^{-s} \right\}$$

$$(47)$$

Imposing stationarity with respect to $\mathbf{u}^{(1)}$ and $\bar{u}^{(2)}$, one finds

$$\frac{\ln Z}{N} = \beta\nu - \tfrac{1}{2}n - \tfrac{1}{2}n \ln \frac{2\beta\nu}{n} + \int_0^\infty \frac{ds}{s} [I_0^d(s) - 1] e^{-s\nu} \quad (48)$$

where ν satisfies

$$\frac{2\beta}{n} = \int_0^\infty ds\, I_0^d(s) e^{-s\nu} \quad (49)$$

The critical value of β corresponds to the lowest possible value of ν, namely $\nu = d$. One verifies that the $1/d$ expansion of β_c coincides with the one of formula (44) when $n \to \infty$.

iii) Mean field equations and systematic corrections in the general case, using irreducible kernels.

Formula (31) can be generalized to arbitrary interactions. However the method is slightly cumbersome, due to the existence of two types of vertices (sites and interactions). Referring to the discussion in section (1.3), let us spell out some details. First let us define irreducibility with respect to interaction vertices. A connected diagram is said to be reducible with respect to a k-vertex if omission of the latter disconnects it into exactly k parts. This is not as general a definition as strong irreducibility, where it was only required that the diagram splits into connected parts. Diagrammatically, we distinguish interaction vertices using black dots, site vertices by white ones. Equation (22)

$$m_i = \text{⊖} = \frac{\partial C\left(\{h_i\}\right)}{\partial h_i} \qquad (50)$$

in which a site vertex i is distinguished, is to be complemented by a corresponding formula for the interaction vertex

$$H_i = \text{⊖⊷⊶⊖} = \left. \frac{\partial S}{\partial \varphi_i} \right|_{\varphi_i = m_i} \qquad (51)$$

Equations (23) and (24) generalize as

$$C_\bullet = \text{⊘} + \text{⊶} + \text{⊶⊷} + \cdots = I\left(\{h_i + H_i\}\right)$$

$$(52)$$

with a similar relation for an interaction vertex. The topological relation

$$v - \sum (k-1)n_k = 1 \qquad (53)$$

where v is the number of blobs and n_k the number of k-interaction vertices, leads to

$$C\left(\{h\}\right) = I\left(\{h + H\}\right) + S\left(\{m\left(i\right)\}\right) - \sum_i m_i H_i \qquad (54)$$

Again (50) and (51) express stationarity with respect to H and m considered as independent variables. The expansion of I involves diagrams such that any interaction vertex belongs at least to one loop. It is natural to organize this series according to the number of loops.

$$I\left(h\right) = \circ + \text{⬡} + \text{⬡} + \cdots \qquad (55)$$

The first term corresponds to mean field, in complete generality (see formulas (6.64)–(6.66)). Again, a diagrammatic approach (as

opposed to a global method) does not allow competing extrema to
be distinguished.

7.2.2   *The role of symmetries*

For models invariant with respect to a group of transformations,
it is possible to achieve simplifications analogous to the use of the
variable $\tanh \beta$ in the Ising case.

A *chiral* model involves fields which are elements of a compact
group G. A natural basis to express any function of φ uses
matrix elements $\mathcal{D}^r_{\alpha\beta}(\varphi)$ of irreducible representations (r labels
such representations). This is the content of the Peter–Weyl
theorem in the case of square integrable functions on a compact
semisimple Lie group. *A fortiori*, this is true for a finite group.
The measure $d\varphi$ is the invariant, or Haar measure on the group.
We assume that it is normalized to one. Then we have the familiar
relations expressing orthogonality and completeness

$$\int d\varphi \; \mathcal{D}^r_{\alpha\beta}(\varphi) \; \overline{\mathcal{D}^s_{\gamma\delta}(\varphi)} = \frac{1}{d_r} \delta_{rs} \delta_{\alpha\gamma} \delta_{\beta\delta} \tag{56}$$

$$\sum_{r,\alpha,\beta} d_r \; \mathcal{D}^r_{\alpha\beta}(\varphi) \; \overline{\mathcal{D}^r_{\alpha\beta}(\varphi')} = \delta(\varphi,\varphi') \tag{57}$$

where d_r is the dimension of the r-representation, and $\delta(\varphi,\varphi')$
the appropriate Dirac measure (δ-function) on G. It is convenient
to expand the Boltzmann weight $\exp S$ in terms of the various
$\mathcal{D}^r(\varphi)$ and to perform the integrations over the φ'_i s using group
theoretic techniques, in particular equation (56), the simplest one
of this kind.

As an illustration, let us return to the n-vector model. We denote in this case the group elements by $\mathcal{R} \in SO(n)$ and, considering the spin manifold as the homogeneous space $SO(n)/SO(n-1)$,
we express the action using group variables as

$$S = \beta \sum_{<ij>} \boldsymbol{\Phi}_0 \mathcal{R}_i^{-1} \mathcal{R}_j \boldsymbol{\Phi}_0 = \sum_{<ij>} S_{ij}(\mathcal{R}_i^{-1}\mathcal{R}_j) \tag{58}$$

with $\boldsymbol{\Phi}_0$ a fixed (isotopic) vector. Using equation (56) one has

$$\exp S = \prod_{<ij>} \sum_r d_r \, \mathrm{Tr} \left(\beta_r \mathcal{D}^r(\mathcal{R}_i^{-1}\mathcal{R}_j) \right) \tag{59}$$

with coefficients β_r ($r \times r$ matrices) given by

$$\beta_r = \int d\mathcal{R} \ \overline{\mathcal{D}^r(\mathcal{R})} \ \exp S_{ij}(\mathcal{R}) \qquad (60)$$

and β_0 is a simple scalar. Each term in the product (59) is represented diagrammatically as follows. For each pair of neighbours (ij), choose a nontrivial representation $r \neq 0$, and draw a link from i to j indexed by r. The corresponding contribution is $d_r \beta_r / \beta_0$ after integration over the group variables. It is understood that in computing the partition function, an overall power of the scalar β_0 is factored out. The graphs will be simple with at most one (oriented) link joining two vertices. Changing the orientations of links amounts to substituting for the representation r its adjoint $\bar{r}$ with $\mathcal{D}^{\bar{r}} = \overline{(\mathcal{D}^r)}$. According to equation (60), we have $\beta_{\bar{r}} = \overline{\beta_r}$.

In the Ising case, the group Z_2 has a single nontrivial representation, isomorphic to the group itself. Therefore $\beta_0 = \cosh \beta$, $\beta_1 = \sinh \beta$ and one recovers the standard rules of the $\tanh \beta$ expansion.

If several links, labelled by $r_1, \cdots, r_n$, are incident on a vertex, the integration over the site field $\mathcal{R}_i$ yields a contribution of the form

$$\int d\mathcal{R} \ \mathcal{D}^{r_1}(\mathcal{R}) \cdots \mathcal{D}^{r_n}(\mathcal{R})$$

The case of a single incident link, $n = 1$, is excluded since the representations r are assumed nontrivial and the corresponding integral vanishes. In the case $n = 2$, reversing one of the orientations, one finds that the two successive links have to carry identical representations. More precisely

$$\int d\mathcal{R}_j \left[d_r \operatorname{Tr} \beta_r \mathcal{D}^r(\mathcal{R}_i^{-1}\mathcal{R}_j) \right] \left[d_s \operatorname{Tr} \beta_s \mathcal{D}^s(\mathcal{R}_j^{-1}\mathcal{R}_k) \right] =$$
$$= \delta_{rs} \left[d_r \operatorname{Tr} \beta_r^2 \mathcal{D}^r(\mathcal{R}_i^{-1}\mathcal{R}_k) \right] \qquad (61)$$

Thus, between sites i and k, we have a contribution similar to the one between two neighbours, with β_r replaced by β_r^2. Repeating the process around a k-rung loop yields a factor

$$d_r \operatorname{Tr} \beta_r^k \qquad (62)$$

When more than two links are incident on a vertex, the computations become more cumbersome and involve coupling coefficients between several representations, as well as the corresponding β'_r s.

If instead of the n-vector model, we consider the chiral case with an action

$$S = \beta \sum_{<i,j>} \mathrm{Tr}(\mathcal{R}_i^{-1}\mathcal{R}_j) \tag{63}$$

a slight simplification arises from the fact that the β'_r s are scalars.

This is also the case for the plaquette action of lattice gauge models

$$S = \beta \sum_p \mathrm{Tr}(\mathcal{R}_{ij}\mathcal{R}_{jk}\mathcal{R}_{kl}\mathcal{R}_{li}) \tag{64}$$

Rather than using the graphical prescriptions of section 2.1, it is more useful to represent the interaction by the plaquette $ijkl$, the sides of which carry the fields (figure 8). The corresponding diagrams have the topology of surfaces.

Fig. 8   Representation of the interaction used in gauge models.

It is also possible in this instance to obtain a restriction to connected diagrams by adapting the cumulant method. Unfortunately this has the disadvantage of generating a proliferation of graphs. For instance, if one enumerates the fourth order graphs arising on a lattice reduced to a single plaquette, one finds 16 distinct topological types including the sphere, the torus, the projective plane, the Klein bottle... Only one of the graphs is not connected. The method is obviously useful only in the case where one attempts to re-sum families of graphs with a fixed topology.

On the other hand, group theoretic integration techniques lead to an economy in the calculations. As above, the expansion

$$\exp S = \sum_p \sum_r \beta_r \chi_r(\mathcal{R}_{ij}\mathcal{R}_{jk}\mathcal{R}_{kl}\mathcal{R}_{li}) \tag{65}$$

in terms of the characters $\chi_r = \operatorname{Tr} \mathcal{D}^r$, allows integration over the $\mathcal{R}$'s using formula (56). Diagrams are expressed as closed surfaces made of plaquettes. The simplest one is the boundary of a three-dimensional cube, with six faces, which may occupy $Nd(d-1)(d-2)/6$ positions on the lattice. This means that the partition functions admits of an expansion, which starts as

$$Z = \beta_0^{Nd}\left(1 + \tfrac{1}{6}Nd(d-1)(d-2)\sum_{r\neq 0} d_r^2\left(\frac{\beta_r}{\beta_0}\right)^6 + \cdots\right) \quad (66)$$

7.2.3 Low temperature expansion – discrete case

In the high temperature expansion, we selected one-by-one a configuration of interactions (depicted by a diagram and summed over field configurations). We now adopt the opposite view to obtain a low temperature series. Starting from a field configuration, chosen in general as an ordered state minimizing the (physical) energy (or maximizing the Boltzmann weight), we modify it successively by taking into account excitations. By necessity, the techniques are different when these excitations have a continuous or discrete spectrum. We consider here the easiest discrete case.

By assumption, the energy (or action) has a discrete spectrum $E_0, E_1, \ldots$, each value with a degeneracy $n_0, n_1, \ldots$, in any finite volume. Let us pretend that the ground state is nondegenerate ($n_0 = 1$) and assume that excited ones arise from a finite modification of the ground state configuration. We return to this point later. The partition function reads

$$Z = \sum n_i e^{-\beta E_i} \quad (67)$$

and the task is to classify states according to increasing energy. Configurations are identified by the set of fields φ_i, taking a different value as compared to the ones $\varphi_i^{(0)}$ in the ground state. Let us split the action as a sum $S_0 + S_1 + S_2 + \cdots$ involving increasing numbers of modified fields

$$S_n \equiv \sum_{i_1,\ldots i_n} S_n(\varphi_{i_1}, \cdots \varphi_{i_n}) \quad (68)$$

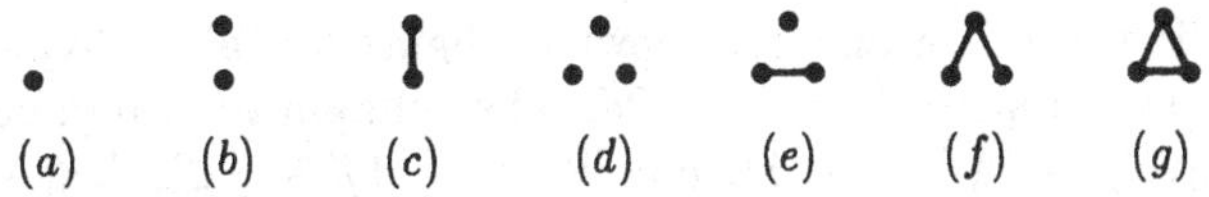

Fig. 9   The first few terms in the low temperature series for the Ising model.

with $S_n\left(\varphi_i^{(0)}, \ldots, \varphi_{i_n}^{(0)}\right) = 0$ for $n > 0$, and $S_0 \equiv S\left(\left\{\varphi_i^{(0)}\right\}\right)$. Thus we have

$$Z = e^{S_0} \sum_{configurations} \prod_i e^{S_1(\varphi_i)} \left[\prod_{ij}(1 + f_{ij})\prod_{ijk}(1 + f_{ijk})\cdots\right] \quad (69)$$

with

$$f_{i_1 i_2 \ldots i_n} = e^{S_n(\varphi_{i_1}, \ldots \varphi_{i_n})} - 1 \quad (70)$$

The quantity in brackets in (69) is then expanded, and represented diagrammatically. Vertices are labelled by the fields $\varphi_i \neq \varphi_i^{(0)}$ which differ from their ground state value. There are also vertices corresponding to "interactions", related to these sites and representing the various terms f_{ij}, f_{ijk}, ... chosen in the expansion of (69). The corresponding graphs are simple, in the sense that n sites are linked at most with one n-interaction vertex. The corresponding expansion is closely related to the low density expansion in the theory of imperfect gases.

As an illustration, let us return once again to the Ising model in the presence of a positive external field which lifts the ground state degeneracy by a quantity $2Nh$. We take the zero field limit $h \rightarrow +0$ only after the thermodynamic one. Only those states differing from the ground state, with all spins aligned parallel to the field, $\sigma_i^{(0)} = +1$, by finitely many spin flips, will contribute to the previous expansion.

In the zero field limit, for a lattice with N sites having q neighbours, we write the action as

$$S = \tfrac{1}{2}\beta N q + \beta q \sum_i (\sigma_i - 1) + \beta \sum_{(i,j)} (\sigma_i - 1)(\sigma_j - 1) \quad (71)$$

The first term yields the ground state contribution $(\exp \tfrac{1}{2}\beta q N)$. The second term yields a factor $\exp(-2\beta q)$ when a spin is reversed. The final term, when expanded as in (69), attributes to every link joining neighbouring sites a factor $(\exp 4\beta - 1)$.

The first few diagrams are shown on figure 9. Take a hypercubic d-dimensional lattice, with $q = 2d$. The configuration numbers are respectively N, $\frac{1}{2}N(N-1)$, Nd, $\frac{1}{6}N(N-1)(N-2)$, $Nd(N-2)$, $Nd(2d-1)$ and 0 (on a cubic lattice). Correspondingly

$$\ln Z/N = \beta d + e^{-4\beta d} + d e^{-8\beta d + 4\beta} - (d + \tfrac{1}{2})e^{-8\beta d}$$
$$+ d(2d-1)e^{-12\beta d + 8\beta} - 4d^2 e^{-12\beta d + 4\beta}$$
$$+ (2d^2 + d + \tfrac{1}{3})e^{-12\beta d} + \cdots \tag{72}$$

The effective expansion parameter is $\exp(-4\beta)$. The first part of the interaction term, namely $\exp(4\beta)$ tends to decrease the order of a given term in the expansion. In fact, it will exactly compensate the contribution of sites located inside a cluster of reversed spin which interact with their q neighbours. Thus only "surface" terms will be unbalanced. Since the surface increases with the size of the cluster, one is assured of being able to arrange the series in increasing powers of $\exp(-4\beta)$, with only a finite number of diagrams contributing to a given order. The case $d = 1$ is exceptional. No matter how large a sequence of reversed spins, its contribution will remain $\exp(-8\beta)$. We find here a phenomenon discussed in chapter 1 (volume 1), with $d = 1$ being the lower critical dimension where the ordered phase disappears. That the low temperature expansion involves a balance between energetic versus entropic contributions from surfaces bounding flipped spins, is the origin of the celebrated analysis of Peierls (1936), leading to a proof of the existence of ordered phases at low temperature. Peierls' argument has been the source of inspiration for many generalizations.

We observe in the series (72) a typical loss of positivity in the coefficients, as compared with the high temperature expansion for the partition function. As a consequence, the analysis of such low temperature series frequently reveals complex singularities which might limit their usefulness.

Low temperature expansions require a knowledge of the ground state which minimizes energy. The latter is in general not unique in the absence of an external field. But in ferromagnetic spin models, an infinitesimal external field is sufficient to lift the degeneracy by a term proportional to the volume. As temperature decreases and the limit of zero field is taken, degenerate ground states are separated by an infinite energy barrier and an expansion around one of them (describing a pure phase) is therefore justified.

Fig. 10   Equivalent low temperature diagrams in a lattice gauge theory.

One can encounter more complex situations. Take for instance an *antiferromagnetic* Ising model, with a reversed sign for the interaction term. At low temperature minimization of energy tends to favor antiparallel neighbouring spins. On certain lattices, which do not admit a bipartition, such configurations are not allowed. A simple example is the triangular two-dimensional lattice. On each elementary triangular face of the lattice, at least one pair of spins cannot be antiparallel. This is the phenomenon of *frustration*: a frustrated link is one for which spins are parallel. The distribution of frustrated links can produce a thermodynamic degeneracy of the ground states and leads to the possibility of transitions among them through excited states.

i) Study the ground state degeneracy of the antiferromagnetic Ising model on a triangular lattice using the expressions given in chapter 2.

ii) Obtain the low temperature expansion of a Z_2-gauge model with an action

$$S = \beta \sum_p \sigma_{ij}\sigma_{jk}\sigma_{kl}\sigma_{li} \qquad (73a)$$

with the field σ_{ij} taking values ± 1. The ground state has all plaquette terms equal to $+1$, meaning that σ_{ij} is a pure gauge

$$\sigma_{ij} = s_i s_j \qquad (s_i = \pm 1) \qquad (73b)$$

whatever the values of the variables s_i are. The corresponding degeneracy is therefore 2^N.

Low temperature diagrams are generated by reversing the link variables σ_{ij}. But if all variables pertaining to links incident on a site are simultaneously reversed, this amounts to a gauge transformation (figure 10). One is consequently led to identify all diagrams corresponding to a reversal of all link variables incident on a site.

7.2.4 *Low temperature expansion – continuous case*

The previous method is to be modified in the continuous case. One has collective excitations (generalized spin waves) and formula (67) is replaced by a genuine functional integral, to which we can only apply the saddle point method and its appended perturbative – possibly renormalized – corrections. This technique is by now familiar to the reader. Let us summarize the main steps. Given a ground state configuration $\{\varphi_i^0\}$ satisfying

$$\left.\frac{\delta S}{\delta \varphi_i}\right|_{\varphi_i=\varphi_i^0} = 0 \tag{74}$$

the action is then expanded around this configuration, starting with a (hopefully) negative definite quadratic form corresponding to a genuine maximum

$$S(\varphi_i) = S(\varphi_i^0) + \tfrac{1}{2}\sum_{i,j}\Delta_{ij}(\varphi_i - \varphi_i^0)(\varphi_j - \varphi_j^0) + V(\{\varphi_i - \varphi_i^0\}) \tag{75}$$

The partition function is expressed as

$$
\begin{aligned}
Z &= \int \prod_i \mathrm{d}\varphi_i \, \exp S(\varphi_i) \\
&= \exp\left(S(\varphi_i^0)\right) \exp\left(V(\{\delta/\delta j_i\})\right) \int \prod_i \mathrm{d}\varphi_i \times \\
&\qquad \left. \exp\left\{\sum_{i,j}(\varphi_i - \varphi_i^0)\Delta_{ij}(\varphi_j - \varphi_j^0) + \sum_i j_i(\varphi_i - \varphi_i^0)\right\}\right|_{j=0} \\
&= \det^{-1/2}\!\Delta \, \exp\left(S(\varphi_i^0)\right) \, \exp\left(V(\{\delta/\delta j_i\})\right) \exp\left(\tfrac{1}{2}j_i(\Delta^{-1})_{ij}j_j\right)\Big|_{j=0}
\end{aligned}
\tag{76}
$$

A diagrammatic expansion follows when $\exp V(\{\delta/\delta j_i\})$ is expanded in a power series. Each monomial in V is associated to a vertex. The effect of the functional derivatives $\delta/\delta j_i$ when we then set the source j to 0 is to contract in all possible ways these vertices $(\Delta^{-1})_{ij}$ (Wick's theorem). A difference with the discrete case, is that one obtains a series in powers of β^{-1}, rather than in $\exp(-\beta)$. Indeed Δ^{-1} is proportional to β^{-1} and, although V contains a factor β, the number of propagators overwhelms the number of interaction vertices. This is the standard counting of the number of loops, with β^{-1} playing the role of $\hbar$. Note that the stationary configuration φ^0 is β-independent.

A crucial difficulty is the possible existence of zero modes, in particular those originating from Goldstone's theorem. As a result, Δ is only semidefinite and cannot be inverted in a straightforward way. It is necessary to isolate these degrees of freedom and to treat them by a direct integration and not by the saddle point method, if possible. An example was worked out in volume 1 (chapter 6, section 2.4) to compute mean field corrections for a lattice continuous gauge field system. These zero modes arise generally from the existence of a continuous transformation group G, leaving the action invariant, and thus producing a continuum of ground state configurations starting from any given one. A general method for dealing with these problems is to introduce a symmetry breaking term which is then averaged out, in the following form. Let a noninvariant *gauge fixing* function $\mathbf{f}(\varphi)$ be introduced. The term *gauge fixing* is borrowed for gauge theories, where the method was introduced. It is not meant here to imply necessarily a local invariance group. Let $^g\varphi$ denote the action of G on the configuration φ, and assume that the equation $\mathbf{f}(^g\varphi) = \mathbf{c}$ has a unique solution $g \in G$. There might not exist such a global gauge fixing, this difficulty is generally bypassed by requiring the above property to hold in a small neighborhood of the identity. For a global invariance group we need only require that $\mathbf{f}(^g\varphi) = \mathbf{c}$ fixes g only up to a transformation of the little group G^o of φ^o (the subgroup of G leaving φ^o invariant, and as a consequence all terms in the action expanded around φ^o). This is an inessential complication in the following, and we leave it to the reader to modify accordingly the formulae.

Write

$$\det\mathcal{M}(\varphi) \int dg\delta(\mathbf{f}(^g\varphi) - \mathbf{c}) = 1 \qquad (77)$$

where

$$\mathcal{M}(\varphi) = \frac{\delta\mathbf{f}(^g\varphi)}{\delta g} \qquad (78)$$

In principle, the Jacobian should be written $|\det\mathcal{M}(\varphi)|$ but according to the assumptions, the absolute value sign, can be omitted since $\det(\mathcal{M}(\varphi))$ does not vanish. Introducing the identity (77) into the functional integral for Z and using the invariance of the action S, the measure, as well as $\det\mathcal{M}(\varphi)$, one gets

$$Z = \int e^{S(\varphi)} \delta(\mathbf{f}(\varphi) - \mathbf{c}) \det \mathcal{M}(\varphi) \, d\varphi \, dg \qquad (79a)$$

One can now integrate over g. Moreover, since the result is $\mathbf{c}$-independent, we can also integrate over $\mathbf{c}$ with an arbitrary (normalized) weight function $\exp F(\mathbf{c})$ (for instance proportional to $\exp -\mathbf{c}^2$), and obtain

$$Z = \int \exp\left[S(\varphi) + F(\mathbf{f}(\varphi)) + \operatorname{Tr} \ln \mathcal{M}(\varphi)\right] d\varphi \qquad (79b)$$

In this highly symbolic form, invariance is apparently lost (due to the presence of $F(\mathbf{f}(\varphi))$) and the saddle point method applies without difficulty. The nonlocal term $\det \mathcal{M}(\varphi)$ can be replaced by an equivalent Grassmannian integral, introducing auxiliary (Faddeev–Popov ghost) fields, $\bar\psi, \psi$

$$Z = \int \exp\left[S(\varphi) + F(\mathbf{f}(\varphi)) + \bar\psi \mathcal{M}(\varphi)\psi\right] d\varphi \, [d\psi d\bar\psi] \qquad (79c)$$

A typical illustration is afforded by the nonlinear σ-model (or $O(n)$ model) in two dimensions (the lower critical dimension) at low temperature. This system with its global symmetry has a behaviour reminiscent of nonabelian gauge fields in four dimensions, namely it is asymptotically ultraviolet free. Since it is much simpler than its gauge counterpart, it has been under very active scrutiny to understand how to obtain reliable information on the spontaneous mass generation from a clever use of perturbative calculations. Moreover, interesting conjectures have been made (Zamolodchikov) concerning an exact expression for the scattering matrix. We refer the reader to the literature for a detailed treatment of the σ-model and various generalizations involving an arbitrary homogeneous manifold instead of a sphere for the target space. Such generalizations have been in particular extremely fruitful in the study of two-dimensional random systems using the replica trick (chapter 10) or in string field theory.

i) Starting from an original G-invariant model, study the remnant symmetry in the form (79c) after gauge fixing.

ii) For a two-dimensional n-vector model, with an action written as

$$S = -\frac{1}{2g_0^2} \int d^2x \, (\partial \mathbf{\Phi})^2 \qquad \mathbf{\Phi} \cdot \mathbf{\Phi} = 1 \qquad (80)$$

the renormalization of the coupling constant is given to lowest order by

$$\frac{1}{g^2} = \frac{1}{g_0^2} + \frac{n-2}{2\pi} \ln \mu a + \cdots$$
$$\mu \frac{\partial}{\partial \mu} g \equiv \beta(g) = -\frac{n-2}{4\pi} g^3 + \cdots$$

(81)

with μ a renormalization scale and a the inverse (ultraviolet) cut-off. This expresses the ultraviolet asymptotic freedom property, stating that, as the mass scale μ gets large, the renormalized coupling g is driven to 0 provided that $n > 2$. The factor $(n-2)$ is in agreement with the fact that the property holds only for $n > 2$, the borderline being the XY-model with $n = 2$. An analytic continuation in n below 2 would imply reversing the conclusions.

7.2.5   *Strong field expansions*

Consider the case where a large external field is present. This does not modify in an essential way the derivation of low temperature series. Rather it suppresses the problem of degenerate extrema. Take for instance the Ising model, the vertex contribution becomes

$$z = \exp(-2H - 2\beta q) \tag{82}$$

where H denotes the external field, while the link contribution remains

$$u = \exp(4\beta) - 1 \tag{83}$$

We get a power series in two variables z and u, and the problem encountered in section 2.3 was to order it in the low temperature variable $(1 + u)^{-1}$. Alternatively, we may look for an expansion valid when z goes to zero, i.e. for strong field. Powers of z count the number k of vertices. For fixed k, the number of links is at most $\frac{1}{2}k(k-1)$, and $\frac{1}{2}qk$ for a finite coordination number q. Each coefficient is a polynomial in u.

Thus equation (72) can be recast in the form

$$\frac{1}{N} \ln Z = H + \beta d + z + z^2 (du - \tfrac{1}{2}) + z^3 [d(2d-1)u^2 - 2du + \tfrac{1}{3}] + \mathcal{O}(z^4) \tag{84}$$

in direct correspondence with the graphs of figure 8.

An application of these series is the study of the Lee–Yang edge singularity, as discussed in Chapter 3 (volume 1).

7.2.6   Fermionic fields

The extension of the diagrammatic rules to Grassmannian integrals is straightforward, with due care paid to antisymmetry properties (chapter 2, volume 1). Let us give an elementary example by rederiving a known result on Gaussian integrals through perturbative means.

Consider the integral over $\{q_i, \bar{q}_i\}$ anticommuting variables with a source term $\{\eta_i, \bar{\eta}_i\}$, itself anticommutative

$$
Z(\bar{\eta}, \eta) = \int \exp\left(m \sum_{i,\alpha} \bar{q}_i^\alpha q_i^\alpha\right) \left[\frac{\mathrm{d}q\,\mathrm{d}\bar{q}}{m}\right]
$$
$$
\times \exp\left(-\bar{q}_i^\alpha \Delta_{ij}^{\alpha\beta} q_j^\beta + \bar{\eta}_\alpha^i q_i^\alpha + \bar{q}_i^\alpha \eta_\alpha^i\right)
$$
$$
= \det(m - \Delta)\exp\bar{\eta}(m - \Delta)^{-1}\eta \tag{85}
$$

To obtain the corresponding expansion, the mass term is essential and is kept as part of the measure. The second term in the exponential is expanded as an infinite series, and the resulting integrals are interpreted diagrammatically. Each term $\Delta_{ij}^{\alpha\beta}$ is represented by an ordered link, from site i to site j say. When integrating over $q_i^\alpha, \bar{q}_i^\beta$ the only nonvanishing terms are those with a factor 1 or a factor $\bar{q}_i^\alpha q_i^\alpha$. As a consequence the diagrams are either closed fermionic oriented loops, or arcs extending from a "source" $\bar{\eta}$ to a "sink" η. A loop $(i_1 i_2 \cdots i_n i_1)$ yields a factor $-\operatorname{Tr}\left(\Delta_{i_1 i_2} \cdots \Delta_{i_n i_1}\right)/nm^n$. The trace is over the internal indices $(\alpha\beta)$ of the matrices $\Delta_{ij}^{\alpha\beta}$. A minus sign is assigned to each fermionic loop, and arises from the anticommutativity of the variables $q, \bar{q}$ when they are reordered in the standard fashion $\bar{q}_i^\alpha q_i^\alpha$ to obtain a unit integral. The additional factor $1/n$ is a consequence of an incomplete cancellation among symmetry factors.

In principle, we have to reject those diagrams which intersect at a site i with the same internal index. This exclusion rule can in fact be ignored using the cumulant method as we saw previously. Due to the identity

$$
\exp m\bar{q}q = 1 + m\bar{q}q
$$

specific to Grassmannian variables, the diagrammatic rules for cumulants are identical to the rules in the (nonconnected) initial expansion.

> Check that the sum over diagrams with more than one line (with the same internal index α) incident on site i vanishes.

Another consequence of the trivial transformation to cumulants is the immediate exponentiation of the expansion. In the absence of sources, $\ln Z$ is the sum over all diagrams with a unique loop. The only connected terms involving sources are those with a single arc from $\bar{\eta}_i$ to η_j.

We have thus reconstructed the expansion in Δ/m of

$$\ln Z(\bar{\eta}, \eta) = \operatorname{Tr} \ln \left(1 - \frac{\Delta}{m}\right) + \frac{1}{m}\bar{\eta}\left(1 - \frac{\Delta}{m}\right)^{-1}\eta \qquad (86)$$

Further examples will be presented in chapter 10. We also quote here that in three dimensions the Ising model admits a nontrivial Grassmannian presentation, with quartic interaction terms, which leads to elaborate expansion rules, loosely referred to as fermionic surfaces.

7.3 Enumeration of graphs

7.3.1 Configuration numbers with exclusion constraint

We have already observed that the notion of free graph enables one to factorize a contribution $n(G)$ involving only the geometry of the lattice. This a function of the lattice size N. If one uses, as is the most frequent case, periodic boundary conditions, a connected diagram will have a configuration number $n(G)$ proportional to N because of translational invariance if N is large enough to exclude special configurations which would take advantage of periodicity to include nontrivial loops. Even though we do not include these terms in the present context, their study is a legitimate part of the theory, and might reveal interesting finite size effects. More generally $n(G)$ will be a polynomial in N, the degree of which is equal to the number of disconnected parts of the graph. The term of degree 1 in N is of particular importance, since it is the only one that survives in the computation of $\ln Z$, an extensive quantity proportional to N.

To compute these configuration numbers, we can restrict ourselves to simple graphs, since a multiple link does not affect the number of configurations. Moreover the computations can be reduced to those pertaining to certain multiply connected diagrams using the following property. Consider the set of graphs $G_1 \cup G_2 \equiv \{G\}$ that can be obtained starting from two given ones G_1 and G_2 by identifying certain of their elements. If $k(G; G_1, G_2)$ denotes the number of ways in which G can be split into G_1 and G_2, the following formula holds

$$n(G_1)n(G_2) = \sum_{G=G_1 \cup G_2} k(G; G_1, G_2)n(G) \qquad (87)$$

We now apply this formula in several instances.

(a) Reduction to connected diagrams

If G_1 and G_2 are free graphs with c_1 and c_2 connected parts respectively, G will have at most $c_1 + c_2$ connected parts, and only one of the graphs in $G_1 \cup G_2$ will have exactly this number, namely the graph where one does not identify any element in both graphs. Hence for instance the equation with two triangular graphs

$$n(\triangle)^2 = 2n(\triangle\,\triangle) + 2n(\triangle\!\triangle) + 2n(\,\diamondsuit\,) + n(\triangle) \qquad (88)$$

allows one to compute $n(\triangle\ \triangle)$ in terms of configuration numbers for connected graphs. Similar formulas are useful for gauge theories where graphs have the topology of surfaces. As an example for two cubes

$$n(\,\boxdot\,)^2 = 2n(\,\boxdot\boxdot\,) + 2n(\ \boxdot\!\boxdot\ ) + n(\boxdot) \qquad (89)$$

(b) Reduction to multiply connected graphs

A similar method applies when G_1 and G_2 have a common root i. In the graphs occurring on the r.h.s. of equation (87), a unique one will have an articulation point i. Thus

$$n(\Delta_i)^2 = 2n(\underset{i}{\bowtie}) + 2n(\underset{i}{\diamondsuit}) + n(\Delta_i) \qquad (90)$$

When one sums the number of configurations of a graph over the position of the root i, one obtains Np times the number of configurations of the free graph, where p is the number of vertices equivalent to the rooted one under automorphisms of the graph.

Table I. Reduced number of configurations $\bar{n}(G)$ for loop graphs on triangular, square, cubic, body centered cubic and face centered cubic lattices.

ring	triangular	square	cubic	b.c.c.	f.c.c.
3	2	—	—	—	8
4	3	1	3	12	33
5	6	—	—	—	168
6	15	2	22	148	970
7	42	—	—	—	6168
8	123	7	207	2736	42069
9	380	—	—	—	301376
10	1212	28	2412	61896	2241420
11	3966	—	—	—	17173224
12	13265	124	31754	1579324	134806948
13	45144	—	—	—	1079802216
14	155955	588	452640	43702920	8798329080
15	545690	—	—	—	$\cdots$
16	1930635	2938	6840774	1282524918	
17	6897210	—	—	$\cdots$	
18	24852576	15268	108088232		
20	$\cdots$	81826	1768560270		
22		449572	$\cdots$		
24		2521270			
26		14385376			

The previous formula in example (90) becomes

$$[3Nn(\Delta)]^2 = 2Nn(\bowtie) + 4Nn(\diamondsuit) + 3Nn(\Delta) \qquad (91)$$

(c) Reduction along a link

One can furthermore envision the case where G_1 and G_2 have a common link ij, which allows a reduction of graphs which can be split along a link. Thus for example

$$n(_i\Delta_j)^2 = 2n(\overset{j}{\underset{i}{\diamondsuit}}) + n(_i\Delta_j) \qquad (92)$$

Since for coordination q, there are $\frac{1}{2}Nq$ links, one obtains

$$[3Nqn(\Delta)]^2 = 2Nqn(\diamondsuit) + 3Nqn(\Delta) \qquad (93)$$

These techniques allow one to derive configuration numbers from those of a smaller set of graphs for which simple and general methods are not available. A case by case study is then unavoidable. In the following tables, we show some of these enumerations. Table I counts closed loops. Table II gives configuration numbers for some irreducible diagrams of a more complex structure on standard lattices (triangular and square two dimensional lattices, simple cubic, body centered cubic, and face centered cubic tridimensional ones). Finally table III gives configuration numbers for lattice gauge theories on a d-dimensional hypercubic lattice.

7.3.2   Multiply connected graphs

The preceding method reveals itself very quickly rather un-tractable because it generates a large number of terms. It is in particular necessary to compute configuration numbers for all graphs, multiply connected or not. It is of course possible to use general techniques such as cumulants or Legendre transforms. But the price is the loss of simplicity for graphs, even if the original ones admitted this restriction. Let us present a third technique, which requires suitable modifications according to the lattice and the model.

Let $A(G)$ and $B(G)$ be two families of weights satisfying

$$A(G) = \sum_{g \subseteq G} B(g) \tag{94}$$

and such that if G is the disjoint union $G_1 \cup G_2$, one has

$$g_1 \subseteq G_1, g_2 \subseteq G_2 \quad \Rightarrow \quad A(g_1 \cup g_2) = A(g_1) + A(g_2) \tag{95}$$

It then follows that the corresponding $B(G) = 0$, for G discon-nected.

The proof relies on the inversion formula

$$(-1)^{p(G)} B(G) = \sum_{g \subseteq G} (-1)^{p(g)} A(g) \tag{96}$$

which is proved by induction on the number of links $p(G)$ of G.

To prove formula (96), one uses relations (94) and (96) for any graph g strictly included in G (as a recursive hypothesis). The coefficient of $B(g)$ is then $\sum_{G \supset g' \supseteq g} (-1)^{p(g')}$. Every pair g', g'' which differ only by a single fixed link not belonging to g, gives no contribution in this sum. The only remaining contribution is the one of the only graph g' equal to G with the fixed link deleted. This proves recursively equation (96).

Table II. Reduced number of configurations $\bar{n}(G)$ for some low order irreducible diagrams.

Diagram	triangular	square	cubic	b.c.c.	f.c.c.
	—	—	—	12	36
	--	--	---	--	2
	—	—	—	—	600
	---	--	--	—	48
	6	—	24	480	3888
	3	—	---	--	3132
	—	—	—	24	
	--	--	--	3	9
	---	--	—	—	192
	—	20	436		

The graphs g contributing to (96) can be subdivided into three categories: $g \subseteq G_1, g \subseteq G_2$ and those of the form $g = g_1 \cup g_2$ with $g_1 \subseteq G_1, g_2 \subseteq G_2$ both nonempty. The proof of the above proposition is then straightforward.

Let us apply this result to a lattice limited to the sites and links of a given graph. Choose $A(G) = \ln Z(G)$. The hypotheses are

Table III. Some graphs in lattice gauge theories and their configuration numbers.

6.1		$\frac{1}{6}d(d-1)(d-2)$
10.1		$\frac{1}{2}d(d-1)(d-2)(2d-5)$
11.1		$\frac{1}{2}d(d-1)(d-2)(2d-5)$
12.1		$-\frac{1}{12}d(d-1)(d-2)(12d-29)$
12.2		$\frac{4}{3}d(d-1)(d-2)(d-3)$
12.3		$\frac{1}{3}d(d-1)(d-2)(d-3)$
14.1		$\frac{1}{2}d(d-1)(d-2)(2d-5)^2$
14.2		$2d(d-1)(d-2)(4d^2-22d+31)$
14.3		$d(d-1)(d-2)(d-3)$
14.4		$4d(d-1)(d-2)(d-3)$
15.1		$\frac{1}{3}d(d-1)(d-2)(d-3)(2d-5)$
15.2		$\frac{1}{3}d(d-1)(d-2)(d-3)$
15.3		$\frac{4}{3}d(d-1)(d-2)(d-3)$

Table III. *(continued)*

15.4		$d(d-1)(d-2)(2d-5)^2$
15.5		$4d(d-1)(d-2)(4d^2-22d+51)$
16.1		$\frac{1}{2}d(d-1)(d-2)(4d^2-24d+37)$
16.2		$\frac{1}{8}d(d-1)(d-2)(d-3)$
16.3		$8d(d-1)(d-2)(d-3)^2$
16.4		$16d(d-1)(d-2)(d-3)^2$
16.5		$16d(d-1)(d-2)(d-3)^2$
16.6		$\frac{1}{3}d(d-1)(d-2)(d-3)(2d-5)$
16.7		$-d(d-1)(d-2)(22d^2-113d+147)$
16.8		$2d(d-1)(d-2)(d-3)$
16.9		$\frac{1}{2}d(d-1)(d-2)(2d-5)^2$
16.10		$2d(d-1)(d-2)(4d^2-22d+31)$
16.11		$4d(d-1)(d-2)(d-3)$

justified when G is disconnected. As a result, for the complete lattice one has

$$\ln Z = \sum_{g \text{ connected}} n(g)B(g) \tag{97}$$

where the contributions $B(g)$ follow directly from equation (94).
To lowest orders with graphs up to three links,

$$
\begin{aligned}
\ln Z(/) &= B(/) \\
\ln Z(\mathsf{V}) &= B(\mathsf{V}) + 2B(/) \\
\ln Z(\mathsf{Z}) &= B(\mathsf{Z}) + 2B(\mathsf{V}) + 3B(/) \\
\ln Z(\mathsf{Y}) &= B(\mathsf{Y}) + 3B(\mathsf{V}) + 3B(/) \\
\ln Z(\triangle) &= B(\triangle) + 3B(\mathsf{V}) + 3B(/)
\end{aligned}
\tag{98}
$$

$$\cdots$$

This procedure is well adapted to automatic implementation on a computer. The inversion formula (96), valid for labelled graphs, is not true for free graphs.

In a second stage one may apply this same proposition, for articulated graphs, where the property indicated in equation (95) remains valid.

7.4   Results and analysis

7.4.1   *Series analysis*

The expansion techniques described in this chapter provide us with truncated series, up to an order n, of the form

$$f(z) = \sum_{k=0}^{n} f_k z^k + \mathcal{O}(z^{n+1}) \tag{99}$$

One is interested in two types of questions.

(1) Find the singularities z_i of the function, and study the behaviour of f in their vicinity. In particular one would like to know whether it is likely that the behaviour is of the form

$$f(z) \sim A_i(z - z_i)^{\gamma_i} \tag{100}$$

and, if this is the case, what is the exponent γ_i, and if possible the amplitude A_i.

(2) Extrapolate the function outside its circle of convergence. For instance, one may be interested by the limiting value as z reaches infinity in a given direction. It is clear that these are *a priori* unsoluble problems, unless one is given further information: regularity of the coefficients f_k, knowledge of the analyticity domain... in which case one will try to exploit it, using a method appropriate to each specific instance. Unfortunately these data are frequently missing, and one can only proceed blindly, using different techniques and comparing their results. We shall only give here a brief survey, referring the reader to specialized treatments.

It must be pointed out parenthetically that a vast knowledge has been accumulated on the large order behaviour of the continuous field theoretic perturbation series, originating in the seminal work of Bender and Wu (for the anharmonic oscillator) and Lipatov (for its transposition to genuine field theories). This information has provided a very important tool in the analysis of such series. Nothing as systematic is known in the case of lattice series, although isolated results exist in some instances, reflecting in particular the nature of the lattice.

The most straightforward technique is the *ratio method* and its refinements. If the coefficients f_k are real and the nearest singularity z_0 is unique, isolated, and "algebraic" (i.e. of the form (100)), the ratio of successive coefficients behaves as

$$\frac{f_k}{f_{k+1}} = z_0 \left[1 + \frac{1+\gamma}{k} + \mathcal{O}(\frac{1}{k^2}) \right] \tag{101}$$

The study of the sequence of ratios yields both z_0 and γ. One can profitably use such accelerators of convergence as the Neville tables in this context.

The Neville tables arise from a practical display of Aitken's iteration formulas. Let $f\,(x\,|x_0, \ldots, x_k)$ denote the unique polynomial of degree k which coincides with $f(x)$ at the points $x_0, \ldots, x_k$, the formulas

$$f(x \mid x_i, \ldots x_{i+n}) = \left| \begin{matrix} f(x \mid x_i, \ldots x_{i+n-1}) & x - x_i \\ f(x \mid x_{i+1}, \ldots, x_{i+n}) & x - x_{i+n} \end{matrix} \right| \Big/ (x_i - x_{i+n}) \tag{102}$$

allow a recursive construction of interpolating expressions on 2, 3,...
points. When this is applied to a sequence $\{u_n\}$ with an asymptotic
behaviour which is assumed (or known) to be polynomial in $1/n$,
one is led to consider the set of sequences

$$
\begin{cases}
u_n^{(1)} = (n+1)u_{n+1} - nu_n = \lim u_n + \mathcal{O}\left(\dfrac{1}{n^2}\right) \\[2mm]
u_n^{(2)} = \frac{1}{2}\left\{(n+2)u_{n+1}^{(1)} - nu_n^{(1)}\right\} = \lim u_n + \mathcal{O}\left(\dfrac{1}{n^3}\right) \\[2mm]
u_n^{(3)} = \frac{1}{3}\left\{(n+3)u_{n+1}^{(2)} - nu_n^{(2)}\right\} = \lim u_n + \mathcal{O}\left(\dfrac{1}{n^4}\right) \\[2mm]
\cdots
\end{cases}
\tag{103}
$$

In practice, one generally observes first an improved convergence,
then, as the order increases, an instability (the correction terms
with an increasing power in $1/n$, might have increasing coefficients).

Problems occur when the number of available terms in the series
is unsufficient for the singularity to impose its behaviour on the
coefficients. Also such a method is tailored to deal with a unique
singularity.

Padé approximants provide a powerful means of investigating
several singularities at once, with larger convergence domains.
The idea is to approximate an analytic function by a sequence of
rational ones. More precisely the Padé approximant $[L/M]\,(z) \equiv
P_L(z)/Q_M(z)$ is the unique rational fraction, a ratio of two
polynomials of degree L and M respectively, defined up to a
common multiple, such that its Taylor series up to degree $L + M$
agrees with the known expansion of $f(z)$

$$
f(z) - P_L(z)/Q_M(z) = \mathcal{O}(z^{L+M+1})
\tag{104}
$$

Equation (104) yields $(L+M+1)$ equations on $(L+1)+(M+1)-1$
unknowns. Padé approximants are by definition particularly suitable for reconstructing meromorphic functions. Experimentally,
one observes that they simulate branch points quite well, by accu-
mulating alternatively zeroes and poles on the branch cut. They
are well suited for obtaining analytic continuation beyond the natural circle of convergence of Taylor series. Difficulties may sometimes occur with an erratic convergence, in addition to the occurrence of spurious singularities.

When one assumes an isolated power law singularity, meaning that $(z - z_0)^{-\gamma} f(z)$ is regular at $z = z_0$, it is better to apply Padé approximants to the logarithmic derivative

$$\frac{1}{f}\frac{df}{dz} = \frac{\gamma}{z - z_0} + \text{regular part}$$

with an isolated pole. In practice, nonfactorizable singularities might occur $(A(z)(z - z_0)^{\gamma} + B(z))$ as well as confluent singularities of the form $(A(z - z_0)^{\gamma} + B(z - z_0)^{\beta} + \cdots$, and even worse. The method looses its accuracy, leading possibly to a reasonable location of z_0, with an unstable estimate of the exponent γ.

Integral approximants generalize the preceding technique. These are solutions of a differential equation of a suitable type, with polynomial coefficients determined in such a way that, up to the order to which it is known, $f(z)$ satisfies the equation. For instance, the first order linear differential equation

$$Q_K(z)f'(z) + P_M(z)f(z) + R_L(z) = \mathcal{O}(z^{K+L+M+2}) \qquad (105)$$

yields standard Padé approximants for $Q_k \equiv 0$ ($k \equiv -1$), and those of the logarithmic derivative, when $R_L \equiv 0$. These approximants are singular at the zeroes of Q, and behave as

$$A(z - z_0)^{\gamma} + B$$

$$\gamma = -\frac{P(z_0)}{Q'(z_0)} \qquad\qquad B = -\frac{R(z_0)}{P(z_0)}$$

These approximants are in general suited to take into account subdominant terms in the vicinity of singularities, and yield better values for the exponents. The trouble is the proliferation in the number of possible approximants, with large deviations between various ones, leading to a difficult choice between several estimates. These integral approximants admit a further generalization to several variables, and hence allow the study of multi-critical points.

Information on the *analytic properties* of the function f, can be usefully combined with the preceding techniques. Let us give an example where it is assumed that at a *known* value z_c, the function $f(z)$ has an accumulation of singularities of the type

$$f(z) = \sum_{n,p=0}^{\infty} C_{np}\left(1 - \frac{z}{z_c}\right)^{\gamma_n} \ln^p\left(1 - \frac{z}{z_c}\right) \qquad (106)$$

Then $f(z)$ admits an integral representation

$$f(z) = \int \left(1 - \frac{z}{z_c}\right)^{\lambda} \sigma(\lambda)\, d\lambda \qquad (107)$$

and its transform

$$g(t) = \int \frac{\sigma(\lambda)}{1 - \lambda t}\, d\lambda = \sum_{n,p=0}^{\infty} C_{np} \frac{\gamma_n^p p!}{(1 - t\gamma_n)^{p+1}} \qquad (108)$$

is a meromorphic function in t. It is not difficult to relate the Taylor coefficients of $g(t)$ to those of $f(z)$. Padé approximants, ill-adapted to the confluent singularities of $f(z)$, are on the contrary very well suited to the analysis of the meromorphic structure of $g(t)$. This is the so-called *Padé–Mellin* method.

In short, it is clear that extracting exponents and amplitudes from a limited perturbative series is an art form. Except in a few simple cases, it requires sophisticated expertise.

7.4.2 An example: the Ising series
on a body centered cubic lattice

We cannot of course show here all the series obtained in a great variety of models. We have given in chapter 6 (volume 1) examples for lattice gauge models. Here we limit ourselves to quoting a few results pertaining to the three-dimensional Ising model on a body centered cubic lattice, a system for which rather long series are known. Other results will be found in the references.

The high temperature series in the variable $t \equiv \tanh \beta$ for the free energy per site is

$$\frac{1}{N} \ln Z = \ln 2 - 4\ln(1+t) + 12t^4 + 148t^6 + 2496t^8 + 52168t^{10} +$$

$$+ 1242078t^{12} + 32262852t^{14} + 892367762t^{16} + \mathcal{O}(t^{18})$$
$$(109)$$

By taking successive derivatives, this series allows the internal energy and the specific heat to be found. One can in this way obtain the exponent α.

The correlation function $\langle \sigma_0 \sigma_{\mathbf{x}} \rangle$ depends on the (vector) variable $\mathbf{x}$. The corresponding high temperature diagrams have odd vertices in $\mathbf{0}$ and $\mathbf{x}$. One is interested in the behaviour at large distance, or, equivalently, for small wavenumber of its Fourier transform,

$$\tilde{\chi}(\mathbf{k}) = \sum e^{i\mathbf{k}\cdot\mathbf{x}} \langle \sigma_0 \sigma_{\mathbf{x}} \rangle = \chi - \frac{\mu_2}{2d}\mathbf{k}^2 + \mathcal{O}(\mathbf{k}^4) \qquad (110)$$

Here χ is the static susceptibility. The lattice anisotropy shows up in fourth order terms in $\mathbf{k}$. If one ignores these anisotropies, one can study alternatively the spherical moments

$$\mu_n = \sum_{\mathbf{x}} |\mathbf{x}|^n \langle \sigma_0 \sigma_{\mathbf{x}} \rangle \qquad (111)$$

which are obviously convergent in the high temperature phase, where the correlation function decreases exponentially. In particular $\chi = \mu_0$ diverges at the critical temperature with an exponent γ. The effective correlation length $\xi_{eff}^2 = 2d\chi/\mu_2$ diverges with the exponent ν. For the body centered cubic lattice, the following series are known

$$
\begin{aligned}
\chi = 1 &+ 8t + 56t^2 + 392t^3 + 2648t^4 + 17864t^5 + 118760t^6 \\
&+ 789032t^7 + 5201048t^8 + 34268104t^9 + 224679864t^{10} \\
&+ 1472595144t^{11} + 9619740648t^{12} + 62823141192t^{13} \\
&+ 409297617672t^{14} + 2665987056200t^{15} + 17333875251192t^{16} \\
&+ 112680746646856t^{17} + 731466943653464t^{18} \\
&+ 4747546469665832t^{19} + 30779106675700312t^{20} \\
&+ 199518218638233896t^{21} + \mathcal{O}(t^{22})
\end{aligned}
$$

$$(112)$$

and

$$
\begin{aligned}
\mu_2 = 8t &+ 128t^2 + 1416t^3 + 13568t^4 + 119240t^5 \\
&+ 992768t^6 + 7948840t^7 + 61865216t^8 + 470875848t^9 \\
&+ 3521954816t^{10} + 25965652936t^{11} + 189180221184t^{12} \\
&+ 1364489291848t^{13} + 9757802417152t^{14} + 69262083278152t^{15} \\
&+ 488463065172736t^{16} + 3425131086090312t^{17} \\
&+ 23896020585393152t^{18} + 165958239005454632t^{19} \\
&+ 1147904794262960384t^{20} + 7910579661767454248t^{21} \\
&+ \mathcal{O}(t^{22})
\end{aligned}
$$

$$(113)$$

Observe that both series have integral positive coefficients, so that we are sure of the existence of a nearest singularity for positive t.

Let us illustrate series analysis using the example of the susceptibility. We face two difficulties. One arises from the symmetries of the lattice which admits a bipartition in two simple cubic sublattices, with the spins on one sublattice interacting only with those of the second. If one changes the sign of all spins on one sublattice and β into $-\beta$ (or t into $-t$), the partition function is invariant, as is clear from equation (109) apart from a normalization term in $\ln(1 + t)$. This symmetry leads to an antiferromagnetic transition at $-t_c$. The susceptibility can be split into even and odd parts, according to the contributions of pairs of spins located on the same or different sublattices. The odd part is related to the even one by increasing each spatial contribution by a unit step. This amounts loosely to multiplying it by the internal energy, with an exponent $\alpha - 1$. One thus expects that the coefficients χ_n in the series (112) behave as

$$\chi_n \approx An^{\gamma-1}t_c^{-n}[1 + B(-1)^n n^{\alpha-\gamma+1}] \qquad (114)$$

It is then natural to split the series into even and odd parts in t, and to consider the quantities

$$y_n = -\left[\ln\left(\frac{\chi_n \chi_{n-4}}{\chi_{n-2}^2}\right)\right]^{-1} \approx \frac{n^2}{\gamma - 1} \qquad (115)$$

The exponent γ is the limit of the sequence

$$\gamma_n = 1 + 2\frac{y_n + y_{n-1}}{(y_n - y_{n-1})^2} \qquad (116)$$

The second difficulty is the existence of a confluent singularity predicted by the renormalization group. The expected correction to formula (114) is of the form $(1 + cn^{-\Delta_1})$, where $\Delta_1 = \omega\nu$. As a consequence, one cannot efficiently use the Neville tables to accelerate the convergence of the sequences. Indeed one observes strong instabilities if one attempts to use these techniques. However, if t_c is sufficiently accurately determined, one can use the Padé–Mellin method. Yet another possibility is to use the approximate value $\Delta_1 \simeq 0.50$ to suppress this singularity. In the analysis carried out by Zinn-Justin, one introduces the average $\bar{y}_n = (y_n + 2y_{n-1} + y_{n-2})/4$ and considers the quantity

$$Y_n = \left(\frac{\bar{y}_n^{1/2\Delta_1} - \bar{y}_{n-2}^{1/2\Delta_1}}{\Delta_1}\right)^{2/(\Delta_1-1)} \qquad (117)$$

which replaces (115) by eliminating entirely the correction due to the first confluent singularity.

From the series, a consistent set of data on the singularities can be extracted in the form

$$t_c^{-1} = 6.3543$$
$$\gamma = 1.239 \pm 0.0025$$
$$\nu = 0.6305 \pm 0.0015 \tag{118}$$
$$\eta = 0.035 \pm 0.003$$

The effective correlation length computed as indicated above using the second moment μ_2, does not coincide with the "true" correlation length defined as

$$\xi(\hat{\mathbf{x}})^{-1} = -\lim_{|\mathbf{x}|\to\infty} \frac{\ln \langle \sigma_0 \sigma_{\mathbf{x}} \rangle}{|\mathbf{x}|} \tag{119}$$

for which one can also obtain high temperature series. The pitfalls in analysing this quantity have been already discussed in chapter 6 (volume 1, section 3.5).

The low temperature series are more irregular on this same b.c.c. lattice, due to the existence of spurious complex singularities, and hence lead to a less trustworthy determination of the critical exponents.

Using as a variable $u = \exp{-4\beta}$, the free energy per site is given by

$$
\begin{aligned}
\ln Z = {} & u^4 + 4u^7 - 4\tfrac{1}{2}u^8 + 28u^{10} - 64u^{11} + 48\tfrac{1}{3}u^{12} + 204u^{13} - 786u^{14} \\
& + 1164u^{15} + 922\tfrac{3}{4}u^{16} - 8760u^{17} + 20032u^{18} - 9164u^{19} \\
& - 84215\tfrac{4}{5}u^{20} + 294677\tfrac{1}{3}u^{21} - 378996u^{22} - 569704u^{23} \\
& - 54012882u^{24} + 112640896u^{25} - 5164464u^{26} \\
& - 694845120u^{27} + 2160781086u^{28} + \mathcal{O}(u^{29})
\end{aligned}
\tag{120}
$$

The pattern of minus signs is suggestive of the complex nearby singularities mentioned above. We use here an abbreviated notation, for instance $4\tfrac{1}{2}$ instead of $(4 + \tfrac{1}{2})$, to avoid huge numerators in the fractions.

The spontaneous magnetization $\langle \sigma \rangle$ is given by the series

$$
\begin{aligned}
\langle \sigma \rangle = {} & 1 - 2u^4 - 16u^7 + 18u^8 - 168u^{10} + 384u^{11} - 314u^{12} - 1632u^{13} \\
& + 6264u^{14} - 9744u^{15} - 10014u^{16} + 86976u^{17} - 205344u^{18} \\
& + 80176u^{19} + 1009338u^{20} - 3579568u^{21} + 4575296u^{22} \\
& + 8301024u^{23} + 3832961\tfrac{1}{2}u^{24} - 7942796u^{25} + 1118118u^{26} \\
& + 43016052u^{27} - 133595088\tfrac{6}{7}u^{28} + \mathcal{O}(u^{29})
\end{aligned}
\tag{121}
$$

Using $\langle\sigma\rangle$, it is possible to compute the connected correlation function $\langle\sigma_0\sigma_x\rangle_c = \langle\sigma_0\sigma_x\rangle - \langle\sigma\rangle^2$, and to obtain the corresponding low temperature susceptibility

$$
\begin{aligned}
\tfrac{1}{4}\chi = {}& u^4 + 16u^7 - 18u^8 + 252u^{10} - 576u^{11} + 519u^{12} + 3264u^{13} \\
& - 12468u^{14} + 20568u^{15} + 26662u^{16} - 215568u^{17} + 528576u^{18} \\
& - 164616u^{19} - 3014889u^{20} + 10894920u^{21} - 13796840u^{22} \\
& - 2990961u^{23} + 190423962u^{24} - 399739840u^{25} - 22768752u^{26} \\
& + 2803402560u^{27} - 8743064909u^{28} + \mathcal{O}(u^{29})
\end{aligned}
\tag{122}
$$

Analyzing the poles and residues of the Padé approximants of the logarithmic derivative of these series allows one to obtain the critical exponents within the ordered phase. However, these results are less reliable than those obtained from high temperature series, due to more important fluctuations in the estimates. One finds that the equality $\gamma' = \gamma$ is approximately satisfied. At this stage the low temperature exponents are given by

$$
\begin{cases}
\alpha' \simeq 0 \\
\beta = 0.312 \pm 0.05 \\
\gamma' = 1.25 \pm 0.05
\end{cases}
\tag{123}
$$

Let us conclude by giving the strong field expansion of the free energy in the variables $\mu = \exp(-2\beta H)$ and $u = \exp(-4\beta)$

$$
\begin{aligned}
\ln Z = {}& u^4\mu + (4u^7 - 4\tfrac{1}{2}u^8)\mu^2 + (28u^{10} - 64u^{11} + 36\tfrac{1}{3}u^{12})\mu^3 + \\
& (12u^{12} + 204u^{13} - 798u^{14} + 948u^{15} - 366\tfrac{1}{4}u^{16})\mu^4 \\
& + (12u^{14} + 216u^{15} + 1262u^{16} - 9072u^{17} + 17592u^{18} \\
& - 14184u^{19} + 4174\tfrac{1}{5}u^{20})\mu^5 + (27u^{16} + 312u^{17} \\
& + 2368u^{18} + 4312u^{19} - 92992u^{20} + 275021\tfrac{1}{3}u^{21} \\
& - 353640u^{22} + 216036u^{23} - 51444\tfrac{1}{2}u^{24})\mu^6 \\
& + (72u^{18} + 704u^{19} + 4404u^{20} + 17616u^{21} - 36348u^{22} \\
& - 833064u^{23} + 3795726u^{24} - 7072736u^{25} + 6798900u^{26} \\
& - 3344712u^{27} + 669438\tfrac{1}{7}u^{28})\mu^7 + \mathcal{O}(\mu^8)
\end{aligned}
\tag{124}
$$

Notes

The subjects treated in this chapter are rather technical and are covered in numerous papers. Some reviews and references can be found in the series *Phase Transitions and Critical Phenomena*, vol. 3 entitled *Series Expansions for Lattice Models*, C. Domb and M.S. Green eds, Academic Press (1974), and more recently in the *Proceedings of the 1980 Cargèse Summer Institute on Phase Transitions*, M. Levy *et al.* eds, Plenum (1982).

General graph theory is treated in C. Berge, *The Theory of Graphs*, Methuen (1962). The graph reduction technique is discussed in F. Englert, *Phys. Rev.* **129** 567 (1963), which contains additional references.

Series expansions, dealing mostly with the Ising model, are presented in C. Domb, *Adv. Phys.* **19**, 339 (1970) and in a sequence of papers by M.F. Sykes and collaborators, *J. Math. Phys.* **6** 283 (1965); *J. Phys.* **A5** 624, 640, 653, 661, 667 (1972); *J. Phys.* **A6** 1517 (1973); *J. Math. Phys.* **14** 1060, 1066, 1071 (1973); *J. Phys.* **A12** L25 (1979). See also S. McKenzie, *J. Phys.* **A8** L102 (1975) and **A12** L185 (1979), and B.G. Nickel in the proceedings of 1980 Cargèse summer institute quoted above. The problems specific to the spin–spin correlation functions are discussed in M.E. Fisher and R.J. Burford, *Phys. Rev.* **156** 583 (1967).

Padé approximants are described in a book by G.A. Baker, Jr, *Essentials of Padé Approximants*, Academic Press (1975). The analysis of confluent singularities using the Padé–Mellin method is due to G.A. Baker and D.L. Hunter, *Phys. Rev.* **B7** 3377 (1973). Examples of other techniques are found in J. Zinn-Justin, *J. Physique* **42** 783 (1981). For a generalization to several variables, see M.E. Fisher and J.H. Chen in the Proceedings of 1980 Cargèse summer institute quoted above, and J.H. Chen, M.E. Fisher and B.G. Nickel, *Phys. Rev. Lett.* **48** 630 (1982).

The work of R. Peierls on the existence of ordered phases at low temperature has been quoted in chapter 2 (volume 1). For the asymptotic freedom of the σ-model, see A.M. Polyakov, *Phys. Lett.* **B59**, 79 (1975), E. Brézin and J. Zinn-Justin, *Phys. Rev.* **B14**, 3110 (1976). Exact S-matrices for two dimensional models are due to A.B. Zamolodchikov and Al.B. Zamolodchikov, *Nucl. Phys.* **B133**, 525 (1978).

8

NUMERICAL SIMULATIONS

A new field opened when modern computers offered the possibility of performing extensive simulations of large systems. This allows known behaviours to be checked and provides an exploratory guide in circumstances where theoretical tools are absent. Measurements of observables can be compared both to existing theoretical expectations – providing a crucial test for their validity – and to experiments – checking the modelling of a physical system –. This chapter presents the background material needed to design simulations on a (relatively) large scale. Some numerical examples have already been presented in previous chapters, and we only give a few further illustrations, pertaining mainly to lattice gauge theory, the usefulness of which relies extensively on this method. We also describe a practical implementation of real space renormalization, known as the *Monte Carlo Renormalization Group* method (Ma, Swendsen, Wilson). Finally, we discuss specific issues relevant to the extension of the simulations to fermionic systems.

8.1   Algorithms

8.1.1   Generalities

Systems with up to 10^6 to 10^7 variables can be handled by computers, and these numbers may soon be significantly increased. The measurements can be sufficiently numerous to allow statistical accuracy. Although simulated systems still have a modest size as compared to macroscopic systems, collective effects already clearly appear and accurate results about critical phenomena emerge from the numerical simulations. It turns out, when investigating more closely the available numerical methods, that one gets a better insight into the foundations of equilibrium statistical physics, the ergodicity problems, the meaning of probabilities and, last but not least, ways to approach equilibrium. Although we seem to deal in this chapter with technical matters, the alert reader will realize that the subject in fact deserves the closest scrutiny and that experience can only be gained through practice.

The large number of variables forbids any *exact* numerical treatment, for instance by enumerating all possible configurations ($\sim 10^{8000}$ for a small Ising system $30 \times 30 \times 30$) and computing their contributions. There exists however a few attemps at deriving explicit expressions for very small lattices (with a $4 \times 4 \times 4$ Ising sytem, the number of configurations is already of macroscopic order 2.10^{19}! Luckily, symmetries simplify considerably the task of enumerating them). In practice, it is therefore necessary to use a representative sampling of the system. On each sample, one performs the desired measurements and the statistical analysis is done exactly as for an ordinary experiment. One has to face two additional difficulties. The first is to simulate nature, which freely provides a representative sampling of a system, and the second is to estimate the biases arising from small size.

A direct method would consist, using a random number generator, in choosing successively and independently the different degrees of freedom of the system and to assign to the resulting configuration its statistical weight. This weight is the product of the Boltzmann factor by the measure on the fields. A well-known application is the roulette wheel which ensures a positive income (without any risk) for casinos at the expense of players who ignore the laws of statistics, or want to ignore them. The application of this idea is at the origin of the name *Monte Carlo method* given to such simulations. The simple sampling is currently used to investigate percolation and fluid flows for instance, and yields accurate results in these cases.

However, simple random sampling is not practical for the type of systems we have in mind, generalized spin systems, gauge theories, random surfaces, etc., for which the Boltzmann weights relate several degrees of freedom and may generate nontrivial collective effects. The energy fluctuations are of order $1/\sqrt{N}$, where N is the system size. Configurations which are not energetically very close to the mean energy (or mean action), have a negligible effect in the computation of mean values. As the simple random sampling does not take into account this fact, it is necessary to introduce a bias. Let $\mathbf{x}$ represent the set of all variables characterizing the configuration of a system. A numerical simulation generates a sequence $\{\mathbf{x}_i\}$ of independent configurations randomly chosen according to the same probability distribution $P(\mathbf{x}_i)$. Using the notation of statistical physics, one replaces the computation of an average observable

$$\langle f \rangle = \frac{\sum_{\mathbf{x}} f(\mathbf{x}) \exp(-\beta \mathcal{H}(\mathbf{x}))}{\sum_{\mathbf{x}} \exp(-\beta \mathcal{H}(\mathbf{x}))} \tag{1}$$

by the evaluation of a sequence

$$f_n = \frac{\sum_{i=1}^{n} f(\mathbf{x}_i) \exp(-\beta \mathcal{H}(\mathbf{x}_i)) P^{-1}(\mathbf{x}_i)}{\sum_{i=1}^{n} \exp(-\beta \mathcal{H}(\mathbf{x}_i)) P^{-1}(\mathbf{x}_i)} \tag{2a}$$

and one checks that, in the infinite "time" limit $n \to \infty$, the random variables f_n tend towards the required quantity

$$\lim_{n \to \infty} f_n = \langle f \rangle \tag{2b}$$

up to errors of order $1/\sqrt{n}$.

Verify that if all configurations $\mathbf{x}_i$ are chosen independently with the same probability law $P(\mathbf{x})$, the time average $(1/n) \sum_{i=1}^{n} f(\mathbf{x}_i)$ tends to $\sum_{\mathbf{x}} P(\mathbf{x}) f(\mathbf{x})$. As a consequence check the validity of the limit (2b).

The Boltzmann factor can vary over several orders of magnitude. If $P(\mathbf{x})$ is almost constant, very few samples contribute to the sum in (2a) and the number n of iterations, i.e. the computing time, becomes enormous if a reasonable accuracy is to be achieved. However, it is possible to correct this misfortune by adjusting the probability $P(\mathbf{x})$. This leads to the idea of *importance sampling*, namely to adapt the bias probability to the quantities to be computed. The optimal choice is obviously the Boltzmann weight itself, $P(\mathbf{x}) = \exp(-\beta \mathcal{H}(\mathbf{x}))$.

Due to the interactions, each field (component of $\mathbf{x}$) cannot be selected independently from the others. A possibility is to use a Markov process which generates $\mathbf{x}_{i+1}$ from the knowledge of $\mathbf{x}_i$. Two successive configurations $\mathbf{x}_i$ and $\mathbf{x}_{i+p}$, separated by a finite "time" interval p, are thus correlated, but this is unimportant if this correlation decreases fast enough with p so that the probability distribution of $\mathbf{x}_{i+p}$ approaches, independently of

$\mathbf{x}_i$, the desired probability $P(\mathbf{x})$. Under these conditions, relation (2b) remains of course valid. Let $W(\mathbf{x}_i, \mathbf{x}_{i+1})$ be the conditional probability to select $\mathbf{x}_{i+1}$ starting from $\mathbf{x}_i$, thus defining the Markov process. It should satisfy the following properties.

i) *Normalization* $\sum_{\mathbf{y}} W(\mathbf{x}, \mathbf{y}) = 1$

ii) *Ergodicity* $W(\mathbf{x}, \mathbf{y}) > 0$. Indeed, one should be able to reach all configurations.

iii) *Limiting probability*

$$\sum_{\mathbf{x}} P(\mathbf{x})W(\mathbf{x}, \mathbf{y}) = P(\mathbf{y}) \tag{3}$$

The distribution W, considered as a matrix in the space spanned by configurations, has a unique maximal eigenvalue equal to 1 (according to conditions i and ii). Equation (3) indicates that the corresponding positive eigenvector is $P(\mathbf{x})$. Hence W^p applied on any state distribution probability transforms it into the equilibrium probability $P(\mathbf{x})$ as p increases indefinitely. The limiting distribution of the estimator f_n of $\langle f \rangle$ can also be easily obtained. For large n, this is of course the normal Gaussian probability $\exp[-(f_n - \langle f \rangle)^2/2\sigma_n^2]$, with a standard deviation

$$\lim_{n \to \infty} n\sigma_n^2 = \lim_{n \to \infty} \left(\frac{1}{n} \sum_{i=1}^{n} (f(\mathbf{x}_i) - \langle f \rangle)^2 \right) \tag{4}$$

Check this result, using Fourier transforms of probabilities.

The three conditions are far from being sufficient to determine in a unique way the conditional probability $W(\mathbf{x}, \mathbf{y})$. In practice, it is convenient to decompose W into elementary steps

$$W(\mathbf{x}, \mathbf{y}) = \sum_{\mathbf{z}_i} W_1(\mathbf{x}, \mathbf{z}_1)W_2(\mathbf{z}_1, \mathbf{z}_2) \cdots W_k(\mathbf{z}_{k-1}, \mathbf{y}) \tag{5}$$

Condition (3) is replaced by the more severe *detailed balance* condition

$$P(\mathbf{x})W_j(\mathbf{x}, \mathbf{y}) = P(\mathbf{y})W_j(\mathbf{y}, \mathbf{x}) \tag{6}$$

Relation (3) is then automatically fulfilled. It is easier to check (6) than (3), and the symmetry of this relation is often useful. Nevertheless, even with this additional condition, the indeterminacy in the choice of W remains large and leads to various possible algorithms.

In practical realizations, one proceeds in elementary steps which affect only a small number of degrees of freedom in the system. For instance, W_j just updates the field at location j and leaves the remaining fields unchanged. This is of interest because, for systems with short-range interactions, the amount of computation needed for this update remains limited and independent of the size of the system. The total computer time for a complete Monte Carlo "sweep" (i.e. a step of the Markov chain affecting the whole configuration) is therefore proportional to the size of the system. To be impartial, one should also mention the main defect of this process. It is difficult to excite long wavelength collective modes at low temperature and near criticality. The thermalization time needed to erase the memory of the initial configuration becomes relatively important. Moreover, in the critical domain, there are important correlations between successive steps (*critical slowing down*). This is equivalent to saying that there exist dynamical critical phenomena in the approach to equilibrium, as a counterpart to the static ones. However, this technique is presently the only well understood approach and one of the most efficient.

8.1.2   *The classical algorithms*

A very simple way to satisfy condition (6) is to make the choice

$$W_j(\mathbf{x}, \mathbf{y}) = P(\mathbf{y}) \propto \exp(-\beta \mathcal{H}(\mathbf{y})) \qquad (7)$$

This choice can be interpreted physically as a thermalization of a spin with a source at the desired temperature, the other parts of the system being frozen. The target spin chooses its final configuration according to the equilibrium temperature, taking into account the interactions with its neighbours which do not evolve. This procedure is known as the *heat bath* algorithm.

Compared to the initial problem of choosing the whole config-uration with a probability proportional to the Boltzmann weight, the process is considerably simplified, since the probability (7) depends only on the neighborhood of one site. In a practical implementation for the Ising model, one computes the local field $h_j = \sum_{k(j)} \sigma_k$ seen by the spin at site j. The spin is then randomly reset to ± 1 (independently of its previous value), with a probability $p_j = [1 + \exp(-2\beta h_j)]^{-1}$ to take the value $+1$ (in practice, one generates a random number uniformly distributed between 0 and 1 and the spin is set to $+1$ if this number is less than p_j, to -1 otherwise). The heat bath algorithm leads also to rapid calculations with more general discrete models.

However, and in particular for continuous systems, generating random numbers according to the law (7) is not straightforward, and may lead to rather long calculations which should be avoided. Let us describe a general method for continuous distributions. Define the function $F(y)$ according to

$$y = \int^{F(y)} P(x) \ dx \qquad (8)$$

One checks easily that, if y is a random variable uniformly distributed on $[0, 1]$, $x = F(y)$ is a random variable with a probability distribution $P(x)$. The function F is unfortunately difficult to obtain numerically, except in some very simple cases such as the normal distribution, the exponential law, etc. The inversion procedure required in computing F can be avoided using a classical algorithm which has numerous refinements. Through a simple transformation, the variable x can be assumed to take its value in the interval $[0, 1]$. The following crudest version of the algorithm generates x according to the probability $P(x)$.

1) Generate two independent random numbers x and y uniformly distributed on $[0, 1]$.
2) If $P(x) < (\mathrm{Sup}P)y$, return to step 1. Otherwise, x is the correctly distributed answer.

   Check the validity of this method and compute the frequency of rejections in step 2.

This algorithm generates two random numbers, computes $P(x)$, makes one comparison and performs these operations in a mean number of steps equal to $\mathrm{Sup}P$. Its efficiency increases as the distribution law becomes flatter. Numerous improvements allow its use with full efficiency for any distribution law; the interested reader will find a general reference in the notes.

A generalization of the above ideas has led Metropolis to propose an alternative algorithm to the heat bath method. The configuration **x** is again updated site by site according to the following rules.

M1) Generate a trial value for the field at a given site, chosen at random, leading to a possible configuration **y**. The probability law $Q(\mathbf{y})$ governing this choice is inessential (except for efficiency considerations) and may even depend on the previous value of the field.

M2) If $\mathcal{H}(\mathbf{y}) \leq \mathcal{H}(\mathbf{x})$, the trial value **y** is accepted and the algorithm terminates.

M3) If on the contrary the energy increases, $\mathcal{H}(\mathbf{y}) > \mathcal{H}(\mathbf{x})$, generate a random number λ uniformly distributed on $]0, 1[$. The trial configuration **y** is accepted if $\lambda < \exp -\beta(\mathcal{H}(\mathbf{y}) - \mathcal{H}(\mathbf{x}))$. Otherwise, the old configuration **x** is kept unchanged.

We observe that step M2 would be the same, if we tried to find the ground state of the system. It is in step M3 that an increase in energy which would be have been rejected is accepted with a temperature dependent probability. Check that the corresponding transition probability

$$W_j(\mathbf{x}, \mathbf{y}) \propto \qquad\qquad (9)$$

$$\begin{cases} Q(\mathbf{y}) & \text{if } \mathcal{H}(\mathbf{y}) \leq \mathcal{H}(\mathbf{x}), \ \mathbf{x} \neq \mathbf{y} \\ Q(\mathbf{y}) \exp -\beta(\mathcal{H}(\mathbf{y}) - \mathcal{H}(\mathbf{x})) & \text{if } \mathcal{H}(\mathbf{y}) > \mathcal{H}(\mathbf{x}) \\ \sum_{\mathbf{z}, \mathcal{H}(\mathbf{z}) > \mathcal{H}(\mathbf{x})} Q(\mathbf{z})(1 - \exp -\beta(\mathcal{H}(\mathbf{z}) - \mathcal{H}(\mathbf{x})) & \text{if } \mathbf{x} \equiv \mathbf{y} \end{cases}$$

obeys the conditions required for W as soon as the choice for $Q(\mathbf{y})$ satisfies the ergodicity condition. A weaker condition is to require that a finite power of W satisfies this criterion.

The probability $Q(\mathbf{y})$ is chosen in order to maximize (empirically) the efficiency of the method. For instance, a continuous field φ will be modified by a random quantity $\delta\varphi$ generated in such a way that the mean value $\langle\delta\varphi\rangle$ remains of the order of the fluctuation of the field in the entire system. If one were to generate frequently too large a shift, one would be led to a large rejection rate. Conversely, too small a shift would lead to a small rejection rate, but the system would only evolve very slowly. In both cases, numerous sweeps of the system would be necessary to obtain independent configurations and thermalization. In practice, a rejection rate of about 50% seems to yield good results.

It is possible to apply the algorithm to the same field several times before going to the next one. If the number of applications is large enough, one creates a secondary Markov chain which converges towards the probability (7). It is thus possible to interpolate between the Metropolis and the heat bath algorithm. This technique may provide a better efficiency.

(1) Computers produce a substitute for random numbers as a deterministic sequence with ergodic properties. In this respect, they deserve the name "pseudorandom" numbers. The classical generators are based on a linear congruential method. They yield a sequence x_n between 0 and 1 using the auxiliary integral dynamical system u_n defined through

$$\begin{cases} x_n = 2^{-p}u_n \\ u_{n+1} = Au_n \qquad [\text{mod } 2^p] \end{cases} \tag{10}$$

The numbers $\{x_n\}$ obtained in this fashion provide a "reasonable" pseudorandom sequence if the odd integer A is appropriately chosen. This algorithm is very easy to implement and is furthermore adapted to the structure of computers. However, some biases might appear. In large-scale simulations, it is a good idea to check the random number generators. For instance, the previous one may produce correlations between three consecutive numbers, so that a three-dimensional configuration may be considerably biased. It is therefore necessary to use a more sophisticated procedure. For instance, one extracts at random (using a sequence such as (10)) a number in a table which is renewed using another independent generator (the sequence (10) can again be used, with different A and u_0). In any case, a serious numerical simulation requires statistical tests checking the independency and the distribution of the configurations. Obviously, the replacement of a generator by another one should not modify the results.

(2) In certain instances, the fields can be cast into families such that two fields in the same family do not interact. As an example, for a model with nearest neighbour interactions on a (hyper)cubic lattice, sites can be classified according to the parity of the sum of their coordinates. It is essential to keep this in mind when programming vectorial computers or array processors. Indeed, these machines run very quickly on iterative loops such that

successive operations are independent. This is the case if one splits the loop over fields into two independent ones over even or odd sites. The organization of the storage should also be carefully studied. For instance, a spin of an Ising model can be represented as a single bit. These can be arranged into words (the unit of information in the computer) in such a way that they can be simultaneously treated by single logical instructions. These considerations are important in simulations, where time, storage and cost are the essential limitations.

(3) Another example of difficulty arises when sweeping the fields in succession leading in certain cases to a spurious stabilization of some states. Let us give one example. Consider a two-dimensional gauge theory with an action

$$S = \beta_1 \sum \sigma_1 \sigma_2 \sigma_3 \sigma_4 + \beta_2 \sum \sigma_1 \sigma_2 \sigma_3 \sigma_4 \sigma_5 \sigma_6 \qquad (11)$$

which uses rectangular loops with two adjacent plaquettes, and assume β_2 large and negative. This model is equivalent to an antiferromagnetic spin system in an external field. A ground state takes a chessboard aspect (figure 1) in which one out of two plaquette variables is set to -1. The reader can convince himself that, starting from a configuration with all plaquette variables equal to $+1$, a regular sweep of the link variables done line by line keeps the configuration unchanged. The same phenomenon appears if one starts from the chessboard configuration with β_2 large and positive (a ferromagnetic system). In both cases, the original state remains stable and it is very difficult to analyze the transition line (a first order one) between the two regions. Such a behaviour also appears in simulations of the eight vertex model (or equivalently, the one-dimensional quantum Heisenberg model, or also the two-dimensional Ising model with a four spin interaction). An improvement uses a random choice of the link to be updated, instead of sweeping regularly through the lattice. The results are, however, not yet satisfactory since the plane is partitioned into the two possible degenerate chessboard (staggered) configurations with boundaries which evolve very slowly. It is necessary to have an understanding of the stabilization process in order to overcome this difficulty. The transition between the pure gauge phase (ordered ferromagnetic) and the staggered phase (ordered antiferromagnetic) is expected to occur near $\beta_1 + 4\beta_2 \approx 0$. For definiteness, we choose β_1 positive and therefore β_2 negative. Flipping one link starting from one of the two ordered states costs respectively the energies $-2\beta_2$ and $-6\beta_2$. The process is therefore very improbable. On the other hand, flipping simultaneously two

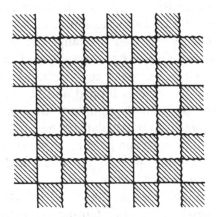

Fig. 1 The fundamental configuration of a gauge system described by the action given in equation (11) for $\beta_2 \ll 0$. Wavy links and shaded plaquettes are set to -1.

parallel neighbouring links is almost energetically costless. This is the fundamental excitation in this domain, and it is wise to build a simulation based on this double flip updating. Unfortunately, such an algorithm would violate the ergodicity condition if one were not to alternate with a link per link sweeping. A simulation taking into account all these considerations will be considerably more efficient than the original one.

8.1.3 Microcanonical simulations

A technique has been proposed which creates pseudorandom sequences of configurations without using the usual random number generators. The idea is to run along a deterministic path in configuration space, with the constraint that the energy lies in a small band of width δE. This restriction ensures that this path is a random one as in the usual analysis of the ergodic theorem. The process should be compared to a random number generator. In both cases, the ergodicity properties are dependent on the specific procedure used to generate the motion. To rely on this microcanonical technique, it is necessary to perform severe statistical tests to check randomness experimentally. The following method seems to satisfy the above criteria.

Consider a *demon* with a reservoir containing positive energy up to δE. As it visits all sites in succession, it flips each field (or spin) *if it can*. If the energy required to change the local configuration is greater than is available in its reservoir, or if this change yields more energy than the reservoir can contain, the field remains unchanged. If not, the field is flipped and the amount of energy in the tank is correspondingly updated. Such a process can be efficiently implemented, in particular for the Ising model, with due attention paid to the remark on parallel processing of independent families. In practice, one uses a "battalion of demons" visiting simultaneously spins which do not interact.

This procedure should be considered as a pseudorandom generator of configurations, on the same footing as the usual algorithms. At present, it seems that Ising systems are well simulated. The interest of the method is its great efficiency with respect to computing time. Empirical improvement tricks are also used to force randomness. For instance, the demon reservoirs are redistributed randomly at fixed time intervals.

In contradistinction with the previous algorithms, in microcanonical simulations, the energy is kept fixed (up to δE) rather than the temperature. It is thus necessary to measure the temperature. The mean energy available to the demons can be used as a thermometer. Indeed, at thermal equilibrium with the system, their energy distribution should follow the Boltzmann law $\exp(-\beta E_{\text{demon}})$. The reservoir has just one degree of freedom and its mean energy is easy to compute exactly as a function of β. For the Ising system, E_{demon} is a multiple of the flipping energy by an integral factor up to δE. Hence

$$\langle E_{\text{demon}} \rangle = \frac{\sum_{j=0}^{\delta E} j e^{-\beta j}}{\sum_{j=0}^{\delta E} e^{-\beta j}} \tag{12}$$

a relation which can be inverted to yield the temperature β.

8.1.4   Practical considerations

As is clear from the previous considerations, Monte Carlo simulations are in fact an art. The specifics of the implementation depend on the goals (and capabilities) of the performer. We collect here further technical details.

Boundary conditions As the tractable lattice sizes are rather small, boundary conditions play an essential role. For instance, a 10^3 three-dimensional lattice with free boundary conditions has 48.8% of its sites lying on its boundary. This effect increases with dimension and the measurement of bulk quantities becomes problematic. The most common way to handle this problem is to use periodic boundary conditions. From the programmer's point of view, this is easily done using a *modulo* operation, an elementary instruction for computers. Moreover, one takes advantage of the binary structure of data in computers by choosing as linear size a power of 2, replacing this operation by an even more rapid *logical and* instruction. There exists an even more efficient trick which simultaneously treats all dimensions, by choosing "twisted" boundary conditions. The identification of sites in a given direction is conveniently shifted in orthogonal directions, wrapping the lattice as a helix. To be specific, consider a two-dimensional system. In the computer memory, sites are arranged into a one-dimensional array $(1, 1)$, $(1, 2)$, $\cdots$, $(1, L)$, $(2, 1)$, $\cdots$, $(2, L)$, $(3, 1)$, $\cdots$, so that the addresses of the neighbours of a given site are obtained by shifting its own address by fixed quantities $(\pm 1$ and $\pm L)$. The twisted boundary conditions extend this rule to the initial and final sites, the addresses being defined modulo L^2. This trick greatly simplifies the programming and is very efficient.

One should be aware that periodicity may introduce spurious effects. In some cases, this might be a disaster if one does not pay sufficient attention to its implications. Let us quote some examples.

i) Mass measurements are performed by fitting two-point correlations $\langle \varphi_0 \varphi_x \rangle$ at large distances, by an exponential falloff Cst exp$(-x/\xi)$ if the medium is infinite. With periodic boundary conditions, the correlation function becomes periodic. Taking into account only the first duplication φ_L of φ_0, the previous formula is replaced by

$$\langle \varphi_0 \varphi_x \rangle \underset{\substack{x \text{ large} \\ 0 < x < L}}{\sim} \text{Cst} \cosh\left(\frac{x - \frac{1}{2}L}{\xi} \right) \qquad (13)$$

A typical example of such a behaviour is shown in figure 2.

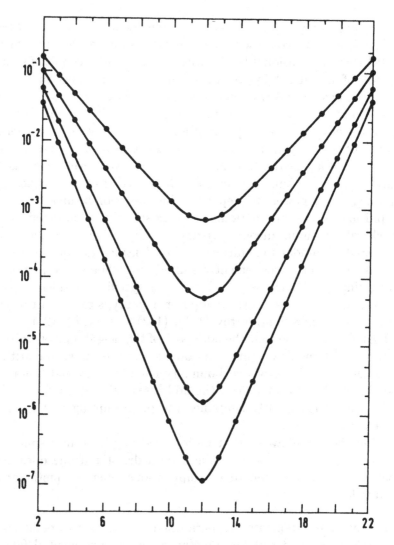

Fig. 2   Typical hyperbolic cosine behaviours in the computation of the pion propagator (from A. Billoire, R. Lacaze, E. Marinari and A. Morel, *Nucl. Phys.* **B251** [**FS13**], 581 (1985)).

ii) In lattice gauge theory, the static quark potential is derived from the observation of the average of a Wilson operator on a loop of size R in the spatial direction multiplied by T in the time direction, in the limit where $T \rightarrow \infty$ at fixed R. If $T = L$ in a periodic box of size L, the loop breaks into two independent closed lines, due to the boundary conditions. These are called Polyakov loops. Their use suppresses the problem of the limit $T \to \infty$ as well as finite size effects arising from the existence of corners in Wilson loops.

iii) When studying field theories at finite temperature, the partition function $Z = \text{Tr} \exp -\mathcal{H}/T$ at temperature T is expressed in terms of a functional integral with an additional Euclidean "temporal" dimension, for instance

$$Z = \int d\Phi \exp \left\{ -\frac{1}{T} \int_0^{1/T} dt \int d^d\mathbf{x} \, \mathcal{L}(\Phi) \right\} \qquad (14)$$

and a periodic constraint $\Phi(0, \mathbf{x}) = \Phi(1/T, \mathbf{x})$ for the field Φ in the simplest case of a scalar bosonic field. Numerical simulations for such problems are easily performed, using an asymmmetric lattice with $N_s^d \times N_t$ sites, $N_s a_s \gg N_t a_t$, periodic boundary conditions and possibly unequal spacings a_s and a_t. In the continuous limit, the temperature is related to $N_t a_t$ through

$$T = \frac{1}{N_t a_t} \qquad (15)$$

iv) Finally, antiperiodic boundary conditions may also lead to interesting effects. For instance, this allows the measurement of interface energies at low temperature.

Lattice size The choice of the lattice linear size L is obviously related to the effective cost of a simulation, especially in high dimensions, and is often the main limitation to numerical accuracy. However, simulations on lattices of rather small size are not irrelevant. The corresponding calculations are rapid and yield exploratory qualitative runs useful in determining the range of parameters to be chosen in a larger scale simulation. Furthermore, the evaluation of some observables (internal energy and some other bulk quantities) are experimentally found to be rather accurate even for small lattice sizes.

On the other hand, large sizes obviously reduce finite size effects. For technical reasons (initialization of loops, vectorization,...), the computing time needed for the upgrading of a single dynamical variable is also reduced with increasing size. Statistical accuracy is of course better for bulk quantities, since the measurements are

made over more lattice sites. Large sizes are in any case needed
to compute correlation functions, but the cost can be reduced by
increasing the size only along the field separation.

Note that a simulation is relevant only if the correlation length is
less than the linear size. The detailed study of critical phenomena
would therefore seem to require huge sizes.

Thermalization time and choice of initial configuration It is
always important to estimate the thermalization time, first to
reach equilibrium, secondly to know the number of sweeps which
should separate two configurations in the generating Markov chain
in order that they can be considered as independent. Let us
recall the definition of relaxation functions. Starting from an
off-equilibrium state characterized by a variation $\delta\theta$ of some
thermodynamic quantity θ (temperature, external field, etc.)
from its equilibrium value, a nonlinear relaxation function for an
observable A can be defined as

$$\phi_A^{\delta\theta}(t) = \frac{\langle A(t)\rangle - \langle A(\infty)\rangle}{\langle A(0)\rangle - \langle A(\infty)\rangle} \tag{16}$$

and allows the introduction of an associated relaxation time

$$\tau_A^{\delta\theta} = \int_0^\infty dt\, \phi_A^{\delta\theta}(t) \tag{17}$$

Similar quantities can be also computed at equilibrium, using
linear response theory

$$\phi_{AB}(t) = \frac{\langle A(0)B(t)\rangle - \langle A\rangle\langle B\rangle}{\langle AB\rangle - \langle A\rangle\langle B\rangle}$$

$$\tau_{AB} = \int_0^\infty dt\, \phi_{AB}(t) \tag{18}$$

The usual assumption is that the two functions are related through

$$\lim_{\delta\theta \to 0} \phi_A^{\delta\theta}(t) = \phi_{AB}(t) \tag{19}$$

where B is here the quantity conjugate to θ (e.g. the internal
energy if θ is the temperature, the magnetization if it is the
external field).

It is expected that the thermalization time is of the order of
equilibrium fluctuation times. This is of course useful only if one
has some *a priori* knowledge of the latter, which is seldom the

case. Similarly, one can also make some guesses when metastable states are present. Assume that an energy barrier ΔE separates an unstable from a stable state. From the theory of thermally activated processes, one may estimate that the relaxation time behaves as

$$\tau \sim \tau_0 \exp(\beta \Delta E) \qquad (\beta \Delta E \gg 1) \qquad (20)$$

where τ_0 is a typical time within the metastable phase. Such phenomena occur in first order transitions and are at the origin of hysteresis loops to be discussed later. The energy barrier is here the energy required for the "nucleation" of the stable phase. Similar energy barriers occur in other problems such as spin glasses and lead to extremely long relaxation times, decreasing the efficiency of the Monte Carlo method.

Close to a second order transition, the relaxation time diverges, since fields are correlated in volumes of order ξ^d. One should thus be aware that a critical slowing down occurs near T_c, with its own dynamical critical exponent

$$\tau \sim \left(1 - \frac{T}{T_c}\right)^{-\zeta} \qquad (21)$$

For the relaxation function ϕ_{AA}, the relaxation time is nearly proportional to the susceptibility $\chi_{AA} = \beta N [\langle A^2 \rangle - \langle A \rangle^2]$. For a finite system, the divergence is of course rounded (see below). In most cases, in the absence of reliable estimates, it is preferable to fit τ by observing, for instance, during thermalization, how fast an observable reaches its (apparent) equilibrium value. Furthermore, one should be aware that relaxation times may vary considerably according to the observable.

Measuring observables The measurement of certain observables may take a significant amount of computing time as compared to the one used to generate equilibrium configurations. We shall see for instance that fermionic averages imply the inversion of huge $L^d \times L^d$ matrices. The magnitude of statistical fluctuations gives rise to another type of limitation, and measurements become meaningless if average values are too small when compared with the background noise. For instance, derivatives cannot be obtained by finite differences between *independent* measurements.

This is the case of the specific heat $C = dE/dT$. We present in the following a technique which permits access to this quantity. In general, bulk quantities are more easily measured, but their interest is restricted since they are less directly related to renormalized observables of the continuous limit. Masses or correlation lengths are physically more interesting, but require much more care and longer simulations since their determination is based on the way a signal decreases with distance, just as it disappears in the noise.

Statistical errors   An estimator $\bar{A}$ for an observable A is the arithmetic average over $\mathcal{N}$ measures. The corresponding statistical error is estimated by

$$\delta A = \sqrt{\frac{1}{\mathcal{N}(\mathcal{N}-1)} \sum_i (A_i - \bar{A})^2} \qquad (22)$$

This formula is meaningful only if the measures A_i are uncorrelated, which is not the case when they are generated by a Markov chain. Hence it is important to perform various tests to check statistical independence of the measurements. An empirical method consists in computing $\mathcal{N}(\delta A)^2$ for various sets of measures and verifying that the results are $\mathcal{N}$-independent.

The value of the relaxation time τ_{AA} (equation (18)) is of course related to this question. One shows that the previous formula should be multiplied by a factor $\sqrt{1 + 2\tau_{AA}}$.

Check this result.

Parametrization of the fields   Fields may take their values in a great variety of domains, such as Lie groups (especially in lattice gauge systems). The nontrivial structure of the manifold lengthens the computing time in a significant way. Moreover, the parametrization may have an influence on the required storage capacity.

As an example, let us consider fields taking their values in the group $SU(2)$. We are faced with the following alternatives.

i) We can parametrize the fields using a 2×2 complex matrix. Storage requires 8 words (floating point numbers) per field. The multiplication is easily implemented and involves 32 elementary floating point multiplications (in fact, this number can be slightly reduced using some programming tricks).

ii) Alternatively one can parametrize the fields as points on the unit sphere S_3 in four-dimensional space as $U = a_0 + \mathrm{i}\mathbf{a}.\vec{\sigma}$, with $a_0^2 + \mathbf{a}^2 = 1$, a_0 and $\mathbf{a}$ real. Now, only 4 words are required. The group multiplication is a bit lengthier to program $(UU' = (a_0 a_0' - \mathbf{a}.\mathbf{a}') + \mathrm{i}(a_0\mathbf{a}' + a_0'\mathbf{a} + \mathbf{a} \wedge \mathbf{a}').\vec{\sigma})$, but requires only 16 elementary multiplications.

The second parametrization is better from all points of view and should be preferred. It has moreover the additional interest (Creutz, 1980) of permitting implementation of the heat bath algorithm. The Boltzmann weight for a field U appears as $\exp \beta \operatorname{Tr} UV$ where V is a sum of $SU(2)$ matrices and admits the same parametrization $V = b_0 + \mathrm{i}\mathbf{b}.\vec{\sigma}$ with no constraint on $b^2 = b_0^2 + \mathbf{b}^2$. The draught of U is taken according to the measure

$$\mathrm{d}P(U' = UV^{-1}) \propto \mathrm{d}\hat{\mathbf{a}}' \sqrt{1 - a_0'^2} \exp(\beta b a_0') \, \mathrm{d}a_0$$

The random choice of a_0' is governed by an exponential and is performed using the usual algorithms, corrected by a rejection procedure to take into account the square root factor. Then the direction of $\mathbf{a}'$ is selected randomly and the group element can be reconstructed. This yields a very efficient implementation of the heat bath algorithm for the $SU(2)$ group.

The structure of higher $SU(n)$ groups ($n \geq 3$) does not allow such simplifications. The manifold structure of the group is more intricate, and it is presently better to use a direct parametrization in terms of $n \times n$ complex matrices, although the required storage capacity is more important.

A sufficiently dense discretization of the domain on which the fields take their values might seem an interesting possibility for decreasing the storage capacity (one integer being sufficient to describe the field) and the computing time (using an internal multiplication table). However, one should be sure that the properties of the system are unaffected by this simplification.

Consider as an example the $U(1)$ group, well represented by Z_n for large n. Spin systems (with an action $\beta \sum_{(i,j)} \cos[2\pi(m_i - m_j)/ n]$, $m_i = 0, \ldots, n-1$) as well as their gauge counterparts with Z_n symmetry have been extensively studied and exhibit two transitions for $n > 4$ (figure 3). The interpretation of the intermediate phase is well understood in terms of massless spin wave excitations, as in the XY-model (chapter 4 in volume 1). The transition between the high temperature and the intermediate phases looks like the $U(1)$ transition and belongs to the same universality class. As the temperature decreases, the discrete structure of Z_n becomes apparent and the system undergoes a second transition. As n increases, the lower critical point is shifted towards lower and lower temperature. As a consequence, the $U(1)$ model can be approximated by a discrete Z_n *clock* model provided the temperature is greater than this lower critical temperature.

This technique would become very interesting if it could be efficiently extended to nonabelian continuous groups. Unfortunately, the corresponding discrete subgroups are only in finite number and might be not sufficiently dense.

Consider again $SU(2)$. From the symmetry groups of regular polyhedra (reduction to $SO(3)$ of the desired discrete subgroups of $SU(2)$ by factoring out the center Z_2), one has the following subgroups (chapter 11).

1) Q quaternionic group, 8 elements. This is the multiplicative group of ± 1, $\pm i\sigma_k$, a double covering of the four group realized as rotations of π around three orthogonal axes.
2) $\bar{T}$ double covering of the tetrahedral group, 24 elements.
3) $\bar{O}$ double covering of the cubic or octahedral group, 48 elements.
4) $\bar{Y}$ double covering of the dodecahedral or icosahedral group, 120 elements.

The groups in this list, which of course omits cyclic and dihedral groups (except for case 1), have been studied in four-dimensional lattice gauge theory and exhibit a unique first order transition at a coupling β respectively equal to 1.23, 2.175, 3.21 and 5.9. Recalling that the crossover region occurs near $\beta \approx 2$ in the $SU(2)$ gauge model, one sees that the last two subgroups are satisfactory in a limited range for simulation of the continuous model.

Turning now to $SU(3)$, among discrete subgroups some are known with 108, 216, 648 and 1080 elements. The respective first order transitions lie at $\beta = 2.5$, 3.2, 3.43 and 3.58. Hence it seems that none of these groups is useful as a substitute for the continuous group, since the interesting crossover region lies near $\beta \approx 5.9$ for the $SU(3)$ system.

8.2   Extraction of results in a simulation

Once a simulation has been programmed, one has at one's disposal
a system where parameters (temperature, external field, etc.)
can be adjusted and on which various measurements can be
made. Beyond the standard treatment of data, we focus on some
peculiarities arising in the study of critical properties.

8.2.1   Determination of transitions

The first task is to determine the phase diagram – if it is not
already known – and to localize the critical regions. A preliminary
exploration is performed by moving back and forth along a path
in the phase diagram, while measuring some observable. As an
example, consider a lattice gauge model with Z_6 symmetry. The
initial temperature is chosen at a rather high value, and the
initial configuration is fully disordered, all fields being chosen
independently at random. The system is then allowed to evolve
during a sufficiently long time to reach thermal equilibrium
(heuristically determined by the fact that observables fluctuate
without drift around their mean value). Then the Markov process
continues as the temperature decreases regularly by a small
amount at each sweep, while measurements are performed on
the plaquette energy $E = 1 - \langle \mathrm{Tr}\, U_p \rangle$. Once a sufficiently low
temperature is reached, the process is reversed to return to the
initial temperature. The resulting curve for the energy is displayed
on figure 3. One observes two hysteresis loops, which reveal the
existence of two transitions and give a rough determination of
their position.

A precise determination of the critical temperatures and of
the order of the transitions requires a more careful study. The
existence of a hysteresis loop is not characteristic of a first order
transition under present circumstances. In the latter case, it is
due to the existence of metastable phases in which the system
remains until a sufficient free energy fluctuation induces a jump
to the stable phase. Observables such as the energy E undergo a
discontinuity. For a second order transition, the hysteresis may be
due to a very long relaxation time as compared to the computing
time. The energy at equilibrium remains continuous as a function
of temperature, and the hysteresis loop should shrink with slower

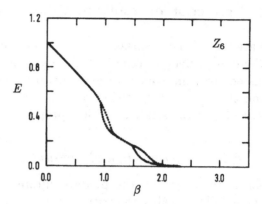

Fig. 3   The plaquette energy in the Z_6 lattice gauge model as the temperature decreases and increases (according to M. Creutz, L. Jacobs and C. Rebbi, *Phys. Rev.* **D20**, 1915 (1979)).

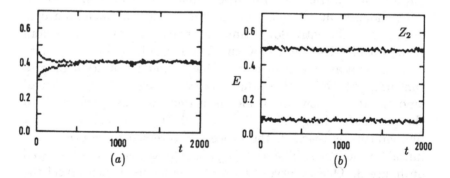

Fig. 4   Evolution of systems at the critical temperature, starting from an ordered or a disordered configuration. (*a*) Z_6 gauge model (second order). (*b*) Z_2 gauge model (first order). Same source as figure 3.

changes in the temperature. However, this is not such a good criterion, since the most important factor is the finite size of the system. Curves are rounded and the two characteristic shapes of observables for first and second order transitions become hardly distinguishable. In any case, there are no transitions in finite systems, and therefore the tests to be presented below should be clear and unambiguous enough for the answer to remain valid in the thermodynamic limit.

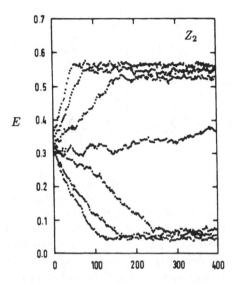

Fig. 5   Time evolution of a Z_2 gauge system initially containing two pure phases as β increases (from top to bottom) from 0.41 to 0.47. Same source as figure 3.

As an example of such a test, one prepares the system in an ordered and a disordered state and let each of these configurations evolve at (or very near) the critical temperature. In the case of a second order transition, both states reach the same equilibrium (figure 4a) while they remain in different phases in the case of a first order transition (figure 4b). Nevertheless, there still remains the possibility that the relaxation time is very long. This can be studied using a sample with an interface separating two pure phases and evolving at various temperatures near the critical one. Figure 5 displays such an experiment for the four-dimensional Z_2 gauge system, with temperatures $\beta = 0.41, 0.42, \ldots, 0.47$. The curves are sufficiently characteristic to ascertain that we have here a first order transition with a critical temperature slightly above 0.44 (duality yields the exact value 0.4407).

In the case of a second order transition, one wants to get a handle on critical exponents and amplitudes, or more generally on all universal quantities. Let us estimate the difficulty of such problems for the Ising model. We measure for instance the susceptibility

$$\chi = \left\langle \left(\sum_i \sigma_i \right)^2 \right\rangle - \left\langle \sum_i \sigma_i \right\rangle^2 \tag{23}$$

or the specific heat

$$C = \left\langle \left(\sum_{(i,j)} \sigma_i \sigma_j \right)^2 \right\rangle - \left\langle \sum_{(i,j)} \sigma_i \sigma_j \right\rangle^2 \tag{24}$$

A technical detail should be mentioned here. As stressed above, these quantities cannot be obtained by numerical differentiation of measured magnetization or energy, due to the statistical uncertainties. This method would not distinguish the signal from the noise. The correct technique is to measure magnetization and energy in various *statistically independent* samples and to estimate their variance as indicated in equations (23) and (24). We insist on the statistical independence, since consecutive configurations of the generating Markov chain are correlated and thus induce an underestimation.

The effect of finite size is illustrated on figure 6 for the three-dimensional Ising system. The critical temperature is determined as the limit of the location of the maximum. The fits of critical exponents are much more delicate and are very sensitive to the determination of T_c. Therefore other methods are needed. A first attempt using additional information from finite size scaling allows the fitting of the measurements by more constrained formulae.

8.2.2   Finite size effects

Scaling laws have been studied in detail in previous chapters. We summarize them here for our present purpose. We consider a system with a finite size L^d depending on only one length L (results can be generalized, e.g. to bars or slices). We have to take into account both the shift in the "critical" temperature $T_c(L)$ and in various exponents $(\gamma(L), \ldots)$. It is assumed that the susceptibility behaves as

$$\chi_L \underset{t_L \to 0}{\sim} A(L) t_L^{-\gamma(L)} \qquad \text{with } t_L = \frac{T - T_c(L)}{T_c(\infty)} \tag{25}$$

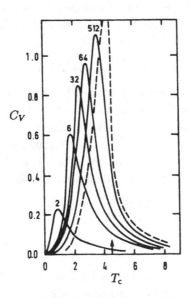

Fig. 6   The three-dimensional Ising model specific heat on finite lattices of various sizes. The dashed curve is the expected thermodynamic limit (from K. Binder, *Physica* **62** 508 (1972)).

For a finite system, all quantities are of course regular and it is *stricto sensu* incorrect to use the term critical temperature. Nevertheless, we define it here as the location of the maximum of χ_L. Hence the corresponding exponent $\gamma(L)$ vanishes. In other cases, such as slices, there exists a nontrivial exponent $\gamma(L)$. For instance, when in the slice $d - 1$ dimensions are large and can be considered in practice infinite as compared to the others, $\gamma(L)$ interpolates between the value $\gamma = \gamma(\infty)$ for the d-dimensional bulk system and $\gamma^* = \gamma(1)$ of the $(d - 1)$-dimensional one.

As the temperature approaches $T_c \equiv T_c(\infty)$, the correlation length ξ increases first according to its expected power law $t_\infty^{-\nu}$ until it reaches the order of magnitude L, where finite size effects make their appearance. Hence, the displacement of the critical temperature is expected to be such that $L \approx \xi$. The behaviour of the temperature shift is thus estimated to be

$$\frac{T_c(L) - T_c}{T_c} \sim L^{-\theta} \qquad \theta = \frac{1}{\nu} \qquad (26)$$

It is natural to expect that the only relevant variable governing the system is, besides the temperature, $L/\xi \sim Lt^{\nu}$. This leads to the scaling behaviour (Fisher 1969)

$$\chi_L(T) \sim L^{\omega} X(L^{\theta} t_L) \tag{27}$$

This hypothesis is confirmed below the upper critical dimension (chapter 4 in volume 1) from a general analysis. It can be proved to hold for some solvable models and is observed in low dimensional precise simulations. As $x \to \infty$, one should recover the critical behaviour obtained in the thermodynamic limit $L \to \infty$. This leads to the following two conditions on the form factor $X(x)$

$$X(x) \underset{x \to \infty}{\sim} A_{\infty} x^{-\gamma} \tag{28}$$

and

$$\omega = \theta\gamma = \frac{\gamma}{\nu} = 2 - \eta \tag{29}$$

Conversely, for fixed L, the behaviour of $X(x)$ near $x = 0$ is governed by equation (25), and one finds

$$X(x) \underset{x \to 0}{\sim} A(L) L^{(\gamma(L)-\gamma)/\nu} x^{-\gamma(L)} \tag{30}$$

In the case of a finite system, $\gamma(L)$ vanishes and $X(0)$ is finite. The height of the peak is therefore predicted to behave as

$$\chi_{L,\max} \sim L^{-\gamma/\nu} \tag{31}$$

Relations (26) and (31) are crucial in obtaining a reasonable accuracy in the determination of critical exponents using data such as those displayed on figure 6.

Logarithmic divergences $\chi \sim \ln t$ are not correctly taken into account in the previous scaling laws. Generalize the above derivation to this case. Formula (27) should be subtracted at some fixed temperature T_0, by writing for instance

$$\chi_L(T) \sim L^{\omega}[X(L^{\theta} t_L) - X(L^{\theta} t_0)] + \chi_L(T_0) \tag{32}$$

As a result, one finds that the height of the peak behaves logarithmically.

The method also allows information about interface energies to be extracted. Assume for instance antiperiodic boundary conditions in a given direction. At low temperature, an interface will appear (a surface crossing frustrated links) with an "area"

$S(L) = L^{d-1}$, of order $1/L$ with respect to the volume $V(L) = L^d$ of the system. The free energy admits an expansion

$$\mathcal{F}(T, L) = V(L)\mathcal{F}_\infty(T) + S(L)\mathcal{F}_s(T) + \mathcal{O}\left(1/L^2\right) \qquad (33)$$

where $\mathcal{F}_\infty(T)$ is the bulk free energy per unit volume in the thermodynamic limit and $\mathcal{F}_s(T)$ is the interface free energy per unit area. The singularities of these functions differ, but are related if the scaling law (27) is valid. Indeed, writing one more term in the large x expansion (28) of the form factor pertaining to the specific heat, with γ replaced by α and ω by α/ν

$$X(x) = X_\infty x^{-\alpha} + Y_\infty x^{-\phi} + \cdots \qquad (34)$$

and inserting into formula (33), one derives the critical exponent α_s for the interface free energy

$$\alpha_s = \phi = \alpha + \nu = 2 - \nu(d-1) \qquad (35)$$

Write the inequalities to be satisfied by the exponents in order that the previous expansions make sense. In particular, one finds $\theta \geq 1$. The limiting logarithmic case $\theta = 1$ can be also studied, allowing an estimate of the corresponding amplitude $A_s(\infty)$. A value $\theta < 1$ would correspond to long-range forces, and the splitting between volume and surface terms is no longer valid.

In practice, when one fits measurements, the most sensitive parameter is the critical temperature, and an error in its determination induces important variations of the critical exponents, which are therefore rather imprecise. The scaling laws have introduced additional constraints which improve the situation. However, a correct determination of the exponents following the previous method implies very precise measurements on samples as large as possible, and hence expansive simulations. An alternative, ingenious and very promising technique has been devised to treat efficiently the critical regime.

8.2.3   *Monte Carlo renormalization group*

Chapter 4 (volume 1) has been devoted to elementary aspects of the renormalization group in real space. The block spin technique seems well adapted to simulation, and we discuss now its practical

implementation. We reproduce once more the essential formulae
needed in the discussion of Ising like systems.

Let us choose a set of operators $\{X_\alpha\}$, which, for practical
reasons, should be as small as possible, leading to fast numerical
computations, and for theoretical reasons, as large as possible to
describe a general action, given the constraints of symmetry and
short-range interactions. The action is expressed as a combination
of the X_α's with a set $\{g_\alpha\}$ of coupling constants. It reads

$$S(\{\sigma\}\,;g_\alpha) = \sum_\alpha g_\alpha X_\alpha(\{\sigma\}) \tag{36}$$

One also defines block variables $\mu_i(\{\sigma\})$, using for instance the
majority rule

$$\mu_i = \text{Sign}\left(\sum_{j\in\text{block }i}\sigma_j\right)$$

which is easily implemented on computers. If a sum in a block
vanishes, μ_i is chosen at random. The system $\{\mu_i\}$ is hence derived
numerically from a given configuration $\{\sigma_i\}$, and both systems
describe the same physics in two different scales, with a ratio λ
given by size of the blocks. Recall, from chapter 4, that the two
corresponding actions are related through

$$\exp S(\{\mu\}\,;g_\alpha^{(1)}) = \sum_{\{\sigma\}}\left(\prod_i \delta(\mu_i - \mu_i(\{\sigma\}))\right)\exp S(\{\sigma\}\,;g_\alpha^{(0)})$$

$$\tag{37}$$

This relation is here only approximate since the set of operators
X_α is by necessity finite. The fixed point g_α^* of the transformation
$g^{(0)} \rightarrow g^{(1)}$ corresponds to the critical point, near which the
linearized version of the transformation reads

$$g_\alpha^{(1)} - g_\alpha^* = \sum_\beta T_{\alpha\beta}(g_\beta^{(0)} - g_\beta^*) \tag{38}$$

The *Monte Carlo Renormalization Group* method (Swendsen,
Wilson) gives numerical access to the matrix T rather than to the
coupling constants of the decimated action. This yields directly
the critical exponents, without the need for a precise *a priori*
knowledge of the critical temperature. Refering to the discussion
in chapter 4 and to figure 4.2, the approach to the critical
point depends on the magnitude of the eigenvalues of T with

respect to unity. In the basis diagonalizing T, relevant operators are those corresponding to eigenvalues of modulus greater than unity. The associated coupling constants will flow away from the fixed point. In particular, the temperature is associated to the largest eigenvalue y_{max} in the even sector, and thus the reduced temperature satisfies

$$t^{(1)} \sim t^{(0)} y_{max} \tag{39}$$

Recalling that the correlation lengths in both configurations differ by a factor λ

$$\xi^{(1)} = \frac{\xi^{(0)}}{\lambda} \tag{40}$$

we derive, from the scaling behaviour $\xi \sim t^{-\nu}$ and from equation (39), the critical exponent ν in the form

$$\nu = \frac{\ln \lambda}{\ln y_{max}} \tag{41}$$

To obtain other exponents such as γ, one needs the introduction of an external field (i.e. more generally odd operators X_α).

In practice, one generates configurations $\{\sigma^{(0)}\}$ very close to the critical temperature. Then the decimated configurations $\left\{\sigma_i^{(k)} = \mu_i(\{\sigma^{(k-1)}\})\right\}$ are determined and one measures the corresponding values $X_\alpha^{(k)}$ of the operators X_α. Performing statistical averages over independent configurations $\{\sigma^{(0)}\}$, one obtains the quantities

$$\frac{\partial}{\partial g_\alpha^{(p)}} \left\langle X_\beta^{(q)} \right\rangle = \left\langle X_\alpha^{(p)} X_\beta^{(q)} \right\rangle - \left\langle X_\alpha^{(p)} \right\rangle \left\langle X_\beta^{(q)} \right\rangle \tag{42}$$

The determination of T immediately follows, since equation (38) can be rewritten as

$$\frac{\partial}{\partial g_\alpha^{(k)}} \left\langle X_\beta^{(k+1)} \right\rangle = \sum_\gamma T_{\alpha\gamma} \frac{\partial}{\partial g_\gamma^{(k+1)}} \left\langle X_\beta^{(k+1)} \right\rangle \tag{43}$$

We emphasize again that an important aspect of this method is that critical exponents are obtained independently from a precise estimate of the critical temperature. The numerical study is to be completed by examining the stability of the results with respect to the number of decimations. One has here a control on the shift between the initial and the critical temperature. If the results are not sufficiently stable, the choice of the operator set $\{X_\alpha\}$ is inadequate and should be modified.

To illustrate the method, we display some of the data obtained by Swendsen for the two-dimensional Ising model. The set of operators are chosen to be

Even operators	Odd operators
1– Nearest neighbours	1– Magnetic field
2– Next nearest neighbour (11)	2– Three spins on a plaquette
3– Four spins coupled on a plaquette	3– Three aligned spins
4– Third neighbour (20)	4– Three spins (00),(10),(21)
5– Fourth neighbour (21)	
6– Four spins along a rectangle	
7– Fifth neighbour (22)	

The symbols such as (10), (11), ..., indicate the relative coordinates of the interacting neighbours. Starting from a 64×64 lattice with only the usual Ising coupling fixed at its (known) critical value, tables Ia, Ib display the leading and subleading critical eigenvalue exponents in the absence of an external field (only even operators are considered). The exact values for these exponents are respectively $2 - \nu = 1$ and -1. The number of couplings introduced increases along the vertical direction. Clearly, a single coupling is insufficient, while the introduction of more than two terms brings little improvement in table Ia. Table Ib pertaining to the second largest eigenvalue requires of course more operators. Note that this example is particularly favorable, although one expects that the method will work with only a few adequately chosen couplings. The number of iterations (by 2×2 blocks) varies horizontally. The instability of the flow is visible on right side of the table and is due here to finite size effects, while, in a more general case, it could also be due to a discrepancy with the critical temperature. Table Ic displays the same analysis for the magnetic exponent (i.e. the dimension of the magnetic field with an exact value $2 - \beta = 1.875$), using odd interactions, and implies similar comments. The study is also completed by using other lattice sizes (the results, not given here, confirm the validity of the method). A further consistency check uses the same data, but considers two consecutive iterations as a single one with scaling factor λ^2. If there is a discrepancy, this means that neglected interactions are important (note that unfortunately agreement does not mean that the method is really correct).

Table I. Monte Carlo Renormalization Group
determination of exponents for a two-dimensional
64×64 Ising system, according to the work of
Swendsen quoted in the notes.

(a)	1	2	3	4
1	0.912(2)	0.963(4)	0.957(2)	0.940(7)
2	0.967(3)	0.999(4)	0.998(2)	0.993(6)
3	0.968(3)	1.001(4)	0.999(2)	0.992(6)
4	0.969(4)	1.002(5)	0.999(2)	0.988(5)
5	0.969(4)	1.001(5)	0.997(2)	0.990(5)
6	0.969(3)	1.001(5)	0.997(2)	0.988(5)
7	0.969(5)	1.000(5)	0.997(2)	0.984(4)

(b)	1	2	3	4
2	$-2.19(3)$	$-2.20(4)$	$-2.20(3)$	$-2.32(8)$
3	$-2.17(3)$	$-2.34(6)$	$-2.56(6)$	$-2.56(22)$
4	$-1.11(3)$	$-1.14(3)$	$-1.16(3)$	$-1.35(20)$
5	$-0.94(3)$	$-0.94(3)$	$-0.91(2)$	$-0.80(8)$
6	$-1.04(3)$	$-1.04(3)$	$-0.98(2)$	$-0.84(9)$
7	$-1.08(21)$	$-1.03(4)$	$-0.99(20)$	$-0.76(20)$

(c)	1	2	3	4
1	1.8810(1)	1.8757(2)	1.8731(4)	1.8706(5)
2	1.8804(1)	1.8758(2)	1.8740(4)	1.8735(7)
3	1.8806(1)	1.8758(2)	1.8740(4)	1.8732(8)
4	1.8808(1)	1.8759(2)	1.8741(4)	1.8737(9)

Errors on the last digit are indicated in parentheses.
(a) Leading thermal exponent. (b) Subleading thermal
exponent. (c) Magnetic exponent.

i) Finite size effects are not included in the previous discussion. A
more complex procedure allows their elimination. This involves
two independent simulations with two lattices differing by a
factor λ in size. Coupling constants are adjusted in such a way
that the $(n + 1)$th decimation of the larger system and the nth
decimation of the smaller one yield identical mean values for
the operators X_α. As the physical sizes have to agree, finite size
effects are identical and disappear in the final result.

ii) Several improvements have been proposed. Let us describe
for instance a version (Gupta and Cordery) in which the
configurations $\{\sigma^{(0)}\}$ are generated with a weight

$$\exp\left[S\left(\left\{\sigma^{(0)}\right\};g_\alpha^{(0)}\right) - S\left(\left\{\sigma_i^{(1)}\right\};g_\alpha^g\right)\right]$$

with

$$\sigma_i^{(1)} = \mu_i \left(\left\{ \sigma^{(0)} \right\} \right) \tag{44}$$

depending on the block variables. The coupling $g_\alpha^{(1)}$ of the decimated action has been approximated by a guess g_α^g. The benefit is that the introduction of both site and block couplings suppresses the long time correlations due to the diverging correlation length. If the guess is perfect, the block spins are totally uncorrelated

$$\left\langle X_\alpha^{(1)} \right\rangle = 0 \qquad \left\langle X_\alpha^{(1)} X_\beta^{(1)} \right\rangle = N_\alpha \delta_{\alpha\beta} \tag{45}$$

If the guess is not perfect, one has to first order

$$\left\langle X_\alpha^{(1)} \right\rangle = \left\langle X_\alpha^{(1)} X_\beta^{(1)} \right\rangle \left(g_\beta^{(1)} - g_\beta^g \right) \tag{46}$$

so that it is possible to determine the renormalized coupling with no truncation error as

$$g_\alpha^{(1)} = g_\alpha^g + \frac{\left\langle X_\alpha^{(1)} \right\rangle}{N_\alpha} \tag{47}$$

The procedure can be iterated. If the irrelevant eigenvalues of T are small, two or three iterations are sufficient to reach the fixed point action, assumed to be of short range. The determination of the exponents proceeds exactly as above.

8.2.4   *Dynamics and the Langevin equation*

We have already mentioned the importance of the underlying dynamics generated by the Markov chain used in the Monte Carlo method. However, we have said very little on this subject, which is not easy to analyze theoretically. Another suggestion has been made of a method to generate configurations with better control of the dynamics. This *stochastic quantization* method uses the classical equations of motion with an additional perturbation generated by white noise. The technique has interesting applications for the determination of correlation functions and in fermionic simulations. It was originally proposed as a substitute for the gauge fixing procedure (Parisi, Wu Yong Shi).

Consider a generic field theory described by an action $S[\varphi]$. The classical evolution of fields with a random, time dependent driving term $\eta(t)$ is decribed by the Langevin equation

$$\dot{\varphi}_i(t) = \frac{\delta S}{\delta \varphi_i(t)} + \eta_i(t) \tag{48}$$

The first term is driving the system towards classical equilibrium. The excitation is chosen to be a random Gaussian variable

$$\langle \eta_i(t) \rangle = 0$$

$$\left\langle \eta_i(t)\eta_j(t') \right\rangle = \delta_{ij}\delta(t - t') \qquad (49)$$

$$\ldots$$

The notation $\langle \cdot \rangle$ denotes here the stochastic average with respect to the η's, while the usual field theoretic expectation values will be denoted by an overbar. The Fokker–Planck equation satisfied by the probability distribution $P(\varphi, t)$ for the field configurations can be easily derived from (48) and reads

$$\frac{\partial}{\partial t}P(\varphi, t) = \frac{\delta}{\delta\varphi_i}\left[-\frac{\delta S}{\delta\varphi_i}P(\varphi, t) + \frac{\delta P(\varphi, t)}{\delta\varphi_i}\right] \qquad (50)$$

In the stationary limit $\dot{P} = 0$, one recovers the usual Boltzmann weight $\exp S$. Hence any observable A satisfies

$$\langle A \rangle \xrightarrow[t \to \infty]{} \bar{A} \qquad (51)$$

If the action is invariant under a symmetry group, such as local invariance in lattice gauge systems, the result is valid for the invariant part of the operators A. Indeed, "longitudinal" components (i.e. corresponding to gauge transforms) have a vanishing driving term in equation (48). As a result, they undergo a random walk, and this yields divergent terms proportional to t (chapter 1). This divergence of course does not affect the measurement of invariant quantities. These properties were the reason for the original proposal and also apply in the context of continuous field theory.

Simulations of the Langevin equation are easy to do. Using random draughts of the noise $\eta_i(t)$, a configuration $\{\varphi\}$ is deterministically followed in time. Mean values of observables are then computed from a time average

$$\bar{A} = \lim_{t \to \infty}\frac{1}{t}\int_0^t dt\, A \qquad (52)$$

However, as in all numerical computations, time should be discretized. Denoting the elementary time step by ε, a discrete version of the Langevin equation reads

$$\varphi_i(\tau+1) = \varphi_i(\tau) + \varepsilon\frac{\delta S}{\delta\varphi_i} + \sqrt{\varepsilon}\eta_{i,\tau} \tag{53}$$

where we have used a reduced integral time variable $\tau = t/\varepsilon$ and rescaled η by a factor $\sqrt{\varepsilon}$, in such a way that the $\eta_{i,\tau}$ are independent normal Gaussian variables

$$\left\langle \eta_{i,\tau}\eta_{i',\tau'} \right\rangle = \delta_{ii'}\delta_{\tau\tau'} \tag{54}$$

The computer running time is proportional to ε^{-1}. It is thus interesting to use ε's as large as possible. We have therefore to discuss the corresponding discretization effects. In particular, the limiting probability distribution of the fields will be shifted from the Boltzmann weight $\exp S$ to $\exp S'$ related to a different effective action S'. Deriving the analog of equation (50) for the discretized case and expanding up to first order in ε, one obtains a continuous Fokker–Planck equation with an effective action given by

$$S' = S + \tfrac{1}{4}\varepsilon\sum_i \left[2\frac{\delta^2 S}{\delta\varphi_i^2} + \left(\frac{\delta S}{\delta\varphi_i}\right)^2 \right] + \mathcal{O}(\varepsilon^2) \tag{55}$$

It is possible to redefine the fields so that the quadratic term in the coefficient of ε disappears

$$\varphi_i' = \varphi_i - \frac{\varepsilon}{4}\frac{\delta S}{\delta\varphi_i} + \cdots$$
$$S'' = S + \frac{\varepsilon}{4}\sum_i \frac{\delta^2 S}{\delta\varphi_i^2} + \cdots \tag{56}$$

In the $\lambda\varphi^4$ theory, the effect is interpreted (to this order) as a shift in the bare square mass $m^2 \to m^2 + \tfrac{1}{4}\lambda\varepsilon$. Irrelevant terms also affect the bare coupling constant.

i) The difficulty in implementing this algorithm for a pure $SU(n)$ lattice gauge theory lies in the nontrivial structure of the manifold on which the fields take their values. The stochastic equation of motion which replaces (48) has to apply to an element in the tangent space, i.e. the Lie algebra. Let us denote by t_α the generators of the Lie algebra, normalized according to $[t_\alpha, t_\beta] = \mathrm{i}f_{\alpha\beta\gamma}t_\gamma$ and $\mathrm{Tr}\, t_\alpha t_\beta = \tfrac{1}{2}\delta_{\alpha\beta}$. If ∂_α stands for the derivative with respect to the group parameter corresponding to t_α, the simplest discretized Langevin equation reads

$$U_{ij}(\tau + 1) = e^{-i\lambda_{ij} \cdot t} U_{ij}(\tau)$$

$$\lambda_\alpha = \varepsilon \partial_\alpha S + \sqrt{\varepsilon} \eta_\alpha \tag{57}$$

For the Wilson action $S = (-\beta/2n) \sum_p \operatorname{Tr}(U_p + U_p^\dagger)$, this leads to

$$i\lambda_{ij} \cdot t = \frac{\varepsilon\beta}{4n} \sum_{kl} (U_{ijkl} - U_{ijkl}^\dagger) - \frac{\varepsilon\beta}{4n^2} \sum_{kl} \operatorname{Tr}(U_{ijkl} - U_{ijkl}^\dagger) + i\sqrt{\varepsilon} H_{ij}(\tau)$$

$$\tag{58}$$

where H is a traceless Hermitian random noise matrix satisfying

$$\left\langle \left(H_{ij}(\tau)\right)_{ab} \left(H_{kl}(\tau')\right)_{cd} \right\rangle = (\delta_{ac}\delta_{bd} - \delta_{ab}\delta_{cd}/n)\delta_{(ij),(kl)}\delta_{\tau,\tau'} \tag{59}$$

After some manipulation, the discretization of the Langevin equation leads to a Fokker–Planck equation, with an effective action of the form

$$S' = \left(1 + \tfrac{1}{12}\varepsilon n\right) S + \tfrac{1}{4}\varepsilon \sum_\alpha \left\{2\partial_\alpha^2 S - (\partial_\alpha S)^2\right\} + \mathcal{O}(\varepsilon^2) \tag{60}$$

This formula can again be interpreted as a rescaling of the fields together with a shift of the coupling, according to

$$U \to e^{-\frac{1}{4}i\varepsilon \partial S \cdot t} U$$

$$\beta \to \beta \left(1 - \varepsilon \frac{5n^2 - 6}{12n}\right) \tag{61}$$

ii) Is it possible to design a more sophisticated discrete Langevin equation leading to errors of order higher than ε?

8.3 Simulating fermions

Grassmannian integrals have been defined in an algebraic way and cannot be computed as usual integrals. The concept of sampling the integration space becomes meaningless, and the existence of "Grassmannian chips" remains a joke. Fermions cannot be simulated using the previous Monte Carlo techniques, and it is necessary to perform the integration over Grassmannian variables in closed form using the properties described in chapter 2 (volume 1). The corresponding numerical simulations therefore have to deal with the estimation of determinants and inverses of huge matrices in the best of all cases where one has reduced the fermionic action to a quadratic one.

Quantum Chromodynamics is one of the domains where fermionic simulations are used extensively. At present, the corresponding programs are extremely heavy and costfull and reach the capacity limits of existing computers. Parts of the following discussion might soon become obsolete.

Using the language and notation of lattice gauge theories, we consider a model with bosonic gauge fields U and fermionic fields $\bar{q}_i$, q_i interacting through the action

$$S = S_{\mathrm{g}}(U) + \sum_{ij} \bar{q}_i D_{ij}(U) q_j \tag{62}$$

with only quadratic fermionic terms. More complicated systems such as the Gross–Neveu model (see chapter 10) can be handled within this scheme at the price of the introduction of additional bosonic fields. The occurence of flavor, spinor,... indices is immaterial for the forthcoming analysis. Their introduction would only complicate the expressions. We omit them and leave it to the reader to restore them for specific applications.

Performing the Grassmannian integrals, one finds that the gauge fields evolve according to an effective action

$$S_{\mathrm{eff}}(U) = S_{\mathrm{g}}(U) + \ln \det D(U) \tag{63}$$

The additional difficulty in the Monte Carlo sampling for the gauge configurations is the computation of the determinant term. Note that it is nonlocal and introduces long-range interactions between gauge fields. Once the configurations have been generated, another difficulty arises in the measurement of the observables involving Fermi fields. We have

$$\langle f(U) \rangle = Z^{-1} \int DU \exp S_{\mathrm{eff}}(U)\, f(U) \tag{64a}$$

$$\left\langle \bar{q}_i q_j f(U) \right\rangle = Z^{-1} \int DU \exp S_{\mathrm{eff}}(U)\, f(U)\, \left(D(U)^{-1}\right)_{ij} \tag{64b}$$

$$\left\langle \bar{q}_i q_j \bar{q}_k q_l f(U) \right\rangle = Z^{-1} \int DU \exp S_{\mathrm{eff}}(U)\, f(U)$$
$$\times \left[\left(D(U)^{-1}\right)_{ij} \left(D(U)^{-1}\right)_{kl} - \left(D(U)^{-1}\right)_{il} \left(D(U)^{-1}\right)_{kj} \right] \tag{64c}$$
$$\cdots$$

and the problem is to estimate the inverse matrix $D(U)^{-1}$, depending on the configuration $\{U\}$ and occuring in the average over fermionic observables. In a first step, we present an approx-

(a) (b)

Fig. 7   The two terms entering in the determination of meson masses.

imation which greatly simplifies the simulation and amounts to neglecting the virtual quark pairs.

8.3.1   *The quenched approximation*

The most drastic assumption is that the term $\ln \det D(U)$ in (63) has little influence in the sampling of gauge configurations, at least in a qualitative, and perhaps even semiquantitative sense (this is called the "quenched approximation"). The practical interest of this assumption is obvious, since the evaluation of the determinant is very long. Another interest is that the effective action becomes independent on the details of the fermionic model (choice for the discretization of fermion fields, number of flavours, bare masses of quarks, etc.), and the same configurations can thus be used while varying the parameters and details of the fermionic part of the model (still necessary to compute averages such as those in equations (64b)–(64c)), allowing an accurate comparison without statistical errors.

An interpretation of the quenched approximation is that it amounts to a decoupling of virtual pairs for large internal quark masses. This is easily understood if one splits the quark propagator into kinetic and mass terms $D(U) = \Delta(U) + m$. Expanding in $1/m$, the effective action reads (up to an unessential additive constant)

$$S_{\text{eff}}(U) = S_{\text{g}}(U) - \sum_{k} \frac{1}{k} \left(\frac{-1}{m} \right)^{k} \text{Tr}\, \Delta^{k}(U) \qquad (65)$$

so that at face value the determinant becomes negligible when m tends to infinity. This naive argument neglects of course renormalization effects and may become totally wrong in the presence of anomalies. The latter corresponds precisely to cases where the combination of a vanishing coefficient (an inverse power of m) and a divergent expression conspire to yield a finite result in one (or several) term(s) in the above expansion. We set aside this possibility. Nevertheless the quenched approximation becomes questionable whenever the creation or annihilation of dynamical quark pairs is expected to play an important role. For instance, let us consider the meson propagator $\left\langle (\bar{q}_i q_i)(\bar{q}_j q_j) \right\rangle$. The two terms entering in the r.h.s. of equation (64c) are graphically represented on figure 7. The diagram (b) generates fermion loops by iteration, so that its contribution is strongly related to the terms neglected in the quenched approximation. Due to selection rules on angular momentum or internal quantum numbers, the contribution of this graph vanishes in most cases of interest, such as when studying the ρ or π meson propagators. It remains however important for states such as the η meson which is a singlet for the flavour group (this case is also related to the above mentioned anomaly). Such states have relatively high masses, and one can therefore expect that the approximation is reasonable in the spectroscopy of light mesons (π, ρ, K, K^*,...) and light baryons (N, Δ, Λ, Σ,...).

Formula (65) yields a procedure to compute systematic corrections in $1/m$ to the quenched approximation. However, the resulting *hopping parameter* expansion seems to be difficult to handle and has not yet provided accurate improvements.

Fermionic simulations using the quenched approximation still have to face the determination of the inverse matrix $G = D(U)^{-1}$. In practice, this task takes an important part of the total computer time.

Two remarks can be used to simplify this matrix inversion. First, it is generally sufficient to compute only a fixed column of the matrix G. Indeed, in many applications, one is interested in determining meson and baryon masses from the decay laws of correlations $\left\langle (\bar{q}_i \gamma q_i)(\bar{q}_j \gamma q_j) \right\rangle$ and $\left\langle (\bar{q}_i \bar{q}_i \bar{q}_i)(q_j q_j q_j) \right\rangle$, working at fixed i and varying j. Secondly, the matrix D has a nearest neighbour structure and multiplication by this matrix is thus not expansive. These facts can be used in iterative procedures. The Gauss–Seidel method, valid for a positive definite matrix D, uses the sequence

$$G^{(n+1)} = G^{(n)} + \varepsilon[1 - DG^n] \qquad (66)$$

These equations decouple for each column of G. One expects to reach the fixed point $G = D^{-1}$ of this sequence if ε is sufficiently small.

Study the convergence properties of the sequence, using a basis diagonalizing D. Show in particular that, for lattice gauge simulations performed at rather high coupling β, convergence is ensured if $\varepsilon < \mathrm{Cst}m$, that is, in practice, for sufficiently massive quarks. The number of iteration steps required to reach a given precision grows as $1/m$ for small quark mass m.

The technique can be improved in several ways. For instance, one can use a second order relaxation procedure

$$G^{(n+1)} = G^{(n)} + \left\{\varepsilon + \rho\varepsilon^2[1 - D]\right\}(1 - DG) \qquad (67)$$

In practice, the values of ρ and ε are empirically adjusted to yield the quickest convergence. The Runge–Kutta value $\rho = \frac{1}{2}$ is theoretically the best one to approach the expected result, but slighly larger values of this parameter yield sequences which overshoot for some time the exact result and provide a better estimate when one uses only a small number of iterations.

Another improvement is to rederive from the sequence $\{G^{(n)} - G^{(n-1)}\}$ an estimate of the errors and thus the best value to be chosen for ε at the next iteration step. This leads to the so-called conjugate gradient method.

8.3.2 Dynamical fermions

Let us now abandon the crude approximation of the previous subsection and analyze the available practical techniques to handle the full nonlocal action (62).

Luckily, in the Monte Carlo updating algorithm, it is not necessary to compute at every stage the complete action (62). Only its variation δS_{eff} is effectively required when varying a field $U \to U + \delta U$. We have

$$\delta S_{\mathrm{eff}}(U) = \delta S_{\mathrm{g}}(U) + \mathrm{Tr}\left[D(U)^{-1}\frac{\delta D}{\delta U}\right] + \mathcal{O}(\delta U^2) \qquad (68)$$

We are therefore facing again the problem of obtaining an estimate for $G = D(U)^{-1}$. This computation occurs now very frequently, not only for the measurements of observables as it did in the quenched approximation. We need it at each step of the updating procedure. The most straighforward procedures are the ones described above. These are however very costly. Alternatively, the pseudofermion method introduces auxiliary complex *bosonic* fields $\{\varphi_i\}$ and uses the identity

$$(D^{-1})_{ij} = \left\langle \varphi_i^\dagger \varphi_j \right\rangle = \frac{\int \prod_k (\mathrm{d}\varphi_k \mathrm{d}\varphi_k^\dagger)\, \varphi_i^\dagger \varphi_j \exp\left(-\varphi^\dagger D\varphi\right)}{\int \prod_k (\mathrm{d}\varphi_k \mathrm{d}\varphi_k^\dagger) \exp\left(-\varphi^\dagger D\varphi\right)} \qquad (69)$$

This expression is computed statistically using a standard Monte Carlo method for the pseudofermions φ.

A difficulty arises from the fact that the matrix D may be non-positive definite, so that the integrals in (69) are meaningless. This feature is corrected by using a pseudofermionic action $\varphi^\dagger D^\dagger D\varphi$ which involves now a next to nearest neighbour interaction. The mean value $\langle \varphi^\dagger \varphi \rangle$ provides the inverse matrix $(D^\dagger D)^{-1}$ from which one extracts the desired quantity $D^{-1} = (D^\dagger D)^{-1} D^\dagger$ by a quick multiplication with a "short-range" matrix.

We deal now with two Monte Carlo simulations, one for the pseudofermions inside the other one for gauge fields. In principle, δS_{eff} has to be computed every time with an updated gauge field and the pseudofermion algorithm could be very slow. At present, one generally computes δS_{eff} only once at the end of a full sweep. This approximation is expected to be accurate enough provided the shift δU remains small.

Extend the method of stochastic quantization to take into account the fermions and generalize formula (58) for the updating of gauge fields to include the effect of the fermion determinant.

8.3.3 Hadron mass calculation in lattice gauge theory

We conclude this chapter with a quick overview of the present attempts to predict the mass spectrum of light "elementary" particles from numerical simulations of lattice quantum chromodynamics. Let us first recall some more details about the model. Strong interactions are described in terms of a $SU(3)$ "colour" gauge group coupling several families of quarks. We restrict ourselves to the light, nonstrange hadrons and thus consider only the lowest mass quarks. These obey a $SU(2)$ isotopic symmetry, with respect to which they form a doublet of "up" and "down" states. Observed particles are bound states which are colour singlets due

Table II. Experimental characteristics of some light particles. We list isospin I, G-parity when relevant, angular momentum J, parity P and charge conjugation C. The quark content is indicated for the mesonic states.

Particle	$I^G(J^P)C$	Mass (MeV)	expected in
π	$1^-(0^-)+$	139	$\bar{q}\hat{\gamma}_5\tau q$
ρ	$1^+(1^-)-$	770	$\bar{q}\gamma_\mu\tau q$
ω,φ	$0^-(1^-)-$	783, 1020	$\bar{q}\gamma_\mu q$
η,η'	$0^+(0^-)+$	549, 958	$\bar{q}\hat{\gamma}_5 q$
A_1	$1^-(1^+)+$	1270	$\bar{q}\hat{\gamma}_5\gamma_\mu\tau q$
B	$1^+(1^+)-$	1235	$\bar{q}[\gamma_\mu,\gamma_\nu]\tau q$
δ	$1^-(0^+)$	980	$\bar{q}\tau q$
S	$0^+(0^+)+$	975	$\bar{q}q$
N	$\frac{1}{2}\left(\frac{1}{2}^+\right)$	939	
Δ	$\frac{3}{2}\left(\frac{3}{2}^+\right)$	1232	

to confinement. The corresponding operators in the quark language are most likely nonlocal. However, as they have a local component involving only quark fields at the same location, one studies for simplicity the correlation lengths of the corresponding operators, $\bar{q}_i q_i$ for mesons, $q_i q_i q_i$ for baryons. Spinorial and isospin indices are arranged to yield a state of definite spin and isospin, taking into account the limitations due to the finite lattice rotation group. For instance, the isospin 1 pseudoscalar π meson is obtained from the operator $\bar{q}_i\hat{\gamma}_5\tau q_i$, τ being the three 2×2 Pauli matrices acting on isospin quark indices. We display in table II the characteristics of the most important light particles. A more complete study has also to include heavy quarks, and in particular the strange quark which appears in rather light strange particles such as the pseudoscalar isotopic doublet K (495 MeV).

In the spectroscopy of light hadrons, chirality plays a very important role although it is only an approximate symmetry. In the limit of massless quarks, the Goldstone theorem predicts a vanishing mass for the pion. Its low value (139 MeV) as compared to the order of magnitude 1 GeV for other particles, as well as the experimental verification of current algebra predictions show the importance of spontaneously broken chiral invariance. Monte Carlo simulations are presently carried out using concurrently

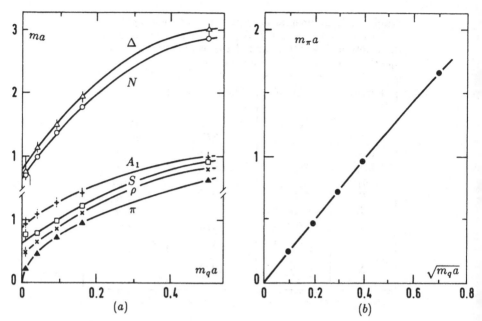

Fig. 8   (*a*) Measured masses of light states in the staggered fermion model as a function of the quark mass. (*b*) Pion mass as a function of the square root of the quark mass, showing the prediction of chiral invariance (from D. Barkai, K.J.M. Moriarty and C. Rebbi, *Phys. Lett.* **156B**, 385 (1985)).

either the staggered fermionic action (6.157), or Wilson's action (6.150). The first one has, as was discussed, a discrete chiral invariance, and it is therefore expected that the pion mass vanishes with the quark mass. Figure 8 shows this behaviour.

On the other hand, using Wilson's action, chiral invariance is explicitly broken by the lattice regularization but one expects to recover it in the continuous limit. With r the parameter occuring in this action, this will however not occur for the naive value $1/2dr$ of the hopping parameter $K = 1/[2(ma + dr)]$ as m goes to zero, due to renormalization effects on the bare quark mass. These effects are not under control in the present case for lack of chiral symmetry. The critical value is displaced to K_c, and chiral invariance will hopefully be restored as the hopping parameter approaches K_c from below. This is observed as expected in the simulations displayed on figure 9, where the critical point corresponds to the vanishing of the pion mass.

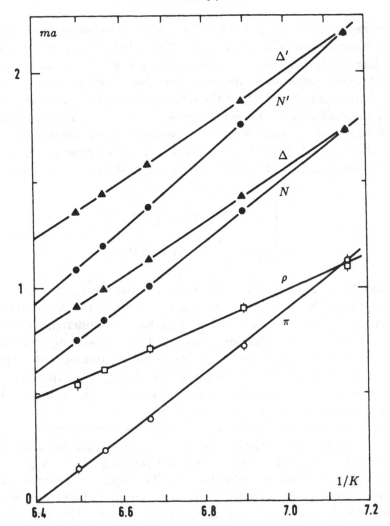

Fig. 9 Hadron masses versus the inverse hopping parameter with the Wilson's fermionic action (from S. Itoh, Y. Iwasaki and T. Yoshie, *Phys. Lett.* **B167**, 443 (1986)).

As the Monte Carlo techniques are in any case extremely expansive, simulations are done extensively only for a fixed gauge coupling constant β_J. In the choice of this parameter, one is squeezed between two opposite constraints. The first one is to have a coupling as low as possible to have a faster convergence rate and to avoid the use of large lattices (the spacing a decreases as β_J

increases due to asymptotic freedom) which are also very onerous. The second constraint is to have a sufficiently high coupling in the scaling window in order to allow a contact with the continuous limit using asymptotic freedom formulae.   Unfortunately, an extensive numerical study of the scaling properties in the presence of fermions has not yet been performed. Let us recall that, in the scaling region, the masses remain constant in the lattice mass scale Λ_L given by (6.31). In the case of a $SU(3)$ colour gauge group, the formulae (6.22), valid for pure gauge theory, have to be modified as

$$
\begin{cases}
b_0 = \dfrac{1}{(4\pi)^2} \left(11 - \tfrac{2}{3} N_f\right) \\[2ex]
b_1 = \dfrac{1}{(4\pi)^4} \left(102 - \tfrac{38}{3} N_f\right)
\end{cases}
\tag{70}
$$

where N_f is the number of flavours, two in the present discussion. Note that this number cannot exceed 16 to ensure asymptotic freedom ($b_0 > 0$). Given the present lack of numerical results about scaling properties, one assumes that the scaling window is the same as in the pure gauge case. String tension data (see figure 16 of chapter 6) yield a lower bound $\beta_J \geq 5.7$. Up to now, simulations have generally been done either at $\beta_J = 5.7$ or 6. These choices are perhaps too optimistic.

It is important to check that the various lattice formulations leading to the same continuous model yield compatible results. Figure 10 displays the ratio m_N/m_ρ as a function of m_π/m_ρ, from a compilation of several simulations. This figure illustrates the state of the art up to 1987, displaying in particular the error bars. With such errors, all data are compatible. However, the simulations done at $\beta_J = 6$ and in the quenched approximation are systematically below the dashed line, while those performed at 5.7 are above this line. This is an evidence for an insufficient scaling for these gauge coupling values.

In conclusion, the small sample of the very numerous results obtained by many groups that we have reported show a remarkable qualitative agreement with the physical spectrum.   It is to be expected that more elaborate simulations incorporating for instance dynamical quarks will improve the present evidence that quantum chromodynamics is indeed the best candidate for a theory of strong interactions.

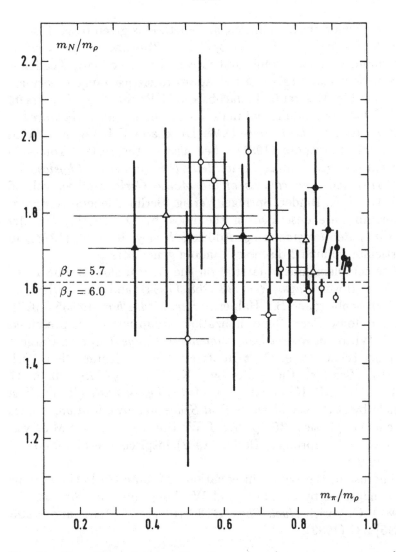

Fig. 10 The relation between the pion and nucleon masses, expressed in units of the ρ mass, according to several sources (from Bowler *et al*, *Phys. Lett.* **162B**, 354 (1985)).

Notes

The first proposition for a practical Monte Carlo simulation was done by N. Metropolis, A.W. Rosenbluth, M.N. Rosenbluth, A.H. Teller and E. Teller, *J. Chem. Phys.* **21**, 1087 (1953). A general review on the Monte Carlo method is given by K. Binder, *Monte Carlo Investigations of Phase Transitions and Critical Phenomena* in the Domb and Green series, vol. 5B, Academic Press, New York (1976). Applications to lattice gauge theory are presented by M. Creutz, L. Jacobs and C. Rebbi, *Phys. Reports* **95** 201 (1983), and additional material can be found in *Advances in Lattice Gauge Theory*, eds D.W. Duke and F.J. Owens, World Scientific, Singapore (1985). See also in the series "Topics in current physics", *Monte Carlo Methods in Statistical Physics*, vol. 7 (1979) and *Applications of the Monte Carlo Method*, vol. 36 (1984), ed. K. Binder, Springer Verlag, Berlin. General computer algorithms are described in the book by D. Knuth, *The Art of Computer Programming*, Addison-Wesley, Reading (1973), in particular volume 2 discusses random generators.

For complementary material on the critical slowing down, see L. Van Hove, *Phys. Rev.* **93** 268 (1954) and the review by P.C. Hohenberg and B.I. Halperin, *Rev. Mod. Phys.* **49** 435 (1977).

The Monte Carlo Renormalization Group method is presented by K. Wilson in *Recent Developments in Gauge Theories*, Cargese summer school 1979, G. 't Hooft *et al* eds., Plenum, New York (1980). See also the works by S.K. Ma, *Phys. Rev. Lett.* **37** 461 (1976), L.P. Kadanoff, *Rev. Mod. Phys.* **49** 267 (1977). The article by R.H. Swendsen in *Real Space Renormalization*, Topics in current physics, **30**, p. 57, T.W. Burkhardt and J.M.J. van Leeuwen eds, Springer, Berlin (1982) inspired the discussion of section 2.3.

The use of the Langevin equation in Monte Carlo simulations was suggested by G. Parisi and Wu Yong Shi, *Sci. Sin.* **24** 483 (1981), G. Parisi, *Nucl. Phys.* **B180 [FS2]** 378 (1981) and **B205 [FS5]** 337 (1982).

9

CONFORMAL INVARIANCE

It was noted by Polyakov and others in the early seventies that critical models implement a global conformal invariance which goes beyond pure scale invariance. The latter affects relative distances by a constant (i.e. space independent) factor, while other conformal transformations involve a space dependent factor. This invariance property enables one to fix not only the form of two-point but also three-point functions at criticality, when they are nonvanishing. However the conformal group is in general a finite dimensional Lie group, so that the resulting constraints are limited in number. In two dimensions, a new phenomenon arises, well known in the theory of analytic functions, namely there exists a plethora of local conformal transformations. As a result, it was tempting to investigate the possible consequences of local scale invariance in two dimensions. This is what was brilliantly undertaken by Belavin, Polyakov and Zamolodchikov in 1983, launching a new wave of applications in statistical physics. As the subject is still in its development, the present chapter will not be as elementary as previous ones, nor will it presumably remain up to date, especially as it is closely related to string field theory, a promising new approach to the quantum description of extended objects, which attempts to embrace all known interactions including gauge theories and gravity.

The ideas of local scale invariance, and possibly of a local renormalization group, are not yet fully appreciated. They prompt us to look at several aspects which were not in the forefront in the study of critical phenomena. One of them is the relevance of finite size effects at criticality where, due to the lack of an intrinsic length scale, there is a maximal sensitivity to geometrical effects like boundaries. A second novel aspect is the meaning of the values obtained for critical exponents, critical correlations, etc. It will appear that for many two-dimensional systems they possess an unexpected group theoretical (and even number theoretical) interpretation, being related to the representation theory of some infinite Lie algebras. A third question, partly answered, but not treated here, is the relation with integrable models. Finally the deepest questions perhaps are to look for suitable generalisations both in the direction of studying the effects of moving off criticality, and of trying to gain a deeper understanding of critical phenomena in higher dimensions. The present success in the two-dimensional case can partly be traced to the use of analytic functions which enables us to reduce two-dimensional problems to one-dimensional ones.

9.1 Energy-momentum tensor – Virasoro algebra

9.1.1 Conformal invariance

Since finite size effects play a major role at criticality, we must first ascertain what we mean by a critical system enclosed in a finite domain. This can be phrased in various ways, either by dilating the box, or by saying that the system has no intrinsic length scale except for those dictated by the geometry, or finally by studying the short-distance behaviour of correlations and making sure that it agrees with the infinite volume one at criticality. This is one of the crucial aspects of what is to follow, namely the local characterization of critical systems, and it will be fully elucidated in two dimensions.

Let us first give an example to demonstrate the phenomenon. Consider the correlation function of Ising spins in the two-dimensional case at criticality. The dimension of the spin field is equal to $\frac{1}{8}$, as we know from chapter 2. In the infinite plane

$$\langle \sigma(\mathbf{r}_1)\sigma(\mathbf{r}_2)\rangle = 1/|\mathbf{r}_{12}|^{1/4} \tag{1}$$

It was also mentioned in chapter 2 that in a strip of width L ($0 \leq x < L$), with periodic boundary conditions, the corresponding correlation reads

$$\langle \sigma(\mathbf{r}_1)\sigma(\mathbf{r}_2)\rangle$$
$$= \left(\frac{\pi}{L}\right)^{\frac{1}{4}}\left[\sinh^2\frac{\pi}{L}y_{12}\cos^2\frac{\pi}{L}x_{12} + \cosh^2\frac{\pi}{L}y_{12}\sin^2\frac{\pi}{L}x_{12}\right]^{-\frac{1}{8}} \tag{2}$$

When $L \to \infty$, or $\mathbf{r}_{12} \to 0$, this expression reduces to (1). Rather than changing the geometry (varying L) it is better to characterize intrinsically the critical model by its short-distance scale invariant behaviour.

A conformal transformation of an Euclidean d-dimensional space (or more generally of a Riemannian manifold) is a point to point transformation which preserves angles but not necessarily distances. These transformations form a group which includes the Euclidean group as a subgroup (translations, d parameters, rotations $\frac{1}{2}d(d-1)$ parameters), dilatations (one parameter), and special conformal transformations (d parameters) obtained by composing an inversion, a translation and a second inversion. This describes the connected part of the noncompact conformal group, $SO(d+1,1)$ with $\frac{1}{2}(d+1)(d+2)$ parameters. To be precise, the Euclidean space is completed by a point at infinity (giving it the topology of a compact sphere) to make the above transformations well-defined everywhere.

To understand globally the conformal group, it is convenient to map the completed Euclidean space $\mathcal{R}^d$, with metric $\mathbf{r}^2 = r_1^2 + \cdots + r_d^2$, onto the unit sphere S_d of the Euclidean space $\mathcal{R}^{d+1}$, $x_-^2 + \mathbf{x}^2 = 1$ using a stereographic (conformal) projection as shown on figure 1. Correspondingly

$$
\begin{aligned}
\mathbf{x} &= \frac{2\mathbf{r}}{1+\mathbf{r}^2} & \mathbf{r} &= \frac{\mathbf{x}}{1-x_-} \\[2mm]
x_- &= \frac{\mathbf{r}^2-1}{\mathbf{r}^2+1} & \mathbf{r}^2 &= \frac{1+x_-}{1-x_-}
\end{aligned}
\tag{3}
$$

Using homogeneous coordinates, S_d may be identified with the section $x_+ = 1$ of the light cone $\mathcal{C}$ of Minkowski space $\mathcal{R}^{d+1,1}$ with metric $x_+^2 - x_-^2 - \mathbf{x}^2$. Rays on $\mathcal{C}$ are in one-to-one correspondence with points of $\mathcal{R}^d \cup \{\infty\}$. Conformal transformations are interpreted as homogeneous Lorentz transformations in $\mathcal{R}^{d+1,1}$. The little group of a lightlike vector can be identified with the Euclidean group in d-dimensions (the point singled out may be thought of as the point at infinity). For instance if the lightlike vector has components $(x_+, x_-, \mathbf{x}) = (1, 1, 0)$ a translation $\Lambda_{\mathbf{a}}$ is represented as

$$
\begin{aligned}
\Lambda_{\mathbf{a}}(1,1,0) &= (1,1,0) \\
\Lambda_{\mathbf{a}}(1,-1,0) &= (1,-1,0) + \mathbf{a}^2(1,1,0) + (0,0,\mathbf{a}) \\
\Lambda_{\mathbf{a}}(0,0,\mathbf{v}) &= \mathbf{a}.\mathbf{v}(1,1,0) + (0,0,\mathbf{v})
\end{aligned}
\tag{4}
$$

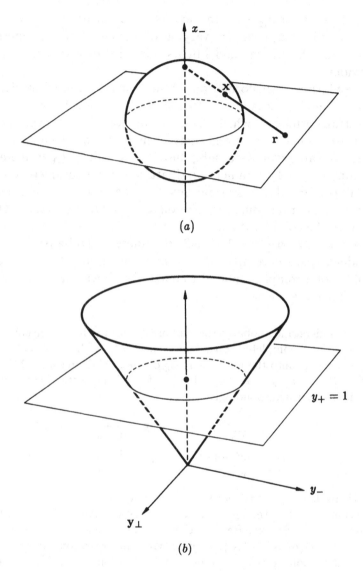

(a)

(b)

Fig. 1   (a) Stereographic projection of S^d onto $\mathcal{R}^d$. (b) The isotropic cone (light cone) in Minkowski space $\mathcal{R}^{d+1,1}$.

An inversion belonging to a disconnected component of $O(d+1,1)$ corresponds to $x_- \rightarrow -x_-$, and a dilatation through λ is represented by an hyperbolic rotation in the x_+, x_- plane through

$$
\begin{aligned}
\Lambda_\lambda(1,1,0) &= \lambda(1,1,0) \\
\Lambda_\lambda(1,-1,0) &= \lambda^{-1}(1,-1,0) \\
\Lambda_\lambda(0,0,\mathbf{v}) &= (0,0,\mathbf{v})
\end{aligned}
\tag{5}
$$

Returning to d-dimensional space a special conformal transformation takes the form

$$
\mathbf{r} \rightarrow \mathbf{r}' = \frac{\mathbf{r} + r^2 \mathbf{b}}{1 + 2\mathbf{b}.\mathbf{r} + b^2 r^2}
\tag{6}
$$

In infinitesimal form

$$
\mathbf{r} \rightarrow \mathbf{r}' = \mathbf{r} + \delta\mathbf{r} \qquad \delta\mathbf{r} = r^2 \left(\delta\mathbf{b} - \frac{2\mathbf{r}(\mathbf{r}.\delta\mathbf{b})}{r^2} \right)
\tag{7}
$$

Euclidean displacements conserve the metric and have therefore a unit Jacobian. This is not the case for dilatations, nor for special transformations, in which case the Jacobian is

$$
\frac{D(\mathbf{r}')}{D(\mathbf{r})} = \frac{1}{(1 + 2\mathbf{b}.\mathbf{r} + b^2 r^2)^d}
\tag{8}
$$

Assume that critical models are not only scale invariant, but also conformal invariant. This hypothesis will receive support as we proceed. Consider for simplicity scalar observables, with corresponding dimension denoted Δ. These assumptions mean that under conformal transformations, the correlation functions satisfy

$$
\langle A_1(\mathbf{r}_1)...A_n(\mathbf{r}_n) \rangle = \prod_{i=1}^{n} \left| \frac{D(r_i')}{D(r_i)} \right|^{\Delta_i/d} \langle A_1(\mathbf{r}_1')...A_n(\mathbf{r}_n') \rangle
\tag{9}
$$

For a two-point function, using translational, rotational and dilatation invariance, we get, with a normalizing constant C_{12}

$$
\langle A_1(\mathbf{r}_1) A_2(\mathbf{r}_2) \rangle = \frac{C_{12}}{r_{12}^{\Delta_1 + \Delta_2}}
\tag{10}
$$

Under a special transformation, we have

$$
r_{12}'^2 = \frac{r_{12}^2}{(1 + 2\mathbf{b}.\mathbf{r}_1 + b^2 r_1^2)(1 + 2\mathbf{b}.\mathbf{r}_2 + b^2 r_2^2)}
\tag{11}
$$

Consequently equations (9) and (10) are only compatible if $\Delta_1 = \Delta_2$

$$\Delta_1 \neq \Delta_2 \qquad \langle A_1(\mathbf{r}_1)A_2(\mathbf{r}_2)\rangle = 0$$

$$\Delta_1 = \Delta_2 = \Delta \qquad \langle A_1(\mathbf{r}_1)A_2(\mathbf{r}_2)\rangle = \frac{C_{12}}{r_{12}^{2\Delta}} \qquad (12)$$

We can define an "orthonormal basis" of observables by choosing the appropriate combinations which diagonalize the symmetric real matrix of coefficients C_{ab}. A single rescaling of the fields will then replace C_{ab} by δ_{ab}. Similar reasoning enables one to obtain the three-point functions

$$\langle A_1(\mathbf{r}_1)A_2(\mathbf{r}_2)A_3(\mathbf{r}_3)\rangle = \frac{C_{123}}{r_{12}^{\Delta_1+\Delta_2-\Delta_3} \, r_{23}^{\Delta_2+\Delta_3-\Delta_1} \, r_{31}^{\Delta_3+\Delta_1-\Delta_2}} \qquad (13)$$

Check that equation (13) is the unique solution to condition (9).

Beyond three-point functions, conformal invariance alone is not sufficient to determine the structure of correlations. This follows from the fact that some combinations such as $r_{13}r_{24}/r_{23}r_{14}$ (where r_{ij} stands for $|\mathbf{r}_i - \mathbf{r}_j|$) are invariant under conformal transformations. More information is required, which will be provided in the two-dimensional case from an assumption of invariance under local conformal transformations. For the time being, let us translate condition (9) in infinitesimal form by writing

$$\mathbf{r}' = \mathbf{r} + \delta\mathbf{r}$$

$$A'(\mathbf{r}) = \left(\frac{D(\mathbf{r}')}{D(\mathbf{r})}\right)^{\Delta/d} A(\mathbf{r}') = A(\mathbf{r}) + \delta A(\mathbf{r}) \qquad (14)$$

$$\delta A(\mathbf{r}) = \left[\delta\mathbf{r}.\boldsymbol{\nabla} + \frac{\Delta}{d}(\boldsymbol{\nabla}.\delta\mathbf{r})\right] A(\mathbf{r})$$

The requirement that $\mathbf{r} \rightarrow \mathbf{r} + \delta\mathbf{r}$ be a conformal map in infinitesimal form means that $\delta\mathbf{r}$ should be at most quadratic in $\mathbf{r}$, i.e. of the form

$$\delta r_i = \delta a_i + \left[\delta\lambda\delta_{ij} + \delta\varepsilon_{ij} + \delta b_i r_j - 2r_i\delta b_j\right] r_j$$

$$\boldsymbol{\nabla}.\delta\mathbf{r} = d(\delta\lambda - 2\mathbf{r}.\delta\mathbf{b}) \qquad (15)$$

where δa_i, $\delta\lambda$, $\delta\varepsilon_{ij} = -\delta\varepsilon_{ji}$ and δb_i are respectively the infinitesimal parameters for translations, dilatation, rotations and special transformations. Explicitly

$$\delta A(\mathbf{r}) = \left[\delta \mathbf{a}.\, \nabla + \delta \lambda \mathbf{r}.\, \nabla + \tfrac{1}{2}\delta\varepsilon_{ij}(r_i \nabla_j - r_j \nabla_i) \right.$$
$$\left. + r^2 \delta \mathbf{b}.\, \nabla - 2(\mathbf{r}.\delta\mathbf{b})(\mathbf{r}.\, \nabla) + \Delta_A(\delta\lambda - 2\mathbf{r}.\delta\mathbf{b}) \right] A(\mathbf{r})$$
$$(16)$$

Therefore, if $\delta A_p(\mathbf{r}_p)$ stands for the corresponding expression pertaining to the (scalar) field A_p, equation (9) states that

$$\sum_{p=1}^{n} \left\langle A_1(\mathbf{r}_1) \ldots \delta A_p(\mathbf{r}_p) \ldots A_n(\mathbf{r}_n) \right\rangle = 0 \qquad (17)$$

Generalize the above expressions in the case of tensorial fields.

9.1.2 Energy–momentum tensor

Critical theories in infinite Euclidean space are obtained from a limiting procedure starting from a sum (or integral) over Boltzmann weights. Local observables generally are expressed in terms of fundamental fields occurring in these weights. Let us denote the latter collectively by A. Correlation functions are averages over the Boltzmann weights with a partition function normalization. If we forget about the limiting procedures, we may think of performing in the numerator a dummy change of variable of the type $A(\mathbf{r}) \to A(\mathbf{r}) + \delta A(\mathbf{r})$.

$$\delta A(\mathbf{r}) = \left[\delta \mathbf{r}.\, \nabla + \frac{\Delta}{d}\, \nabla .\delta \mathbf{r} \right] A(\mathbf{r}) \qquad (18)$$

where now $\delta\mathbf{r}(\mathbf{r})$ is an arbitrary infinitesimal vector field, i.e. not necessarily of the form (15). We further assume that the observables A_p transform also as (18), in which case they are likely to belong to a restricted class (with corresponding dimensions Δ_i). At any rate we denote by δA_p the corresponding variation of the observables A_p. Then the total variation of the correlation functions will not only contain a contribution analogous to (17), but will also include a term arising from the noninvariance of the measure (including the Boltzmann weight) under the change of variable (18). This term will be of first order in $\delta\mathbf{r}$ and should obviously vanish when $\delta\mathbf{r}$ is a constant (corresponding to a translation). It therefore only depends on derivatives of $\delta\mathbf{r}$. Since we are working with a local theory we write the simplest possible form, as a postulate defining the (local) energy–momentum tensor density

$$\sum_{p=1}^{n} \Big\langle A_1(\mathbf{r}_1) \ldots \delta A_p(\mathbf{r}_p) \ldots A_n(\mathbf{r}_n) \Big\rangle$$
$$+ \int d^d \mathbf{r} \Big\langle A_1(\mathbf{r}_1) \ldots A_n(\mathbf{r}_n) T_{ij}(\mathbf{r}) \Big\rangle \partial_i \delta r_j(\mathbf{r}) = 0 \tag{19}$$

This is analogous to the corresponding definition in Minkowskian local field theory, where this construction provides a Noether current for the conserved quantities, energy and momentum. This explains the origin of the name. For a given Lagrangian, one obtains a natural "classical" candidate for T_{ij}, but there might be difficulties in the way, which have their origin in the behaviour of the remaining integration measure. Therefore it is at this point more honest to assume (19) as a postulate, referred to as the basic Ward identity. In a more general setting, the energy–momentum tensor appears as the response to an infinitesimal change of metric, which insures that it is symmetric. We refrain from proceeding along these lines, although most two-dimensional applications suggest that there is much to learn from quantum field theory in curved space.

Returning to equation (19), let us look at the constraints implied by assuming invariance under Euclidean transformations and global dilatations. We substitute the corresponding expressions for δA_p and $\delta \mathbf{r}$. Then we have

i) Invariance under translations, $\delta \mathbf{r}$ constant, equation (19) reduces to (17).

ii) Invariance under rotations, $\partial_i \delta r_j = \delta \varepsilon_{ji} = -\delta \varepsilon_{ij}$ constant. For equation (19) to reduce to (17), it is required that T_{ij} be symmetric

$$T_{ij}(\mathbf{r}) = T_{ji}(\mathbf{r}) \tag{20}$$

iii) Invariance under dilatations, $\delta_i \delta r_j = \delta_{ij} \delta \lambda$. To recover (17) the energy–momentum must be traceless

$$\sum_i T_{ii}(\mathbf{r}) = 0 \tag{21}$$

If one now considers an infinitesimal special transformation, with $\partial_i \delta r_j = 2\left(r_i \delta b_j - r_j \delta b_i\right) - \delta_{ij}(\delta \mathbf{b}.\mathbf{r})$, the second term in (19) vanishes identically if we take (20) and (21) into account. In other words in a local theory, where the Ward identity (19) holds, Euclidean invariance and scale invariance alone insure at the formal level invariance under the full conformal group. This is the important original observation.

If an infinitesimal coordinate transformation is interpreted in an active sense, then the infinitesimal square length between neighbouring points is changed according to $\left(g_{ij}^{(0)} = \delta_{ij}\right)$

$$\mathrm{d}s^2 = \mathrm{d}\mathbf{r}^2 \to \mathrm{d}(\mathbf{r} + \delta \mathbf{r})^2 = \left(g_{ij}^{(0)} + \delta g_{ij}\right) \mathrm{d}r_i\,\mathrm{d}r_j$$
$$\delta g_{ij} = \partial_i \delta r_j + \partial_j \delta r_i \qquad\qquad (22)$$

If we substitute $\frac{1}{2}T_{ij}\delta g_{ij}$ for $T_{ij}\partial_i \delta r_j$ we find that $\int \mathrm{d}^d\mathbf{r}T_{ij}(\mathbf{r})\delta g_{ij}(\mathbf{r})$ appears as the response of the functional measure (including the Boltzmann weight) to a change in the metric. Since here everything is understood as relative to the flat metric $g_{ij}^{(0)}$, we did not bother to distinguish between covariant and contravariant indices.

Let us assume that for a general coordinate variation the variations δA_p are local, of the form (18). Furthermore consider the case where $\delta \mathbf{r}$ is arbitrary and of compact support. Then after a partial integration, we can rewrite equation (19), by identifying the coefficient of $\delta \mathbf{r}$, as

$$\partial_i \left\langle T_{ij}(\mathbf{r})A_1(\mathbf{r}_1)\ldots A_n(\mathbf{r}_n)\right\rangle =$$
$$= \sum_{p=1}^{n}\left\{\delta(\mathbf{r} - \mathbf{r}_p)\partial_j^{(p)} - \frac{\Delta_p}{d}\left(\partial_j\delta(\mathbf{r} - r_p)\right)\right\}\langle A_1(\mathbf{r}_1)\ldots A_n(r_n)\rangle$$
$$(23)$$

This is what is meant by saying that the energy–momentum is conserved. Indeed the right-hand side vanishes for $\mathbf{r}$ in any domain omitting a neighbourhood of the points $\mathbf{r}_1,\ldots \mathbf{r}_n$, so does therefore the left-hand side with the insertion of the divergence of T.

It is not possible for the time being to go much further in the general case. We turn therefore to the special features occurring in two dimensions.

9.1.3 Two-dimensional conformal transformations

In two dimensions, the global conformal group is the familiar
Lorentz group $SO(3,1)$. It may be realized by acting on the
complex plane (completed by a point at infinity) through linear
rational transformations (or Möbius transformations) exhibiting
the isomorphism of $SO(3,1)$ with $SL(2,\mathcal{C})/\{\pm 1\}$. Instead of
using two real independent coordinates x and y, it is convenient to
introduce the two independent complex combinations $z = x + iy$,
$\bar{z} = x - iy$. Using this notation, the flat metric becomes
antidiagonal, and care must be exercised with tensors. We write

$$ds^2 = g_{zz}\,dz^2 + (g_{z\bar{z}} + g_{\bar{z}z})\,dz d\bar{z} + g_{\bar{z}\bar{z}}d\bar{z}^2 = dz d\bar{z}$$
$$dx \wedge dy = d\bar{z} \wedge dz/2i \qquad \partial_x^2 + \partial_y^2 = 4\partial_z \partial_{\bar{z}}$$
$$\partial_z = \tfrac{1}{2}(\partial_x - i\partial_y) \qquad\qquad \partial_{\bar{z}} = \tfrac{1}{2}(\partial_x + i\partial_y) \tag{24}$$

For short we sometimes abreviate ∂_z as ∂ and $\partial_{\bar{z}}$ as $\bar{\partial}$. It follows
that

$$g_{zz} = g_{\bar{z}\bar{z}} = 0 \qquad\qquad g_{z\bar{z}} = g_{\bar{z}z} = \tfrac{1}{2}$$
$$g^{zz} = g^{\bar{z}\bar{z}} = 0 \qquad\qquad g^{z\bar{z}} = g^{\bar{z}z} = 2 \tag{25}$$

A symmetric traceless tensor such as T is obtained in the $z, \bar{z}$ basis
by forming the scalar

$$T_{ij}\xi^i\xi^j = \tfrac{1}{2}\left(T_{11} - iT_{12}\right)\xi^2 + \tfrac{1}{2}\left(T_{11} + iT_{12}\right)\bar{\xi}^2$$

where $\xi = \xi^1 + i\xi^2$, $\bar{\xi} = \xi^1 - i\xi^2$. Thus

$$T_{zz} = \tfrac{1}{2}\left(T_{11} - iT_{12}\right) \qquad T_{\bar{z}\bar{z}} = \tfrac{1}{2}\left(T_{11} + iT_{12}\right) \qquad T_{z\bar{z}} = T_{\bar{z}z} = 0 \tag{26}$$

With a, b, c, d complex satisfying $ad - bc = 1$, the global
conformal transformations read

$$z, \bar{z} \to z', \bar{z}' \quad z' = f(z) = \frac{az+b}{cz+d} \quad \bar{z}' = \overline{f(z)} = \frac{\bar{a}\bar{z}+\bar{b}}{\bar{c}\bar{z}+\bar{d}} \tag{27}$$

$$\frac{D(z',\bar{z}')}{D(z,\bar{z})} = f'(z)\overline{f'(z)} = \frac{1}{(cz+d)^2(\bar{c}\bar{z}+\bar{d})^2} \tag{28}$$

We can also envision other global conformal transformations which
change the orientation and correspond to a disconnected part of
the group. They are generated by complex conjugation and will
not play a crucial role, since we will mostly be concerned with
infinitesimal transformations.

Tensor fields transform with integral powers of f' and $\bar{f}'$. But here we encounter a specificity of two dimensions, in that it makes perfect sense to think of a generalization of the notion of tensor with a fractional number of z or $\bar{z}$ indices. More precisely we define a field with conformal weights $(h, \bar{h})$ as one which transforms according to

$$A(z, \bar{z}) \to A'(z, \bar{z}) = f'(z)^h \overline{f'(z)}^{\bar{h}} A(f(z), \overline{f(z)}) \qquad (29)$$

where $\bar{h}$ is not meant as the complex conjugate of h, but the notation $\bar{h}$ is suggestive of the fact that it relates to the behaviour associated to the $\bar{z}$ variable. The quantity $h - \bar{h}$ plays the role of angular momentum or spin, indicating the behaviour under rotations $z \to e^{i\varphi} z$ with A' picking a factor $e^{i(h-\bar{h})\varphi}$. For most (but not all) of what follows, we shall consider integral spin fields ($h - \bar{h}$ an integer). Indeed these are the only observable ones. However it is natural to envision also cases involving Fermi fields with half integral spins, and further generalizations have been investigated. The scaling dimension Δ is the sum

$$\Delta = h + \bar{h} \qquad (30)$$

and for a real scalar field we have

$$h = \bar{h} = \tfrac{1}{2}\Delta \qquad (31)$$

We can interpret equation (29) by saying that, in infinitesimal form

$$(A + \delta A)\, dz^h \, d\bar{z}^{\bar{h}} = A(z + \delta z, \bar{z} + \delta \bar{z})[d(z + \delta z)]^h [d(\bar{z} + \delta \bar{z})]^{\bar{h}} \quad (32)$$

Therefore, with

$$\delta z = \varepsilon(z) \qquad \qquad \delta A = \delta_z A + \delta_{\bar{z}} A \qquad (33)$$

we have

$$\begin{aligned}
\delta_z A(z, \bar{z}) &= \left\{ \varepsilon(z)\partial_z + h\varepsilon'(z) \right\} A(z, \bar{z}) \\
\delta_{\bar{z}} A(z, \bar{z}) &= \left\{ \overline{\varepsilon(z)}\partial_{\bar{z}} + \bar{h}\overline{\varepsilon'(z)} \right\} A(z, \bar{z})
\end{aligned} \qquad (34)$$

The novel feature arising in two dimensions is the existence of an infinite class of conformal transformations generated by an arbitrary analytic function $f(z)$ instead of a rational linear function as was used until now (we forget here the possibility that f might be an analytic function of $\bar{z}$). That these transformations are conformal follows immediately from the fact that the metric changes into

$$d z d\bar{z} \rightarrow d f d\bar{f} = f'(z)\overline{f'(z)}\, d z d\bar{z} \qquad (35)$$

so that angles are preserved, but lengths are locally multiplied by a dilatation factor $|f'(z)|$. The interpretation of such transformations is not straightforward. The point is that in general

i) $f(z)$ might not be defined throughout the complex plane
ii) and even when this is the case (admitting isolated poles) it will not be a one-to-one map.

The only one-to-one analytic maps of the (completed) complex plane are the linear fractional transformations, i.e. those corresponding to the global group. There are two possible interpretations, both of them interesting. In the first one, we use the assumption of local invariance to obtain quantities pertaining to the transformed domain. Shortly we shall look at such a map from the punctured plane onto a periodic strip. In the second interpretation, we study the effect of infinitesimal local conformal transformations such as (33) with $\varepsilon(z)$ considered of first order but otherwise an arbitrary function of z. It is good to keep in mind that analytic functions are rigid, i.e. $\varepsilon(z)$ cannot be made to be smooth and vanishing outside a bounded domain while keeping it analytic.

Under an arbitrary "infinitesimal" holomorphic change of coordinates

$$\begin{aligned} \delta z &= \delta\varepsilon g(z) \\ \delta\bar{z} &= \delta\varepsilon\bar{g}(\bar{z}) \end{aligned} \qquad (36)$$

fields which transform as

$$\delta A(z,\bar{z}) = \delta\varepsilon \left[g\partial_z + h(\partial_z g) + \bar{g}\partial_{\bar{z}} + \bar{h}(\partial_{\bar{z}}\bar{g}) \right] A(z,\bar{z}) \qquad (37)$$

are called primary ones. Here the emphasis is on the fact that g is arbitrary, and not restricted to be a second degree polynomial in z (in which case A could be called quasiprimary). We shall deal almost exclusively with primary fields, the most notable exception being, surprisingly, the energy–momentum tensor. For a general $g(z,\bar{z})$, we have

$$T_{ij}\partial_i\delta\mathbf{r}_j = 2\delta\varepsilon \left[T_{zz}\partial_{\bar{z}}g + T_{\bar{z}\bar{z}}\partial_z\bar{g} \right] \qquad (38)$$

This quantity vanishes for g ($\bar{g}$) holomorphic (antiholomorphic). Let us insert this expression into equation (19). Of course the second term vanishes in those (finite) regions of the $z, \bar{z}$ plane where g and $\bar{g}$ have the above analyticity property, suggesting a covariance property of correlations of primary fields under general analytic transformations. Let us make this statement more precise. It is convenient to rescale the energy–momentum tensor in order to insert a factor $1/2\pi$ into the integral defining T. Choosing a pair of functions $g, \bar{g}$, vanishing sufficiently fast at infinity to allow integrations by parts, holomorphic for g and antiholomorphic for $\bar{g}$ in the vicinity of the points $z_1, \ldots, z_n$, so that we can use equation (37), and setting $g_p \equiv g(z_p, \bar{z}_p), \partial_p \equiv \partial_{z_p}$, we obtain

$$\sum_{p=1}^{n} \left[g_p \partial_p + h_p(\partial_p g_p) + \bar{g}_p \bar{\partial}_p + \bar{h}_p(\bar{\partial}_p \bar{g}_p) \right] \langle A_1 \ldots A_n \rangle$$

$$= \int \frac{\mathrm{d}\bar{z} \wedge \mathrm{d}z}{2i\pi} [g(z,\bar{z}) \partial_{\bar{z}} \langle T_{zz}(z,\bar{z}) A_1 \ldots A_n \rangle$$
$$+ \bar{g}(z,\bar{z}) \partial_z \langle T_{\bar{z}\bar{z}}(z,\bar{z}) A_1 \ldots A_n \rangle] \qquad (39)$$

Comparing both sides, it follows that $\partial_{\bar{z}} \langle T_{zz}(z,\bar{z}) A_1 \ldots A_n \rangle$ vanishes except at the points $z_1 \ldots z_n$. A similar conclusion holds for $\partial_z \langle T_{\bar{z}\bar{z}}(z,\bar{z}) A_1 \ldots A_n \rangle$. Indeed this equation shows that the first of these functions is a meromorphic one in z with single and double poles located at $z_1 \ldots z_n$, and the corresponding conjugate property for the second, so that we are entitled to write the Ward identity in the useful form

$$\langle T_{zz}(z) A_1(z_1,\bar{z}_1) \ldots A_n(z_n,\bar{z}_n) \rangle =$$
$$\sum_{p=1}^{n} \left[\frac{h_p}{(z-z_p)^2} + \frac{1}{(z-z_p)} \partial_{z_p} \right] \langle A_1(z_1,\bar{z}_1) \ldots A_n(z_n,\bar{z}_n) \rangle$$
$$\langle T_{\bar{z}\bar{z}}(\bar{z}) A_1(z_1,\bar{z}_1) \ldots A_n(z_n,\bar{z}_n) \rangle =$$
$$\sum_{p=1}^{n} \left[\frac{\bar{h}_p}{(\bar{z}-\bar{z}_p)^2} + \frac{1}{(\bar{z}-\bar{z}_p)} \partial_{\bar{z}_p} \right] \langle A_1(z_1,\bar{z}_1) \ldots A_n(z_n,\bar{z}_n) \rangle \quad (40)$$

Given the parallel treatment of the z and $\bar{z}$ variables, we shall in the following not duplicate the equations, giving only the z part. However, the reader should be aware that some fields, such as the A's above, might have both a z and $\bar{z}$ dependence, while it follows from equation (40) that T_{zz} depends only on z at least in the sense that all correlation with T_{zz} insertions are meromorphic functions of z. Henceforth we shall denote it $T(z)$, similarly for $T_{\bar{z}\bar{z}}$ denoted $\bar{T}(\bar{z})$.

If now $g(z)$ is analytic in a domain enclosing the points $z_1,...,$ z_n and if C is a curve in this domain encircling these points once in the positive sense, the integral form of (40) reads

$$\sum_{p=1}^{n} \left[g(z_p)\partial_{z_p} + h_p g'(z_p) \right] \langle A_1(z_1, \bar{z}_1) \dots A_n(z_n, \bar{z}_n) \rangle =$$
$$\int_C \frac{\mathrm{d}z}{2\mathrm{i}\pi} g(z) \langle T(z) A_1(z_1, \bar{z}_1) \dots A_n(z_n, \bar{z}_n) \rangle \tag{41}$$

and a similar conjugate equation is obtained with $\bar{T}$. The generators of global conformal transformations correspond to functions $g(z)$ which are second degree polynomials in z, in which case the left-hand side of the above equation must vanish. As a result, a correlation function with a $T(z)$ insertion vanishes like z^{-4} for large z

$$\langle T(z) A_1(z_1, \bar{z}_1) \dots A_n(z_n, \bar{z}_n) \rangle \underset{z \to \infty}{\sim} z^{-4} \Gamma(z_1, \bar{z}_1; \cdots; z_n, \bar{z}_n) \tag{42}$$

with a conjugate relation for $\bar{T}$. The canonical dimension of T is 2 (and d in d dimensions) since its integral must be dimensionless (as is the action). A naive idea would have been that the left-hand side of (42) vanishes like z^{-2} at infinity. The present statement is stronger, implying global conformal invariance. A second degree polynomial depends on three complex parameters, which is of course the (complex) dimension of the group $SL(2, C)$. Combining (42) with (40), we find by an expansion around $z = \infty$, that global conformal transformations imply

$$\sum_{p=1}^{n} \partial_{z_p} \langle A_1(z_1, \bar{z}_1) \dots A_n(z_n, \bar{z}_n) \rangle = 0$$

translations

$$\sum_{p=1}^{n} \left(z_p \partial_{z_p} + h_p \right) \langle A_1(z_1, \bar{z}_1) \dots A_n(z_n, \bar{z}_n) \rangle = 0$$

(complex) dilatations

$$\sum_{p=1}^{n} \left(z_p^2 \partial_{z_p} + 2 h_p z_p \right) \langle A_1(z_1, \bar{z}_1) \dots A_n(z_n, \bar{z}_n) \rangle = 0$$

special transformations   (43)

In the same way that we do not write explicitly the conjugate equations, we shall sometimes omit to indicate the $\bar{z}$ variables in the z-equations. In two dimensions, rotations and real dilatations are conjugate transformations. In the complex formalism they combine into complex dilatations as indicated in equation (43).

Relations (41) express a generalized form of covariance under infinitesimal local conformal transformations. To proceed further, we need some properties of correlations involving the energy–momentum tensor, and the first task is to study its behaviour under the above mentioned transformations.

9.1.4 Central charge

From translational invariance in the plane, the average value $\langle T(z) \rangle$ is a constant. It is implicit in equation (42) that this constant has been chosen equal to zero

$$\langle T(z) \rangle = \langle \bar{T}(\bar{z}) \rangle = 0 \tag{44}$$

Similarly translational and global scale invariance imply that the corresponding two-point correlation be a homogeneous function of z_{12}. From equation (42) we should therefore have

$$\langle T(z_1)T(z_2) \rangle = \frac{\frac{1}{2}c}{(z_{12})^4} \qquad \langle \bar{T}(\bar{z}_1)\bar{T}(\bar{z}_2) \rangle = \frac{\frac{1}{2}c}{(\bar{z}_{12})^4}$$
$$\langle T(z_1)\bar{T}(\bar{z}_2) \rangle = 0 \tag{45}$$

from which we learn that T has conformal weights (2,0) ($\bar{T}$ has weights (0,2)) as expected. The choice of the factor $\frac{1}{2}$ in the normalization of the real constant c common to T and $\bar{T}$ is conventional. It is designed to produce the value $c = 1$ for the scalar massless field (see below). Since the scale of T is already fixed by relations such as equation (40), we cannot dispose of it. If the theory admits an underlying Hamiltonian quantum mechanical interpretation, with a positive definite Hilbert space of states and energies bounded from below, then c has to be positive, as we shall see later. But this is by no means necessary for a consistent statistical interpretation, and we shall find interesting cases with $c < 0$.

The dimensionless constant c is called the central charge for reasons which will appear below. One of the main conclusions of this chapter will be that knowledge of c is (almost) sufficient to characterize a critical model in many interesting cases. At any rate, it gives essential information from which much can be derived.

The nonvanishing of c corresponds to an interesting anomaly. Such anomalies occur when a classical symmetry cannot be implemented quantum mechanically as a consequence of renormalization effects. In the context of path integrals, this can also be interpreted by saying that the complete functional measure cannot be made invariant. Such instances arise frequently when dealing with local symmetries. One can also say that the renormalization group equations express an anomalous behaviour under global dilatations off the fixed points.

We just stated that $\langle T(z) \rangle$ vanishes in the infinite plane. If we had invariance under infinitesimal local conformal transformations, we would expect that this mean value would remain equal to zero. But except for those transformations generated by a second degree polynomial in z, we know that one cannot, in a strict sense, speak of an invariance, since the transformations are not one-to-one mappings of the (completed) complex plane onto itself. As a result, we do not expect $\langle \delta T(z) \rangle$ to vanish. Indeed combining equations (41) and (45), we have

$$z \rightarrow z' = z + \delta\varepsilon g(z)$$
$$T \rightarrow T' = T + \delta T \tag{46a}$$
$$\langle \delta T(z) \rangle = \delta\varepsilon \int_C \frac{dz'}{2i\pi} g(z') \langle T(z')T(z) \rangle = \tfrac{1}{2}c\delta\varepsilon \int_C \frac{dz'}{2i\pi} \frac{g(z')}{(z'-z)^4}$$

and as a result

$$\langle \delta T(z) \rangle = \tfrac{1}{12}cg'''(z)\,\delta\varepsilon \tag{46b}$$

with a similar expression for $\bar{T}$. When g is a second degree polynomial, g''' vanishes, and we have the expected invariance. Equation (46) is a first expression of the anomaly. Given that $\langle T(z) \rangle = 0$, and that the weights of T are $(2,0)$, it is natural to postulate that the minimal transformation law incorporating the above anomaly is

$$\delta T(z) = \delta\varepsilon \left[(2g'(z) + g(z)\partial_z)T(z) + \tfrac{1}{12}cg'''(z)\right] \tag{47}$$

with a similar formula for $\bar{T}$. This behaviour should be contrasted with the one of primary fields, given by equation (33). Equation (47) teaches us that T is not a primary field. It exhibits an inhomogeneous behaviour, with an extra contribution proportional to the identity. This suggests that the anomaly may be derived from the corresponding behaviour of the free energy, or partition function.

Let us transcribe here the expressions (12) and (13) for arbitrary two- and three-point functions, omitting the $\bar{z}$ variables,

$$\langle A_h(z_1)A_h(z_2)\rangle = \frac{\alpha_{12}}{(z_{12})^{2h}}$$

$$\left\langle A_{h_1}(z_1)A_{h_2}(z_2)A_{h_3}(z_3)\right\rangle = \frac{\alpha_{123}}{z_{12}^{h_1+h_2-h_3}\,z_{23}^{h_2+h_3-h_1}\,z_{31}^{h_3+h_1-h_2}} \tag{48}$$

Consequently

$$\langle T(z_1)T(z_2)T(z_3)\rangle = \frac{c}{(z_{12}z_{23}z_{31})^2} \tag{49}$$

Check that this is in agreement with

$$\langle \delta T(z_1)T(z_2)\rangle + \langle T(z_1)\delta T(z_2)\rangle = \delta\varepsilon \int_C \frac{\mathrm{d}z_3}{2\mathrm{i}\pi} g(z_3)\,\langle T(z_1)T(z_2)T(z_3)\rangle$$

It is interesting to know the generalization of equation (47) to a finite conformal transformation mapping the plane onto a possibly different domain. We now have $z' = f(z)$. To $T(z)$ will correspond $T'(z')$. With the active interpretation, the corresponding formula reads

$$T(z)\,\mathrm{d}z^2 = T'(z')\,\mathrm{d}z'^2 + \tfrac{1}{12}c\,\{z',z\}\,\mathrm{d}z^2 \tag{50}$$

where the symbol $\{z',z\}$ stands for the Schwarzian derivative of the map $z \to z' = f(z)$, namely

$$\{z',z\} = \frac{f'''}{f'} - \tfrac{3}{2}\left(\frac{f''}{f'}\right)^2 \tag{51}$$

To first order in $\delta\varepsilon$, this reduces to the third derivative of $\delta\varepsilon g(z)$ for $f(z) = z + \delta\varepsilon\,g(z)$.

It is not obvious how, from the composition of infinitesimal maps, one recovers the Schwarzian derivative. We shall present a proof using free field theory in section 2.1. Let us digress a little to give a simple interpretation of the Schwarzian derivative. It is related to the global conformal group very much as the ordinary derivative is related to translations. As will shortly be clear, if we use the notation

$$[z', z] = \{z', z\}\, dz^2 \tag{52a}$$

for the corresponding quadratic differential (recall that z' and z are analytically related), the compatibility of equation (50) requires that

$$[z_1, z_2] + [z_2, z_1] = 0 \tag{52b}$$

$$[z_1, z_2] + [z_2, z_3] + [z_3, z_1] = 0 \tag{52c}$$

$$[z_1, z_2] = \left[\frac{az_1 + b}{cz_1 + d}, z_2\right] = \left[z_1, \frac{az_2 + b}{cz_2 + d}\right] \tag{52d}$$

In equation (52c), it is of course assumed that $z_1 = f_{12}(z_2)$, $z_2 = f_{23}(z_3)$ and $z_3 = f_{31}(z_1)$ with $f_{12} \circ f_{23} \circ f_{31}$ the identity transformation. Finally, if z_1 and z_2 are related by a linear fractional transformation, the corresponding quadratic differential vanishes, as a consequence of (52b) and (52d)

$$az_1 z_2 + bz_1 + cz_2 + d = 0 \quad \Leftrightarrow \quad [z_1, z_2] = 0 \tag{52e}$$

The anharmonic or cross ratio of four complex numbers is invariant under Möbius transformations. To measure how a given transformation $y = f(x)$ departs locally from a fractional linear transformation, let us compare the anharmonic ratios of four points x_i and of their images. Setting $x_{ij} = x_i - x_j$, we compute the quantity

$$Q(x, y) = \frac{x_{13} x_{42}}{x_{12} x_{43}} - \frac{y_{13} y_{42}}{y_{12} y_{43}}$$

for values x_i close to a point x, of the form $x_i = x + t\varepsilon_i$. Correspondingly we expand y_i to third order in t. A short calculation shows that the leading term in Q is of order t^2, of the form

$$Q(x, y) = \frac{\varepsilon_{13} \varepsilon_{42}}{6 \varepsilon_{12} \varepsilon_{43}} \left\{ \frac{f'''(x)}{f'(x)} - \frac{3}{2}\left(\frac{f''(x)}{f'(x)}\right)^2 \right\} \varepsilon_{14} \varepsilon_{23} t^2 + \cdots$$

The properties (52b)–(52e) follow immediately. Let us finally observe that

$$\{f(x), x\} = -2(f')^{1/2} \frac{d^2}{dx^2}(f')^{-1/2} = \frac{d^2}{dx^2} \ln f' - \frac{1}{2}\left(\frac{d}{dx} \ln f'\right)^2 \tag{53}$$

We can now turn to an interpretation of equation (50) and hence of the anomaly. For that purpose, let us map conformally

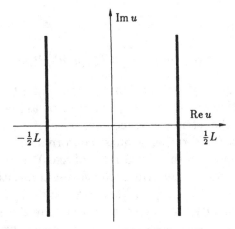

Fig. 2   A periodic strip of width L in the u-plane

the complex z-plane, with the point at the origin deleted, onto a periodic strip in the u-plane, of width L (figure 2). The corresponding map is exponential

$$z = \exp(2i\pi u/L) \tag{54}$$

Therefore

$$T_{\text{strip}}(u) = -(2\pi/L)^2 \left[T_{\text{plane}}(z)z^2 - \tfrac{1}{24}c \right] \tag{55}$$

and since $\left\langle T_{\text{plane}}(z) \right\rangle = 0$, we find

$$\left\langle T_{\text{strip}}(u) \right\rangle = \tfrac{1}{24}c(2\pi/L)^2 \tag{56}$$

This is the basis of an understanding of the anomaly as a Casimir effect, i.e. a shift in free energy due to the finite geometry. This effect was originally derived, then measured, in quantum electrodynamics where it gives rise to forces between neutral conductors. The very fact that it is measurable, although very tiny, is evidence of the long-range character of electromagnetic interactions. Similarly we expect in the present case an L-dependent shift in the free energy due to the confining boundary conditions. For the purpose of normalization, we assume that the free energy per unit area at criticality, in the infinite plane, is equal to zero. Including the extra factor $1/2\pi$, introduced above, the variation of the total free energy is given by a formula analogous to (19), namely

$$\delta \ln Z = - \int_{\mathcal{D}} \frac{d^2 u}{2\pi} \left\langle T_{\mu\nu}(u, \bar{u}) \right\rangle \partial_\mu \delta r_\nu \tag{57}$$

where the domain of integration $\mathcal{D}$ is the infinite strip. Let us choose in particular $\delta r_1 = \delta\varepsilon u_1$, $\delta r_2 = 0$, meaning an horizontal dilatation of the strip. Of course, this is not a conformal transformation (it is rather a quasi conformal one, transforming circles into ellipses, with a fixed ratio of axes). But the definition of $T_{\mu\nu}$ does not imply conformal transformations either. The only nonvanishing term in $\partial_\mu \delta r_\nu$ is $\partial_1 \delta r_1 = \delta\varepsilon$, and $T_{11} = T(u) + \bar{T}(\bar{u})$. We face the difficulty that for an infinite strip $\ln Z$ is infinite. From translational invariance, we expect that there is a well-defined free energy per unit longitudinal length, in other words that

$$F(L) = \lim_{M \to \infty} \frac{1}{M} \ln Z(L, M) \tag{58}$$

exists, provided that an extra boundary condition at a distance M is imposed in the longitudinal direction. Assuming that the limit is independent of the longitudinal boundary condition, we conclude from equations (56)–(58), that

$$\delta F(L) = \delta\varepsilon L \frac{dF(L)}{dL} = -\tfrac{1}{12} c \frac{2\pi}{L} \delta\varepsilon$$

Hence

$$F(L) = \tfrac{1}{6} c \frac{\pi}{L} \tag{59a}$$

We have used the fact that the free energy per unit area in the infinite plane, F_0, vanishes. Otherwise it would contribute an extra term $F_0 L$ to the r.h.s. of (59a). An equivalent statement if one uses a transfer matrix method, is that, at criticality, the logarithm of the largest eigenvalues λ_0 behaves as

$$\ln \lambda_0 = F_0 L + \tfrac{1}{6} c \frac{\pi}{L} \tag{59b}$$

as L tends to infinity. We shall discuss in section 2 the validity of equation (59) with respect to the assumption of a smooth limit starting from a finite domain. Clearly if some correlations grow with distance, as can occur in nonunitary models, the result is expected to be invalid. For the time being, assuming that all correlations decrease with distance, we see that a relation such as (59) enables one to have access to the value of the central charge, as has been exemplified in numerical studies. For instance for polymers ($O(n)$ model $n \to 0$) or percolation (Q-state Potts model with $Q \to 1$), one finds as expected $c = 0$, while we shall see that the Ising model corresponds to $c = \frac{1}{2}$, the three-state Potts model to $c = \frac{4}{5}$ and the free Bose field (Gaussian model) to $c = 1$, to quote just a few of these values.

9.1.5   Virasoro algebra

The above mapping on a strip suggests the use of an operator formalism with a transfer matrix $\mathcal{T}$. Since it corresponds to the use of polar coordinates in the plane, this formalism deserves the name of radial quantization. Time evolution amounts to dilatations and the eigenvalues of the corresponding generator are the conformal weights. To simplify notations, we absorb the previous factor $2i\pi/L$ in the definition of the variable u, so that

$$z = e^u = \rho e^{i\theta} \qquad \ln \rho = \operatorname{Re} u \qquad \theta = \operatorname{Im} u$$

$$T_{\text{strip}}(u) = T_{\text{plane}}(z)z^2 - \tfrac{1}{24}c \qquad (60)$$

According to (60), the propagation axis is now along the real u-axis. Mean values in the strip are identified with traces of ordered products of operators (the ordering being along the "time" axis $\operatorname{Re} u$). To an observable $A(u)$ this procedure associates an operator $\hat{A}(u)$. The transfer matrix is the analog of the exponential of a Hamiltonian, and relates operators at different values of $\operatorname{Re} u$ according to

$$\hat{A}(\operatorname{Re} u, \operatorname{Im} u) = \mathcal{T}^{-\operatorname{Re} u} \hat{A}(0, \operatorname{Im} u) \mathcal{T}^{\operatorname{Re} u} \qquad (61)$$

Correlations are written for a decreasing sequence of values of $\operatorname{Re} u$ as

$$\langle A_1(u_1) \cdots A_n(u_n) \rangle_{\text{strip}}$$
$$= \lim_{M \to \infty} \left[\operatorname{Tr} \left(\mathcal{T}^{\frac{1}{2}M - \operatorname{Re} u_1} \hat{A}_1(0, \operatorname{Im} u_1) \mathcal{T}^{\operatorname{Re}(u_1 - u_2)} \hat{A}_2(0, \operatorname{Im} u_2) \cdots \right. \right.$$
$$\left. \left. \cdots \hat{A}_n(0, \operatorname{Im} u_n) \mathcal{T}^{\operatorname{Re} u_n + \frac{1}{2}M} \right) \right] / \operatorname{Tr} \mathcal{T}^M$$
$$= \lim_{M \to \infty} \frac{\operatorname{Tr} \mathcal{T}^M \hat{A}_1(u_1) \ldots \hat{A}_n(u_n)}{\operatorname{Tr} \mathcal{T}^M} \qquad (62)$$

If required, necessary subtractions are understood to convert these expressions to connected ones. In a quantum mechanical context, this is the Gell-Mann–Low formula. Let us assume for simplicity the model to be unitary. In Euclidean space, this means $T = \exp -H$, where H is Hermitian bounded from below. Let its unique ground state be denoted $|0\rangle$, with $\langle 0|$ the adjoint bra. For a unitary model, the limit $M \to \infty$ projects on the ground state if we assume a finite gap (in energy) to excited states. As the situation under consideration involves a strip with finite width, these assumptions look realistic. Therefore we find

$$\langle A_1(u_1)\dots A_n(u_n)\rangle = \left\langle 0 \left| \hat{A}_1(u_1)\dots\hat{A}_n(u_n) \right| 0\right\rangle \qquad (63)$$
$$\mathrm{Re}\, u_1 > \mathrm{Re}\, u_2 > \cdots > \mathrm{Re}\, u_n$$

For a real observable A, the corresponding operator is Hermitian

$$\hat{A}^\dagger(0, \mathrm{Im}\, u) = \hat{A}(0, \mathrm{Im}\, u)$$
$$\hat{A}^\dagger(u) = \hat{A}(-\bar{u}) \qquad (64)$$

and the correlations satisfy the reflection property

$$\left\langle 0 \left| \hat{A}_1(u_1)\dots\hat{A}_n(u_n) \right| 0\right\rangle^* = \left\langle 0 \left| \hat{A}_n(-\bar{u}_n)\dots\hat{A}_1(-\bar{u}_1) \right| 0\right\rangle \qquad (65)$$

The second argument $\bar{u}$ has not been written explicitly, but is understood. With the help of the exponential map, we can view time translations in the u-plane as dilatations in the (punctured) z-plane, provided we keep in mind the correspondence between operators

$$\hat{A}_{\mathrm{strip}}(u, \bar{u}) = \hat{A}_{\mathrm{plane}}(z, \bar{z}) z^h \bar{z}^{\bar{h}} \qquad (66)$$

As a result

$$\langle A_1(z_1)\dots A_n(z_n)\rangle = \left\langle 0 \left| \hat{A}_1(z_1)\dots\hat{A}_n(z_n) \right| 0\right\rangle \qquad (67)$$
$$|z_1| > |z_2| > \cdots > |z_n|$$

Under a local conformal map $z \to z' = z + \delta\varepsilon g(z)$, with $g(z)$ analytic, a primary field behaves as

$$\delta\hat{A}(z) = \delta\varepsilon \left[h\frac{\mathrm{d}g(z)}{\mathrm{d}z} + g(z)\frac{\mathrm{d}}{\mathrm{d}z} \right] \hat{A}(z) \qquad (68)$$

and equation (39) implies the existence of an operator $\hat{T}(z)$ such that for $|z_1| > |z_2| > \cdots > |z_n|$,

$$\sum_{p=1}^{n} \left\langle 0 \left| \hat{A}_1(z_1) \ldots \delta\hat{A}_p(z_p) \ldots \hat{A}_n(z_n) \right| 0 \right\rangle$$

$$= \delta\varepsilon \left\{ \int_C \frac{\mathrm{d}z}{2\mathrm{i}\pi} g(z) \left\langle 0 \left| \hat{T}(z)\hat{A}_1(z_1) \ldots \hat{A}_n(z_n) \right| 0 \right\rangle \right.$$

$$\left. - \int_{C'} \frac{\mathrm{d}z}{2\mathrm{i}\pi} g(z) \left\langle 0 \left| \hat{A}_1(z_1) \ldots \hat{A}_n(z_n)\hat{T}(z) \right| 0 \right\rangle \right\}$$

where C encircles all points z_i while C' leaves all these points outside. Introducing intermediate contours, we see that this is equivalent to the operator statement

$$\delta\hat{A}(z) = \delta\varepsilon \left\{ \oint_{|z'|>|z|} \frac{\mathrm{d}z'}{2\mathrm{i}\pi} g(z')\hat{T}(z')\hat{A}(z) \right.$$

$$\left. - \oint_{|z|>|z'|} \frac{\mathrm{d}z'}{2\mathrm{i}\pi} g(z')\hat{A}(z)\hat{T}(z') \right\}$$

As a particular case

$$\delta\hat{T}(z) = \delta\varepsilon \left\{ \left[2g'(z) + g(z)\frac{\mathrm{d}}{\mathrm{d}z} \right] \hat{T}(z) + \tfrac{1}{12}cg'''(z) \right\} \tag{70}$$

$$= \delta\varepsilon \left\{ \oint_{|z'|>|z|} \frac{\mathrm{d}z'}{2\mathrm{i}\pi} g(z')\hat{T}(z')\hat{T}(z) - \oint_{|z|>|z'|} \frac{\mathrm{d}z'}{2\mathrm{i}\pi} g(z')\hat{T}(z)\hat{T}(z') \right\}$$

Expand now $\hat{T}$ as well as $\hat{\bar{T}}$ in Laurent series

$$\hat{T}(z) = \sum_{n=-\infty}^{+\infty} \frac{L_n}{z^{n+2}} \qquad \hat{\bar{T}}(\bar{z}) = \sum_{n=-\infty}^{+\infty} \frac{\bar{L}_n}{\bar{z}^{n+2}} \tag{71}$$

The extra factor z^{-2} has its origin in relation (60). We substitute (71) into (70), with $g(z)$ also expanded in a Laurent series with arbitrary coefficients. Identifying the factor of each of those on both sides, we obtain a set of commutation relations defining two infinite Lie algebras, called Virasoro algebras

$$\left[L_j, L_k \right] = (j-k)L_{j+k} + \tfrac{1}{12}cj(j^2-1)\delta_{j+k,0}$$

$$\left[\bar{L}_j, \bar{L}_k \right] = (j-k)\bar{L}_{j+k} + \tfrac{1}{12}cj(j^2-1)\delta_{j+k,0} \tag{72}$$

$$\left[L_j, \bar{L}_k \right] = 0$$

These algebras were originally introduced by Virasoro in the study of the dual model of particle physics, an ancestor of string theory, as a means of enforcing what we now interpret as reparametrization invariance. Viewed in the z-plane, the origin plays no special role, and $\hat{T}$ and $\hat{\bar{T}}$ could be expanded around any other point obtained as the image of the origin through a Möbius transformation.

The formulas (72) can be understood as follows. Suppose first that $c = 0$, and look at one of the two equivalent algebras. One observes that the operators can be realized as first order differential operators of the form

$$\ell_m = -z^{1+m}\frac{\partial}{\partial z} \qquad \left[\ell_j, \ell_k\right] = (j - k)\ell_{j+k} \qquad (73)$$

If we restrict z to be of unit modulus, the ℓ_m's act on functions defined on the circle, and are generators for the diffeomorphisms of the unit circle, meaning one-to-one infinitely differentiable maps (here connected to the identity). The extra term in (72) proportional to c commutes with all elements in the algebra. It therefore indicates an extension of the Lie algebra (73) by a one-dimensional algebra. Since the latter commutes with all other elements, one speaks of a central extension, and hence of a central charge. The situation is the analog at the Lie algebra level of the one which prevails in quantum mechanics when one looks for representations of symmetries, up to a phase. For instance, half integral spins exhibit a central extension of $SO(3)$ by Z_2, in the form of SU_2. A simple calculation shows that, up to a change of basis, the form indicated in equation (72) is the only central extension of the algebra (73).

The operators L_0, $L_{\pm 1}$, or $\bar{L}_0$, $\bar{L}_{\pm 1}$, or ℓ_0, $\ell_{\pm 1}$, generate in each case a closed three-dimensional algebra, which is nothing but the complex Lie algebra of the global conformal group $SL(2,\mathcal{C})$. Note that correspondingly, the coefficient of the anomaly vanishes. Furthermore, L_0, the generator of dilatations, is diagonal in the adjoint representation $\left[L_0, L_{-p}\right] = pL_{-p}$, with integral eigenvalues. It serves to define a grading of the enveloping algebra generated by the L's, i.e. a splitting in homogeneous subspaces with a definite eigenvalue for L_0. Said otherwise, L_{-p} acting on an eigenstate of L_0 increases the eigenvalue by p.

The vacuum state is invariant under global conformal transformations. As a result, it is annihilated by $L_0, L_{\pm 1}$. But since $T(z)\,|0\rangle$ remains regular as $z \to 0$ (as well as $\langle 0|\, T(z)$ as z goes to infinity) it follows that one has the stronger properties

$$\begin{aligned} p \geq -1 && L_p\,|0\rangle = 0 \\ p \leq 1 && \langle 0|\, L_p = 0 \end{aligned} \tag{74}$$

with similar equations for $\bar{L}_p$.

i) Show that in order to obtain (74), it is sufficient to require that beyond $L_0, L_{\pm 1}$, L_2 annihilates the state $|0\rangle$ to the right, or L_{-2} annihilates $\langle 0|$ to the left.

ii) As a consequence of (74), $\langle 0\,|L_p|\,0\rangle = 0$, for any p. Rederive from the commutation relation (72) alone that, for $|z_1| > |z_2|$ say

$$\left\langle 0 \left| \hat{T}(z_1)\hat{T}(z_2) \right| 0 \right\rangle = \frac{\frac{1}{2}c}{z_{12}^4}$$

In a unitary model, a Hermitian observable satisfies the relation

$$\hat{A}(z, \bar{z})^\dagger = \bar{z}^{-2h} z^{-2\bar{h}} \hat{A}(\bar{z}^{-1}, z^{-1}) \tag{75}$$

When applied to T or $\bar{T}$, this entails that a necessary condition for unitarity of a representation is that

$$L_p^\dagger = L_{-p} \quad , \qquad \bar{L}_p^\dagger = \bar{L}_{-p} \tag{76}$$

Equation (40) implies that, as $z_1 \to z_2$, $\hat{T}(z_1)\hat{A}(z_2)\,|0\rangle$ behaves as $h_A z_{12}^{-2} \hat{A}(z_2)\,|0\rangle$. Let us define the state $|h, \bar{h}\rangle$ through

$$\lim_{z \to 0} \hat{A}(z, \bar{z})\,|0\rangle = |h, \bar{h}\rangle$$

$$\langle h, \bar{h}| = \lim_{z \to 0} \langle 0|\, \hat{A}^\dagger(z, \bar{z}) = \lim_{z \to \infty} \langle 0|\, \hat{A}(z, \bar{z}) z^{2h} \bar{z}^{2\bar{h}} \tag{77a}$$

It then follows that

$$\begin{aligned} & L_0\,|h, \bar{h}\rangle = h\,|h, \bar{h}\rangle && \bar{L}_0\,|h, \bar{h}\rangle = \bar{h}\,|h, \bar{h}\rangle \\ p > 0 \quad & L_p\,|h, \bar{h}\rangle = \bar{L}_p\,|h, \bar{h}\rangle = 0 \end{aligned}$$
$$\tag{77b}$$

and similar conjugate relations

$$\begin{aligned} & \langle h, \bar{h}|\, L_0 = \langle h, \bar{h}|\, h && \langle h, \bar{h}|\, \bar{L}_0 = \langle h, \bar{h}|\, \bar{h} \\ p < 0 \quad & \langle h, \bar{h}|\, L_p = \langle h, \bar{h}|\, \bar{L}_p = 0 \end{aligned}$$

In a unitary representation the generator of (real) dilatations, $L_0 + \bar{L}_0$, and the one of rotations, $L_0 - \bar{L}_0$, are both Hermitian.

We conclude that the space of states in a critical model carries a representation of the product of algebras $\left\{ L_p \right\} \times \left\{ \bar{L}_p \right\}$, which we denote $V \times \bar{V}$. Such a representation will be in general reducible, containing states such as $|h, \bar{h}\rangle$ and their "descendants" i.e. states generated by applying a product of L_p and $\bar{L}_p$'s.

Let us focus our attention on one of these algebras, V say. By isomorphism the reasoning applies to $\bar{V}$ as well. Let us consider a representation with a so-called highest weight vector, i.e. a state $|h\rangle$ fulfilling the requirements (77). We henceforth omit the label $\bar{h}$ pertaining to $\bar{V}$. The name highest weight vector derives from the terminology used in the representation theory of Lie algebras. In the present context, it is rather a ground state in a given sector. Consider the so-called *Verma module* generated by $|h\rangle$, i.e. the infinite-dimensional vector space spanned by linear combinations of monomials in the L's applied to $|h\rangle$. For our purpose, we may think of finite linear combinations, so that we need not worry about convergence properties. This space splits into homogeneous subspaces characterized by the eigenvalues $h+n$ of L_0, where n is a non-negative integer, which will be called the level. The quantity $h + n$ is sometimes also called the weight of the corresponding states. Here it is nothing but a conformal weight of a nonprimary field which creates this state when applied to the vacuum state $|0\rangle$. When considering the states of level n, we may use the commutation relations (72) as well as the properties of the state $|h\rangle$ annihilated by $L_p, p > 0$, to show that they all reduce to combinations of some basic ones, involving only $L_p, p < 0$, and ordered conveniently, for instance states of the form

$$|\{p\}, h\rangle \equiv L_{-p_1} L_{-p_2} \dots L_{-p_k} |h\rangle \qquad (78a)$$
$$0 < p_1 \le p_2 \le \dots \le p_k, \qquad k \le n, \qquad p_1 + p_2 + \dots + p_k = n$$

Similarly in the conjugate space

$$\langle h, \{p\}| \equiv \langle h| L_{p_k} \dots L_{p_1} \qquad (78b)$$

with the same constraints on the indices. Such infinite-dimensional spaces carry obviously a representation of the algebra.

The Virasoro algebra is closely akin to the Kac–Moody algebras defined as infinite-dimensional extensions of semisimple Lie alge-

bras (appendix C) which occur in the study of two-dimensional
current algebras as well as in infinite-dimensional integrable systems. In both cases, we deal with graded Lie algebras, and the
techniques for handling them are closely related.

Applying an operator $L_{p'}, p' > 0$ to a state such as $|\{p\}\,h\rangle$
decreases its level by p'. Therefore if a product of such operators
has a degree equal to n, it will produce a multiple of $|h\rangle$ when
applied to $|\{p\}\,,h\rangle$. We write

$$L_{p'_q} \cdots L_{p'_1} L_{-p_1} \cdots L_{-p_k} |h\rangle = (\{p'\}\,,\{p\})_{c,h}\, |h\rangle \qquad (79a)$$

$$0 < p_1 \leq p_2 \leq \cdots \leq p_k \qquad\qquad 0 < p'_1 \leq p'_2 \cdots \leq p'_q$$

$$\Sigma p'_i = \Sigma p_j = n$$

The coefficients $(\{p'\}\,,\{p\})_{c,h}$ for fixed c,h and level n can be
computed solely from the structure of the Virasoro algebra and
the properties of the state $|h\rangle$. They form a matrix indexed by
partitions of the integer n, called the contragredient form. In
the unitary case, this contragredient form is the matrix of scalar
products at level n,

$$(\{p'\}\,,\{p\})_{c,h} = \frac{\langle h, \{p'\}|\,\{p\}\,, h\rangle}{\langle h\,|h\rangle} \qquad\qquad (79b)$$

where we recall that

$$\begin{aligned}
L_0\,|\{p\}\,,h\rangle &= (h + \Sigma p_i)\,|\{p\}\,,h\rangle \\
\langle h, \{p\}|\,L_0 &= \langle h, \{p\}|\,(h + \Sigma p_i)
\end{aligned} \qquad (80)$$

To achieve unitarity, we require that at each level n, the contragredient form is positive semidefinite. If it is definite positive,
the Verma module itself yields a unitary irreducible representation. The remaining option, when the contragredient form is only
semidefinite positive, is that there exist "null states" generating
invariant subspaces.

Since for any $p > 0$

$$L_p L_{-p}\,|h\rangle = \left[L_p, L_{-p}\right]|h\rangle = [2ph + \tfrac{1}{12}cp(p^2 - 1)]\,|h\rangle \qquad (81)$$

a necessary condition for unitarity is $h \geq 0$ and $c \geq 0$. We
therefore find only positive weights in unitary models. But
negative ones, which correspond to correlations growing with
increasing distance, are *a priori* not excluded in a nonunitary
situation. Thus we have

$$\text{unitarity} \quad \Longrightarrow \quad c \geq 0, \qquad h \geq 0 \qquad (82)$$

In any representation the number of linearly independent states at level n cannot exceed $p(n)$, the number of partitions of the integer n. If one agrees to define $p(0) = p(1) = 1$, Euler's generating function for partitions is

$$\sum_0^\infty p(n)q^n = \frac{1}{\prod_1^\infty (1 - q^n)} \equiv \frac{1}{P(q)} \tag{83}$$

where for convergence $|q| < 1$. Dedekind's function $\eta(\tau) = q^{1/24}P(q)$, where $q = \exp 2i\pi\tau$, will play a prominent role in the following discussion.

A means of studying the contragredient form is to compute its determinant at each level n (at level 0 it is 1 by definition). These determinants, first obtained by Kac

$$\det{}_n(c, h) = \det\left(\{p'\}, \{p\}\right)_{c,h} \tag{84}$$

are obviously polynomials in c and h. They enjoy the property that $\det_n(c, h)$ vanishes if and only if at some level $n' \leq n$ there exists a linear combination

$$|s\rangle = \sum \alpha_{\{p\}} L_{-p_1} \cdots L_{-p_k} |h\rangle \qquad \sum p_i = n' \leq n$$

which is annihilated by all $L_p, p > 0$. If such a vector exists at level $n' \leq n$, it generates at level n a subspace of vectors annihilated by the L_p, $p > 0$. Conversely if the determinant vanishes, some of its columns satisfy linear relations. The corresponding vectors of the form $\sum \alpha_{\{p\}} L_{-p_1} \cdots L_{-p_k} |h\rangle$, $\sum p_i = n$ are annihilated by all products of $L_{p'}$, $\sum p_i' = n$. A recursive reasoning shows that there must exist a singular vector of degree $n' \leq n$. If such singular (or "null") vectors exist with $n' > 0$, in order that the representation of V be reduced, the invariant subspace that they generate must be eliminated using an equivalence relation. This means that the field that creates these singular vectors when applied to the vacuum state $|0\rangle$ must be set equal to zero. Since the Kac determinants are polynomials in h, this will enable us to single out for fixed c special values for the conformal weights.

These determinants can at first be computed by merely using the commutation relations. Beyond the trivial cases

$$\det{}_0(c, h) = 1 \qquad\qquad \det{}_1(c, h) = 2h \tag{85}$$

where the vanishing of $\det_1$ for $h = 0$ reflects the fact that $|0\rangle$ itself is a null vector, let us look at the instructive case $n = 2$ with

$$(\{p'\}, \{p\})_{c,h} = \begin{pmatrix} 4h + \frac{1}{2}c & 6h \\ 6h & 4h(2h + 1) \end{pmatrix} \qquad (86a)$$

where the lines or columns are indexed respectively by the partitions $\{2\}$ and $\{1, 1\}$. Therefore

$$\det_2(c, h) = 2h \left[16h^2 + 2(c - 5)h + c \right] \qquad (86b)$$

We note that $\det_2$ contains $\det_1$ as a factor. The new information is the existence of two new zeroes, occurring for

$$h_\pm = \frac{5 - c \pm \sqrt{(1 - c)(25 - c)}}{16} \qquad (87)$$

These zeroes are complex conjugate for c real between 1 and 25, and real otherwise. We exclude the former case from consideration and look at the corresponding singular vectors of level two, i.e. with weight $h_\pm + 2$. We can check directly that

$$|s_\pm\rangle = \left(\frac{3}{2(2h_\pm + 1)} L_{-1}^2 - L_{-2} \right) |h\rangle \qquad (88a)$$

satisfy

$$L_0 |s_\pm\rangle = (h_\pm + 2) |s_\pm\rangle \qquad L_p |s_\pm\rangle = 0 \qquad p > 0 \ (88b)$$

To verify (88b), we observe that L_1 and L_2 generate, through commutation, all the L_p, $p > 0$. Let us introduce the following useful parametrization of the central charge, for any real or complex, finite or infinite m

$$c = 1 - \frac{6}{m(m + 1)} \qquad (89)$$

The above values $h_\pm$ take the form

$$h_+ = \frac{m + 3}{4m} = \frac{[2(m + 1) - m]^2 - 1}{4m(m + 1)}$$

$$h_- = \frac{m - 2}{4(m + 1)} = \frac{[m + 1 - 2m]^2 - 1}{4m(m + 1)} \qquad (90)$$

All this is stated in operator language. We can relate it also to short-distance expansions of a primary field A (of weight h) with a product of fields $T(z)$. For short, we omit again the argument

$\bar{z}$ of A. Equations (40) can be reinterpreted by saying that, when inserted in correlation functions, one has the short-distance expansion

$$T(z)A(u) = \frac{A^{(0)}(u)}{(z-u)^2} + \frac{A^{(-1)}(u)}{(z-u)} + A^{(-2)}(u) + (z-u)A^{(-3)} + \cdots$$

$$A^{(0)}(u) = hA(u)$$

$$A^{(-1)}(u) = \frac{\partial A(u)}{\partial u}$$

$$\cdots$$

$$(91a)$$

as well as

$$T(z)T(u) = \frac{\frac{1}{2}c}{(z-u)^4} + \frac{2T(u)}{(z-u)^2} + \frac{1}{(z-u)}\frac{\partial T(u)}{\partial u} + \cdots \quad (91b)$$

We have therefore a correspondence $A(0) \leftrightarrow |h\rangle$, $A^{(0)}(0) \leftrightarrow L_0|h\rangle$, $A_{(0)}^{(-p)} \leftrightarrow L_{-p}|h\rangle$. One can repeat with the derived fields $A^{(-p)}$ the above short-distance expansion defining $A^{(-p_1,-p_2)}$ in correspondence with $L_{-p_1}L_{-p_2}|h\rangle$, and so on.

The primary field A or $A^{(0)} = hA(z)$ has known transformation properties. Similarly for $A^{(-1)}(u) = \partial_u A(u)$. It is instructive to pursue the idea, in order to obtain an alternative interpretation of singular vectors in terms of the corresponding fields. Recall that if A is primary

$$A(z) = A'(\xi)\left(\frac{d\xi}{dz}\right)^h$$

$$T(z) = T'(\xi)\left(\frac{d\xi}{dz}\right)^2 + \frac{1}{12}c\{\xi,z\}$$

$$(92)$$

It is obvious that $A^{(-1)}$ does not transform as a primary field. To find the behaviour of $A^{(-2)}$, we compare the short-distance expansions for $T(z')A(z)$ and $T'(\xi')A'(\xi)$, using the transformation laws (92). Thus

$$hA(z) + (z'-z)\partial A(z) + (z'-z)^2 A^{(-2)}(z) + \cdots$$

$$= hA'(\xi)\left(\frac{z'-z}{\xi'-\xi}\right)^2\left(\frac{d\xi}{dz}\right)^h\left(\frac{d\xi'}{dz'}\right)^2$$

$$+ (z'-z)\frac{z'-z}{\xi'-\xi}\left(\frac{d\xi}{dz}\right)^h\left(\frac{d\xi'}{dz'}\right)^2\partial_\xi A'(\xi)$$

$$+ (z' - z)A'^{(-2)}(\xi)\left(\frac{d\xi}{dz}\right)^{h}\left(\frac{d\xi'}{dz'}\right)^{2}$$

$$+ \tfrac{1}{12}c(z' - z)\{\xi', z'\}\, A'(\xi)\left(\frac{d\xi}{dz}\right)^{h} + \cdots$$

After some rather laborious calculations, we find

$$A^{(-2)}(z) = A'^{(-2)}(\xi)\left(\frac{d\xi}{dz}\right)^{h+2}$$

$$+ hA'(\xi)\left(\frac{d\xi}{dz}\right)^{h}\left[\tfrac{2}{3}\frac{d^{3}\xi/dz^{3}}{d\xi/dz} - \tfrac{1}{4}\left(\frac{d^{2}\xi/dz^{2}}{d\xi/dz}\right)^{2}\right]$$

$$+ \tfrac{3}{2}\frac{dA'(\xi)}{d\xi}\left(\frac{d\xi}{dz}\right)^{h+1}\frac{d^{2}\xi/dz^{2}}{d\xi/dz}$$

$$+ \tfrac{1}{12}cA'(\xi)\left(\frac{d\xi}{dz}\right)^{h}\left[\frac{d^{3}\xi/dz^{3}}{d\xi/dz} - \tfrac{3}{2}\left(\frac{d^{2}\xi/dz^{2}}{d\xi/dz}\right)^{2}\right]$$

$$(93a)$$

We let the courageous reader find the corresponding formula for $A^{(-p_1,-p_2,\ldots)}$! To say the least, the derived fields have rather complicated transformation laws. But let us compare $(93a)$ with the behaviour of $\partial_z^2 A(z)$ corresponding to $L_{-1}^2 |h\rangle$ we have

$$\partial_z^2 A(z) = \partial_\xi^2 A'(\xi)\left(\frac{d\xi}{dz}\right)^{h+2} + \partial_\xi A'(\xi)(2h + 1)\left(\frac{d\xi}{dz}\right)^{h}\frac{d^{2}\xi}{dz^{2}}$$

$$+ hA'(\xi)\left(\frac{d\xi}{dz}\right)^{h}\left[\frac{d^{3}\xi/dz^{3}}{d\xi/dz} + (h - 1)\left(\frac{d^{2}\xi/dz^{2}}{d\xi/dz}\right)^{2}\right]$$

$$(93b)$$

If we now form the combination

$$\chi_\pm(z) = \frac{3}{2(2h_\pm + 1)}\partial_z^2 A(z) - A^{(-2)}(z) \qquad (94a)$$

for either value $h_\pm$ given by (87) or (90), it is straightforward to verify that

$$\chi_\pm(z) = \chi'_\pm(\xi)\left(\frac{d\xi}{dz}\right)^{h_\pm + 2} \qquad (94b)$$

In other words, singular vectors are associated to combinations of derived fields, arising from the short-distance expansion with the energy–momentum tensor, which transform as primary fields with an incremented weight. This is a useful equivalence. It shows that if we call χ as above such a combination, requiring χ to vanish is a consistent (covariant) condition. This amounts to factoring out invariant subspaces leaving irreducible representations. Finally, such conditions lead to partial differential equations satisfied by correlation functions involving the original field. Of course, its weight h has to be selected as a zero of a Kac determinant.

Let us consider a field A_1 with $h_1 \equiv h_\pm$. Requiring $\chi_\pm = 0$, leads to second order partial differential equations of the form

$$\frac{3}{2(2h_1 + 1)} \frac{\partial^2}{\partial z_1^2} \langle A_1(z_1) \dots A_n(z_n) \rangle$$

$$= \sum_{p=2}^{n} \left[\frac{h_p}{(z_1 - z_p)^2} + \frac{1}{(z_1 - z_p)} \partial_{z_p} \right] \langle A_1(z_1) \dots A_n(z_n) \rangle \quad (95)$$

obtained from equation (40) by using the short-distance expansion of T with A_1. Higher order equations will correspond to singular vectors of higher level. We see that the theory offers the possibility of being able to discuss interesting circumstances involving both the selection of specific values of the conformal weights (or critical exponents) and the determination of the corresponding correlation functions. This appears an amazing achievement. It is therefore quite useful that an explicit expression is known for the Kac determinant at any level.

Show that at level 3

$$\det_3(c, h) = 48h^2 \left[16h^2 + 2(c - 5)h + c \right] \left[3h^2 + (c - 7)h + c + 2 \right]$$

and obtain the corresponding singular vectors. Show that in general, $\det_{n-1}$ is a factor of $\det_n$.

9.1.6   The Kac determinant

The following remarkable expression was found by Kac

$$\det_n(c, h) = \text{cst} \times \prod_{\substack{r,s=1 \\ 1 \leq rs \leq n}}^{n} (h - h_{r,s})^{p(n-rs)} \quad (96)$$

where the central charge is parametrized as in (89), the zeroes $h_{r,s}$ are indexed by two positive integers, r and s, and read

$$h_{r,s} = \frac{[r(m+1) - sm]^2 - 1}{4m(m+1)} \quad (97)$$

A given zero occurs first at level $n' = \text{Inf}(rs)$ and propagates with a multiplicity $p(n - n')$ at level n (the infimum is over the set of pairs r, s giving the same value for $h_{r,s}$). One can check (96) for small values of n by direct computation. The parametrization $c = 1 - 6/m(m+1)$ is singular for $c = 1$ or 25. These cases require special consideration.

The proof given by Feigin and Fuchs of the fundamental result (96), (97) is involved but interesting, so we shall reproduce it below. As the values obtained for the weights solve in essence the question of finding two-dimensional critical exponents, it rightly deserves to be presented in some detail, but can, however, be skipped without harm, should the reader wish to do so.

It is not difficult to obtain the degree of $\det_n(c, h)$ as a polynomial in h. One readily convinces oneself that the term of highest degree arises from the product of diagonal terms in the contragredient form, each of which is the coefficient of $|h\rangle$ in

$$L_n^{\alpha_n} \cdots L_1^{\alpha_1} L_{-1}^{\alpha_1} \cdots L_{-n}^{\alpha_n} |h\rangle \qquad \sum_1^n k\alpha_k = n \qquad \alpha_k \geq 0$$

of degree $\sum_1^n \alpha_k$ in h. We therefore have the following formula for the degree ρ_n

$$\rho_n = \sum_{\alpha_k \geq 0, \sum_1^n k\alpha_k = n} \left(\sum_{k=1}^n \alpha_k \right) \tag{98}$$

where the first sum is over partitions of n, with α_k the number of times a particular integer $1 \leq k \leq n$ occurs in such a partition. We set $\rho_0 \equiv 1$ and write a generating function, for $|q| < 1$, as

$$\sum_0^\infty \rho_n q^n = t \frac{\partial}{\partial t} \sum_{\alpha_1, \alpha_2 \ldots = 0}^\infty t^{\sum_1^\infty \alpha_k} q^{\sum_1^\infty k\alpha_k} \bigg|_{t=1}$$

$$= t \frac{\partial}{\partial t} \frac{1}{\prod_1^\infty (1 - tq^k)} \bigg|_{t=1} = \frac{1}{P(q)} \sum_{r=1}^\infty \frac{q^r}{1 - q^r}$$

$$= \sum_{n'=0}^\infty p(n') q^{n'} \sum_{r,s=1}^\infty q^{rs} = \sum_{n=0}^\infty q^n \left[\sum_{\substack{r,s \geq 1 \\ rs \leq n}} p(n - rs) \right] \tag{99}$$

which is the expected result.

As a side remark, Coste has obtained the numerical, c-independent coefficient, omitted in (96). This prefactor is, with ρ_n as in equation (98)

$$2^{\rho_n} \prod_{\text{partitions of } n} \alpha_1! \cdots \alpha_n! 1^{\alpha_1} \cdots n^{\alpha_n}$$

Using their notations, let us now turn to the proof of Feigin and Fuchs. It involves four stages.

i) Introduce the set $F_{\lambda,\mu}$ of Laurent series (in the sequel we think of $z = e^u$, allowing for arbitrary powers in z)

$$z^\mu f(z) = \sum_{-\infty}^{+\infty} f_k z^{k+\mu} \tag{100}$$

of weight $-\lambda$, i.e. transforming as $f(z) \leftrightarrow f'(\xi)$, with

$$z^\mu f(z) = \xi^\mu f'(\xi) \left(\frac{\mathrm{d}\xi}{\mathrm{d}z} \right)^{-\lambda} \tag{101}$$

in an analytic map $z \leftrightarrow \xi$. Of course, the prime on $f'(\xi)$ does not mean a derivative. Taking a basis $f_{(k)}(z) = z^k$, we have for an infinitesimal transformation $\xi = z - \delta\varepsilon z^{1-j}$ for j integer

$$f_{(k)} \rightarrow f'_{(k)} = f_{(k)} + \delta f_{(k)} \qquad \delta f_{(k)} = \delta\varepsilon(\ell_j f_{(k)})$$
$$\ell_j f_{(k)} = [\mu + k + \lambda(j-1)] f_{(k-j)} \qquad [\ell_j, \ell_k] = (j-k)\ell_{j+k} \tag{102}$$

showing that the set $\{\ell_j\}$ gives a representation of the Lie algebra of diffeomorphisms of the unit circle, generalizing (73). Provided that $\lambda + \lambda' + 1 = \mu + \mu' + p + 1 = 0$, with p an integer, the integral

$$\int \frac{\mathrm{d}z}{2\mathrm{i}\pi z^{1+p}} g_{\lambda'\mu'}(z) f_{\lambda\mu}(z) \tag{103a}$$

gives an invariant bilinear form on $F_{\lambda'\mu'} \otimes F_{\lambda\mu}$. Similarly when $\lambda + \lambda' + 1 = \mu - \mu' + 2\lambda' + p + 1 = 0$, the contragredient form

$$\int \frac{\mathrm{d}z}{2\mathrm{i}\pi z^{1+p}} g_{\lambda'\mu'}(z^{-1}) f_{\lambda\mu}(z) \tag{103b}$$

is also invariant. Consider now the nth exterior product $\bigwedge^n F_{\lambda\mu}$. Let us show that the determinant

$$\varphi(z) = \begin{vmatrix} f_0 & \frac{\mathrm{d}f_0}{\mathrm{d}z} & \cdots & \frac{\mathrm{d}^{n-1}}{\mathrm{d}z^{n-1}} f_0 \\ \vdots & \vdots & \ddots & \vdots \\ f_{n-1} & \frac{\mathrm{d}f_{n-1}}{\mathrm{d}z} & \cdots & \frac{\mathrm{d}^{n-1}}{\mathrm{d}z^{n-1}} f_{n-1} \end{vmatrix}$$

defines a map $\bigwedge^n F_{\lambda\mu} \to F_{n\lambda-\frac{1}{2}n(n-1),n\mu}$. For that purpose we associate to z the n-component vector $f_0(z),..., f_{n-1}(z)$. Consider now n points $z_0,..., z_{n-1}$. Under a map $z \leftrightarrow \xi$, by definition

$$\det f_i'(\xi_j) = \prod_{k=1}^{n} \left(\frac{z_k}{\xi_k}\right)^\mu \left(\frac{\mathrm{d}z_k}{\mathrm{d}\xi_k}\right)^{-\lambda} \det f_i(z_j)$$

If $z_i = z + \varepsilon_i$, where all ε_i tend to zero and are of the same order, we have

$$\det f_i(z + \varepsilon_j) \sim \frac{1}{\prod_0^{n-1} k!} \prod_{0\leq s<r\leq n-1} (\varepsilon_r - \varepsilon_s)\varphi(z)$$

Similarly $\xi_i = \xi + \varepsilon_i\, \mathrm{d}\xi/\,\mathrm{d}z + \cdots$, so that, comparing the leading terms,

$$\varphi'(\xi) = \left(\frac{z}{\xi}\right)^{n\mu} \left(\frac{\mathrm{d}z}{\mathrm{d}\xi}\right)^{\frac{1}{2}n(n-1)-n\lambda} \varphi(z)$$

which is the required property.

The constant function belongs to $F_{0,0}$. As a result, using (103), if

$$\lambda = (n-2)(n+1)/2n \qquad\qquad \mu = -1 - p/n \qquad (104a)$$

we conclude that whenever n functions f_j belong to $F_{\lambda,\mu}$ the integral over the unit circle

$$\gamma = \oint \frac{\mathrm{d}z}{2i\pi z^{p+n}} \det \left[\frac{\mathrm{d}^k f_j}{\mathrm{d}z^k}\right]_{0\leq k,j\leq n-1} \qquad (104b)$$

is an invariant quantity. The definition (103a) with $p = 0$ enables one to assert the existence of an element of $\bigwedge^n F_{-1-\lambda,-1-\mu} \equiv \bigwedge^n F_{-(n-1)(n+2)/2n,p/n}$, the dual of $\bigwedge^n F_{\lambda,\mu}$, corresponding to the invariant γ. All what we have to do is to substitute in (104b) the expansion

$$\frac{\mathrm{d}^k}{\mathrm{d}z^k} f_j(z) = \sum_{q=-\infty}^{+\infty} q(q-1)\cdots(q-k+1)z^{q-k} \oint \frac{\mathrm{d}z'}{2i\pi z'^{q+1}} f_j(z')$$

Consequently with P denoting a permutation on n symbols, $(-1)^P$ the signature of P, we find

$$\gamma = \oint \frac{\mathrm{d}z}{2i\pi z^{p+n}} \sum_P (-)^P \sum_{q_0,...,q_{n-1}=-\infty}^{\infty} \int \frac{1}{(2i\pi)^n} \frac{\mathrm{d}z_0 \cdots \mathrm{d}z_{n-1}}{z_0^{1+q_0} \cdots z_{n-1}^{1+q_{n-1}}}$$

$$\times f_{P_0}(z_0)\cdots f_{P_{n-1}}(z_{n-1}) z^{\sum_{i=0}^{n-1} q_i - \frac{1}{2}n(n-1)}$$

$$\times q_1 q_2(q_2 - 1) \cdots q_{n-1}(q_{n-1} - 1) \cdots (q_{n-1} - n + 2)$$

$$= \sum_{q_0,\ldots,q_{n-1}=-\infty}^{+\infty} \delta_{\Sigma_{i=0}^{n-1} q_i, \frac{1}{2}n(n-1)+n-1+p} \oint \prod_{r=0}^{n-1} \left(\frac{dz_r f_r(z_r)}{2i\pi z_r^{1+q_r}} \right)$$

$$\times \det \begin{vmatrix} 1 & q_0 & \cdots & q_0^{n-1} \\ \vdots & \vdots & \ddots & \vdots \\ 1 & q_{n-1} & \cdots & q_{n-1}^{n-1} \end{vmatrix}$$

The last determinant follows from Vandermonde's formula. Let us change all q_i into $-q_i$, and recall that the basis vectors z^j were denoted $f_{(j)}$. We then reach the conclusion that

$$\gamma = \frac{1}{n!} \sum_{\substack{q_0,\ldots,q_{n-1}=-\infty \\ \Sigma q_i = -\frac{1}{2}n(n+1)+1-p}}^{+\infty} \prod_{0 \le r < s \le n-1} (q_r - q_s) \det \left(f_{(q_i)}, f_j \right)_0$$

where $(\ ,\ )_0$ is the bilinear form (103a) for $p = 0$. We read from this expansion that the corresponding invariant element $\varphi_{n,p}$, belonging to $\bigwedge^n F_{-(n-1)(n+2)/2n,p/n}$ is given by

$$\varphi_{n,p} = \sum_{\substack{q_0,\ldots,q_{n-1}=-\infty \\ \Sigma q_i = -\frac{1}{2}n(n+1)+1-p}}^{+\infty} \left[\prod_{0 \le r < s \le n-1} (q_r - q_s) \right] f_{(q_0)} \wedge \cdots \wedge f_{(q_{n-1})}$$

(105)

ii) In the second stage, we associate to $F_{\lambda,\mu}$ a fermionic Fock space and a corresponding representation of the Virasoro algebra. To each integer i, positive or negative, we associate a pair $a_i, a_i^\dagger$ of operators satisfying the usual canonical anticommutation relations

$$\left\{ a_i, a_j^\dagger \right\} = \delta_{ij} \qquad \left\{ a_i, a_j \right\} = \left\{ a_i^\dagger, a_j^\dagger \right\} = 0 \qquad (106)$$

The reference "vacuum" state $|\Omega\rangle$ is taken by convention to satisfy

$$\begin{aligned} i \ge 0 \qquad & a_i |\Omega\rangle = 0 \\ i < 0 \qquad & a_i^\dagger |\Omega\rangle = 0 \end{aligned} \qquad (107)$$

meaning that all "particle" states such that $i < 0$ are occupied, and all those with $i \ge 0$ empty. Corresponding to the choice (107) we have a natural Wick ordering : :, where the operators a_i $(i \ge 0)$ or $a_i^\dagger$ $(i < 0)$ are on the right of the $a_i(i < 0)$ or $a_i^\dagger(i \ge 0)$, taking into account the sign of the permutation required to bring a given monomial in this order. With this construction, we can associate two quadratic operators, a charge Q and a Hamiltonian H, defined as follows. First

$$Q = \sum_{-\infty}^{+\infty} : a_k^\dagger a_k := \sum_{k \geq 0} a_k^\dagger a_k - \sum_{k < 0} a_k a_k^\dagger \qquad (108)$$

annihilates $|\Omega\rangle$ and will commute with the representation of the Virasoro algebra. In a sector of charge $Q = n$, it is useful to introduce a new reference state $|\Omega, n\rangle$ (such that $|\Omega, 0\rangle \equiv |\Omega\rangle$). This state is annihilated by $a_i (i \geq n)$ and $a_i^\dagger (i < n)$ and corresponds to filling one-particle states up to $n - 1$. If the index is thought of as an "energy", then the Hamiltonian H reads

$$H = \sum_{-\infty}^{+\infty} k : a_k^\dagger a_k := \sum_{k \geq 0} k a_k^\dagger a_k - \sum_{k < 0} k a_k a_k^\dagger \qquad (109)$$

Of course, the one-particle state with $k = 0$ does not contribute to H, which is a positive operator and satisfies

$$H |\Omega, n\rangle = \tfrac{1}{2} n(n - 1) |\Omega, n\rangle \qquad (110)$$

valid for n positive or negative. Observe that $|\Omega, 0\rangle$ and $|\Omega, 1\rangle$ have the same zero energy. For any integer j different from zero, we define the second quantized form of (102), namely

$$L_j = \sum_k a_{k-j}^\dagger \{ \mu + k + \lambda(j - 1) \} a_k \qquad (111)$$

$$j \neq 0 \qquad \left[L_j, a_k^\dagger \right] = \{ \mu + k + \lambda(j - 1) \} a_{k-j}^\dagger \qquad [L_j, Q] = 0$$

The Wick ordering is irrelevant here since k and $k - j$ are distinct. In computing commutators, we use

$$[AB, C] = A\{B, C\} - \{A, C\} B$$
$$[AB, CD] = A\{B, C\} D - AC\{B, D\} + \{A, C\} DB - C\{A, D\} B \qquad (112)$$

The above formulas for Q, H and L_j assume a more compact form if we introduce a pair of complex Fermi fields

$$\psi(z) = \sum_k a_k z^{-k} \qquad \bar{\psi}(z) = \sum_k a_k^\dagger z^k \qquad (113)$$

Then

$$Q = \oint \frac{dz}{2i\pi z} : \bar{\psi}(z)\psi(z) :$$

$$H = \frac{1}{2} \oint \frac{dz}{2i\pi} : \frac{\partial \bar{\psi}}{\partial z} \psi - \bar{\psi} \frac{\partial \psi}{\partial z} : \qquad (114)$$

$$j \neq 0 \qquad L_j = \oint \frac{dz}{2i\pi} z^j \left\{ \frac{(\mu - \lambda)}{z} \bar{\psi}\psi - \lambda \frac{\partial \bar{\psi}}{\partial z}\psi - (1 + \lambda)\bar{\psi}\frac{\partial \psi}{\partial z} \right\}$$

If now j, k and $j + k$ are all distinct from zero

$$j, k, j + k \neq 0 \qquad [L_j, L_k] = (j - k)L_{j+k} \qquad (115)$$

To obtain an expression of L_0 and of the central charge, we compute for $j \neq 0$ the following commutator

$$[L_j, L_{-j}]$$

$$= \sum_k \{\mu + k + j + \lambda(j - 1)\} \{\mu + k - \lambda(j + 1)\} \left(a_k^\dagger a_k - a_{k+j}^\dagger a_{k+j} \right)$$

We reach here a delicate point where care must be exercised. In such a singular expression, it is not advisable to perform a translation on indices. In a subspace of fixed charge Q, and acting on a vector with a finite energy, the operator $a_k^\dagger a_k - a_{k+j}^\dagger a_{k+j}$ will vanish for $|k|$ large enough. Therefore, as it stands, the previous expression is well-defined. Assuming j is positive, we observe that

$$a_k^\dagger a_k - a_{k+j}^\dagger a_{k+j} =: a_k^\dagger a_k : - : a_{k+j}^\dagger a_{k+j} : + \begin{cases} 0 \text{ if } k \geq 0 \\ 1 \text{ if } -j \leq k < 0 \\ 0 \text{ if } k < -j \end{cases}$$

Thus

$$[L_j, L_{-j}] = 2j \sum_k (\mu - \lambda + k) : a_k^\dagger a_k :$$

$$+ \sum_{k=0}^{j-1} \{\mu + k + \lambda(j - 1)\} \{\mu + k - j - \lambda(j + 1)\}$$

With

$$\sum_{k=0}^{j-1} 1 = j \qquad \sum_{k=0}^{j-1} k = \tfrac{1}{2}j(j - 1) \qquad \sum_{k=0}^{j-1} k^2 = \tfrac{1}{6}j(j - 1)(2j - 1)$$

we define in the sector of charge $Q = n$

$$L_0 = H - \tfrac{1}{2}n(n-1) + h_n \qquad h_n = \tfrac{1}{2}(n+\mu)(n+\mu-2\lambda-1) \quad (116)$$

such that

$$[L_j, L_{-j}] = 2jL_0 + \tfrac{1}{12}cj(j^2 - 1) \qquad (117)$$

$$c = -2\left[6\lambda(\lambda + 1) + 1\right] \qquad (118)$$

Obviously equation (117) also holds for negative j. Finally combining the definitions (111) and (116), we can check the last commutation relations of the Virasoro algebra

$$[L_0, L_j] = -jL_j \qquad (119)$$

This is a typical example of a representation, which we denote $\mathcal{F}_{\lambda,\mu,n}$ in the sector with charge n, exhibiting the central charge as a quantum anomaly. Here c is a function of λ only, satisfying

$$c(\lambda) = c(-1 - \lambda) \qquad (120)$$

The adjoint of $L_j(\lambda, \mu)$ for real λ and μ is such that

$$L^\dagger_{-j}(\lambda, \mu) = L_j(-1 - \lambda, \mu - 1 - 2\lambda) \qquad (121)$$

The correspondence $\lambda \to \lambda' = -1 - \lambda$, $\mu \to \mu' = \mu - 1 - 2\lambda$, which leaves both the central charge c and the c-number part of L_0 invariant, is recognized to be the same as occurring in the contragredient form (103b) for $p = 0$. Thus $\mathcal{F}_{\lambda,\mu}$ and $\mathcal{F}_{-1-\lambda,\mu-1-2\lambda}$ are contragredient.

Parenthetically we observe that when $\lambda = 1$, corresponding in the present notation to vector fields (or dually when $\lambda' = -1 - \lambda = -2$ to quadratic differentials) the value (118) of the central charge is $c = -26$. This is the significant value in the quantization of the bosonic string theory, giving the anomaly from the Faddeev-Popov ghost system required to compensate for reparametrization invariance. This anomaly has to be cancelled by the coordinate contribution associated to free fields, each of central charge unity, and leads to a requirement of 26 dimensions for a consistent model (chapter 11).

There was nothing special in our choice of the reference state $|\Omega\rangle$ and the corresponding definition of the charge zero states. The shift operator defines a canonical transformation

$$\begin{aligned} (a_i, a_i^\dagger) &\to (a_{i-n}, a_{i-n}^\dagger) \\ |\Omega, n\rangle &\to |\Omega, 0\rangle \end{aligned} \qquad (122)$$

which exhibits an isomorphism between the charge n and charge 0 sectors. Of course, its effect on Wick ordering is not trivial. Consequently, the representations $\mathcal{F}_{\lambda,\mu;n}$ and $\mathcal{F}_{\lambda,\mu+n;0}$, denoted simply $\mathcal{F}_{\lambda,\mu+n}$, are isomorphic. Indeed μ was only appearing in the combination $\mu + n$ in (116), and the central charge is independent both from μ and n. We verify that

$$\begin{aligned} L_0 |\Omega, n\rangle &= h_n |\Omega, n\rangle \\ L_j |\Omega, n\rangle &= 0 \qquad j > 0 \end{aligned} \qquad (123)$$

Therefore $\mathcal{F}_{\lambda,\mu;n}$ admits $|\Omega, n\rangle$ as a vector corresponding to a dominant weight h_n.

As an example, the simplest case of this construction corresponds to $\lambda = \mu = -\frac{1}{2}$. In the charge zero sector, for any j

$$L_j = \tfrac{1}{2} \oint \frac{\mathrm{d}z}{2\mathrm{i}\pi} z^j \left[: \partial_z \bar{\psi} \psi : - : \bar{\psi} \partial_z \psi : \right]$$

$$c = 1 \tag{124}$$

This sector involves two types of fermions with opposite charge and $c = 1$. One therefore suspects that there exists a "real" fermionic theory (Majorana fermions), with only one type of fermions, which will be studied in a later part of this chapter, corresponding to $c = \frac{1}{2}$. For instance, if $\mathcal{Z}$ denotes the set of integers, and if we define

$$\psi(z) = \sum_{k \in \mathcal{Z}+\frac{1}{2}} a_k z^{-k} \qquad \{a_k, a_{k'}\} = \delta_{k+k',0} \tag{125}$$

the field ψ is double valued, but bilinears are well-defined on the unit circle. We define

$$L_j = \tfrac{1}{4} \sum_{k \in \mathcal{Z}+\frac{1}{2}} (2k - j) : a_{j-k} a_k := \tfrac{1}{4} \oint \frac{\mathrm{d}z}{2\mathrm{i}\pi} z^j \left(: \partial_z \psi \psi : - : \psi \partial_z \psi : \right)$$

$$[L_j, L_k] = (j - k) L_{j+k} + \tfrac{1}{24} j(j^2 - 1) \delta_{j+k,0} I \tag{126}$$

In the expression above, the : : symbol requires the a_k's, $k > 0$, to be ordered to the right of those with $k < 0$, keeping track of the permutation sign. The reference state is annihilated by the operators a_k, $k > 0$ and $L_0 = \sum_{k \geq 1/2} k a_{-k} a_k$ is a positive operator.

iii) We now set λ equal to the value

$$\lambda_0 = -\frac{(n-1)(n+2)}{2n} \tag{127}$$

with n an integer larger or equal to 2 in order to construct a function $\varphi_{n,p}$ as in (105) with $\mu = p/n$. The operator

$$\phi' = \sum_{\substack{q_0,\dots,q_{n-1}=-\infty \\ \Sigma q_i = -\frac{1}{2}n(n+1)+1-p}}^{+\infty} \prod_{0 \leq r < s \leq n-1} (q_r - q_s) a_{q_0}^\dagger \cdots a_{q_{n-1}}^\dagger \tag{128}$$

is such that

$$[Q, \phi'] = n\phi' \qquad [H, \phi'] = \left(1 - p - \tfrac{1}{2}n(n+1)\right)\phi' \tag{129}$$

Hence ϕ'^k increases the charge by nk, and by construction satisfies

$$L_j \left(\lambda_0, \mu = \frac{p}{n}\right) \phi'^k \,|\Omega\rangle = 0 \qquad j > 0 \tag{130}$$

which means that it generates, using (122), a singular vector in $\mathcal{F}_{\lambda_0,\mu=p/n,nk} \sim \mathcal{F}_{\lambda_0,\mu=p/n+nk}$. In $\mathcal{F}_{\lambda_0,p/n+nk}$ this state may be obtained by acting on $|\Omega, -nk\rangle$ with the operator ϕ^k (this was the reason for the prime above), where

$$\phi = \sum_{\substack{q_0,\dots,q_{n-1}=-\infty \\ \Sigma q_i = -\frac{1}{2}n(n+1)-n^2k+1-p}}^{+\infty} \prod_{0 \le r < s \le n-1} (q_r - q_s) a_{q_0}^\dagger \cdots a_{q_{n-1}}^\dagger \quad (131a)$$

satisfying

$$[Q, \phi^k] = nk\phi^k \qquad [H, \phi^k] = k\left(1 - p - \tfrac{1}{2}n(n+1) - n^2k\right)\phi^k \tag{131b}$$

Set

$$\mu_0 = \frac{p}{n} + nk \tag{132}$$

Since $H|\Omega, -nk\rangle = \frac{1}{2}nk(nk+1)|\Omega, -nk\rangle$, the singular vector in the zero charge sector

$$|s\rangle = \phi^k |\Omega, -nk\rangle \tag{133}$$

obeys

$$\begin{aligned} Q|s\rangle &= 0 & L_0(\lambda_0, \mu_0)|s\rangle &= (h_0 + \ell k)|s\rangle \\ j &> 0 & L_j(\lambda_0, \mu_0)|s\rangle &= 0 \end{aligned} \tag{134}$$

with λ_0 and μ_0 given by (127) and (132), and the remaining symbols ℓ and h_0 stand for

$$\ell = 1 - p - \tfrac{1}{2}n^2(1+k) \qquad h_0 = \tfrac{1}{2}\mu_0(\mu_0 - 2\lambda_0 - 1) \tag{135}$$

Finally the central charge c is given by (118), with λ_0 substituted for λ. Since

$$L_0(\lambda_0, \mu_0)|\Omega\rangle = h_0|\Omega\rangle \tag{136}$$

the construction of the singular vector makes sense (i.e. $|s\rangle \ne 0$) provided ℓk is positive, i.e.

$$p < 1 - \tfrac{1}{2}n^2(k+1) \tag{137}$$

The whole procedure involves three arbitrary integers n, p, k and is such that ℓk is always an integer. As a result ℓ is an integer except when n is odd and k even, in which case ℓ is a half integer. To avoid this possibility we assume n even.

iv) We return to the proof of the Kac formula. Consider an abstract Verma module with central charge $c(\lambda_0)$ and dominant weight h_0, with corresponding vector $|h_0\rangle$. Let $\{L_{-j}\}$ denote an ordered string of operators L_{-j}. To the vector $\{L_{-j}\}\,|h_0\rangle$ we can let correspond one of the two kets

$$
\begin{aligned}
\{L_{-j}\}\,|h_0\rangle &\xrightarrow{f} \{L_{-j}(\lambda_0,\mu_0)\}\,|\Omega\rangle \\
\{L_{-j}\}\,|h_0\rangle &\xrightarrow{g} \{L_{-j}(-1-\lambda_0,\mu_0-1-2\lambda_0)\}\,|\Omega\rangle
\end{aligned}
\tag{138}
$$

The second ket is dual to the bra $\langle\Omega|\,\{L_j^\dagger(\lambda_0,\mu_0)\}$ according to equation (121), with h_0 invariant in the corresponding substitution.

Assume that the Verma module possesses a singular vector at a level smaller or equal to r. If the mapping f, which preserves the level sequence, is surjective, $\mathcal{F}_{\lambda_0,\mu_0}$ will also possess such a singular vector. If the application is not surjective, there must exist in the zero charge sector a ket $|\sigma\rangle$ with an excitation energy smaller or equal to r which is orthogonal to the image of f. Hence $\langle\sigma|\,\{L_{-j}(\lambda_0,\mu_0)\}\,|\Omega\rangle = 0$ for all strings of L_{-j} such that $\Sigma j = r$. If one takes complex conjugates, one concludes that $|\sigma\rangle$ is a singular vector for $\mathcal{F}_{-1-\lambda_0,\mu_0-1-2\lambda_0}$. Conversely, if the image under f or g admits a singular vector at a level $\leq r$, one readily sees that this must also be the case for the abstract module.

For λ and μ arbitrary, consider the quantity

$$
\begin{aligned}
b_{\alpha,\beta}(\lambda,\mu) &= b_{\beta,\alpha}(\lambda,\mu) = b_{-\alpha,-\beta}(-1-\lambda,\mu-1-2\lambda) \\
&= \left\{(\alpha+1)\left(\lambda+\tfrac{1}{2}\right)-\mu\right\}\left\{(\beta+1)\left(\lambda+\tfrac{1}{2}\right)-\mu\right\} - \tfrac{(\alpha-\beta)^2}{2}
\end{aligned}
\tag{139}
$$

The product $b_{\alpha\beta}(\lambda,\mu)b_{-\alpha,-\beta}(\lambda,\mu)$, symmetric in α, β, is also invariant in the correspondence $\lambda \to -1-\lambda$, $\mu \to \mu-1-2\lambda$. Hence it is only a function of h (given by equation (116) with $n = 0$), and c (given by (118)). We find

$$
\begin{aligned}
B_{\alpha,\beta}(c,h) = B_{-\alpha,-\beta}(c,h) &= \tfrac{1}{4}b_{\alpha\beta}(\lambda,\mu)b_{-\alpha,-\beta}(\lambda,\mu) \\
&= \left[h + \tfrac{1}{24}(\alpha^2-1)(c-13) + \tfrac{1}{2}(\alpha\beta-1)\right] \\
&\times \left[h + \tfrac{1}{24}(\beta^2-1)(c-13) + \tfrac{1}{2}(\alpha\beta-1)\right] + \tfrac{1}{16}(\alpha^2-\beta^2)^2
\end{aligned}
\tag{140}
$$

For $\lambda = \lambda_0 = -(n-1)(n+2)/2n$, $\mu = \mu_0 = nk + p/n$, n even, hence $\ell = 1 - p - n^2(k+1)/2$ integer, we just saw that the Kac determinant of degree $k\ell$ has to vanish. One verifies that

$$
b_{-k,-\ell}(\lambda_0,\mu_0) = 0
\tag{141}
$$

As a consequence, the determinant of level $k\ell$ has to vanish on an infinite number of points of the curve $B_{k,\ell}(c,h) = 0$. The latter is irreducible if $k \neq \ell$. Hence $\det_{k\ell}(c,h)$ is divisible by $B_{k,\ell}(c,h)$ or its root when $k = \ell$, and $B_{k,k}$ is a square. As a result, $\det^2_{k\ell}(c, h)$ admits $B_{k,\ell}(c, h)$ as a factor. Since a singular vector at level r generates a zero of order $p(s - r)$ at level $s \geq r$, the quantity

$$\prod_{\substack{k,\ell \\ k\ell \leq r}} B_{k,\ell}(c, h)^{p(r-k\ell)}$$

divides $\det_r(c, h)^2$. One checks that both expressions have the same degree in h. As a consequence, up to an h-independent factor,

$$[\det_r(c, h)]^2 = \text{cst} \prod_{\substack{1 \leq k,\ell \\ k\ell \leq r}} B_{k,\ell}(c, h)^{p(r-k\ell)} \tag{142}$$

Finally $B_{k,\ell}$ is symmetric in k and ℓ, hence occurs twice in the product (142) for $k \neq \ell$, while $B_{k,k}$ is a square

$$B_{k,k}(c, h) = \left[h - \frac{(1 - c)}{24}(k^2 - 1) \right]^2$$

Furthermore, when we use the parametrization $c = 1 - 6/m(m+1)$,

$$B_{k,\ell} = (h - h_{k,\ell})(h - h_{\ell,k}) \tag{143}$$

with $h_{k,\ell}$ given by equation (97). This allows the square root in (142) to be taken and produces Kac's formula (96).

9.1.7   Unitary and minimal representations

Let us complete the discussion on the conditions on c and h under which a representation is unitary, i.e.

$$L_j^\dagger = L_{-j} \tag{144}$$

For the module (or vector space) generated by a highest weight h to be a Hilbert space with its scalar product giving rise to the contragredient form (79a), it is necessary that (i) either the latter is positive definite (ii) or else if there exists singular vectors, that the invariant subspaces generated by the latter be factored out, leaving a quotient space in which the form is again positive. These conditions are also sufficient.

We have already seen that a necessary condition for unitarity is

$$c \geq 0, \qquad h \geq 0$$

In the range $0 < c < 1$, Friedan, Qiu and Shenker have succeeded in showing that the only allowed values required by unitarity are such that (i) c has the form (89) with m integer, $m \geq 2$ ($m = 2$ corresponds to $c = 0$ and the trivial identity representation) (ii) the weights h are given by the zeroes of the Kac determinant with the two integers r and s ranging over a finite set of values, $1 \leq s \leq r \leq m - 1$. In summary

$$c = 1 - 6/m(m + 1) \qquad m \text{ integer} \geq 2$$

$$h \equiv h_{r,s} = \frac{[r(m + 1) - sm]^2 - 1}{4m(m + 1)} \qquad 1 \leq s \leq r \leq m - 1$$

$$(145)$$

If we use only the restriction $1 \leq s \leq m, 1 \leq r \leq m - 1$, each weight $h_{r,s}$ occurs twice, since $h_{r,s} \equiv h_{m-r,m+1-s}$.

Truly interesting unitary representations of this sort are therefore characterized by an integer $m \geq 3$ and a finite set of possible weights, hence a finite set of possible scaling dimensions for the fundamental or primary observables. In this sense, we call these representations minimal. The corresponding Verma modules possess singular vectors, i.e. invariant subspaces. For the same value of c, it is also possible to consider fields belonging to representations with a weight $h_{r,s}$ with r and/or s outside the above range. Those might enter into the construction of other types of models which will not principally be dealt with here. Nevertheless, it is useful to introduce a table of these inclusions of invariant subspaces, because they enable one to understand the counting of the remaining linearly independent states at a given level, once the invariant subspaces are factored out. Fix a value of c as in (145). Feigin and Fuchs have established the set of inclusions between Verma modules, depicted on figure 3, arising from the Kac formula. On the picture, an arrow points from the larger space to the subspace. The corresponding weights differ by products of integers. Moreover five subspaces such as M, N', N'', P', P'' are such that M contains the sum (not direct sum) $N' \oplus N''$, while the intersection $N' \cap N''$ is equal to the sum $P' \oplus P''$. To obtain the irreducible representation at the top, with r and s in the restricted range given in equation (145), one has to factor out the sum of its first two "descendents". We shall see in the next section applications of these sets of inclusions.

In the range 1 to 25, there is not constraint on the values of c due to unitarity. At $c = 1$, there exists again a (simpler) form

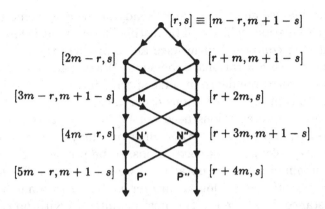

Fig. 3   Inclusions of Verma modules $[r', s']$ corresponding to a central charge $c = 1 - 6/m(m+1)$, $m \geq 3$, and a succession of weights $h_{r', s'} = [(r'(m+1) - s'm)^2 - 1]/4m(m+1)$.

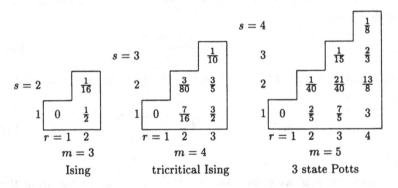

Fig. 4   Grid of conformal weights for the first three instances of unitary minimal representations.

of reducibility (and unitarity) whenever $h = \frac{1}{4}n^2$ with n integer.

As we do not claim here to give an exhaustive presentation, we shall limit ourselves to values of c in the range $c \leq 1$. Such cases are in essence the simplest instances. They include some well-known examples, with at most discrete symmetries for $c < 1$.

For the minimal unitary series, we illustrate the above considerations by displaying in figure 4 the first three examples of the grid of conformal weights. They are already labelled by some models, with qualifications which will be justified later on. It is fascinating to find that conformal invariance alone predicts the existence of systems with a few rational exponents. What is to be found is the key to the correspondence in some specific instance, as well as the physical meaning of the various observables.

Minimal representations need not be unitary, as shown by the simplest example below. Of course one may admit of certain pathologies, like growing correlations, to be properly interpreted. These minimal cases are obtained if we substitute in the formula $c = 1 - 6/m(m + 1)$ not an integer m but a rational fraction. For instance, if $p > p'$ are two positive integers with no common factors (coprimes) such that

$$m = \frac{p'}{p - p'} \qquad m + 1 = \frac{p}{p - p'} \qquad (146a)$$

the corresponding value of c (no longer restricted to be positive) is given by

$$c = 1 - \frac{6(p - p')^2}{pp'} \qquad (146b)$$

A conformal grid of minimal representations and their weights, is given by

$$h_{r,s} = h_{p'-r,p-s} = \frac{(rp - sp')^2 - (p - p')^2}{4pp'} \qquad (147)$$
$$1 \leq r \leq p' - 1 \qquad 1 \leq s \leq p - 1$$

Since for p, p' as above we can find r_0 and s_0 in the indicated range such that $(r_0 p - s_0 p')^2 = 1$, it is clear that, for $p - p' \geq 2$, there will exist negative weights in the table, and therefore the corresponding scalar field $A_{h_0,\bar{h}_0}$ will have correlations increasing with distance. This was excluded in the unitary minimal case. The consistency of these minimal models in the sense that at most a finite set of conformal weights appear for the primary operators will be studied in section 3.

To show that this is not a totally academic possibility, let us present the following example due to Cardy. Let us look at the most elementary case of the above family with a unique nontrivial scalar field, apart for the unit operator (and its descendents like the energy-momentum tensor). Concentrating on the V algebra, we look for a grid with two entries (taking into account the symmetry indicated in (147)). With duplication this means four entries in $1 \leq r \leq p' - 1$ $1 \leq s \leq p - 1$ with $p > p'$ and coprimes. The only possibility is a unique column ($r = 1$, $p' = 2$) of four boxes ($p = 5$, $1 \leq s \leq 4$). Consequently we have two weights, and the values for h and c are

$$h = 0, -\tfrac{1}{5} \qquad c = -\tfrac{22}{5} \qquad (148)$$

The unique nontrivial scalar field $A_{-\frac{1}{5}, -\frac{1}{5}}$ has a scaling dimension

$$\Delta_A = -\tfrac{2}{5} \qquad (149)$$

Now let us try to guess which scalar interacting field theory (at criticality) this could possibly correspond to. The only candidate seems to be the $i\varphi^3$ theory with imaginary cubic coupling, and coincides with the effective continuous model for the Lee–Yang edge singularity. In this instance no symmetry is broken at criticality (but the correlation length diverges) and the model has no chance to be unitary. Let, as usual, η denote the critical exponent for the two-point correlation. It follows in this two-dimensional case that $\eta = 2\Delta = -\tfrac{4}{5}$, indeed negative (hence correlations grow !). Furthermore we recall that we defined the exponent σ such that the singular part of the magnetization behaves as

$$m_s \sim h^\sigma \qquad (150)$$

where h denotes the deviation from the critical magnetic (external) field. Scaling relations imply that

$$\sigma = \frac{d - 2 + \eta}{d + 2 - \eta} \underset{d=2}{\rightarrow} \frac{\eta}{4 - \eta} = -\tfrac{1}{6} \qquad (151)$$

in excellent agreement with numerical simulations. There is a relation between this singular growing part of the magnetization and the fact that after subtraction we are left at criticality with a connected correlation growing at large distance.

9.1.8   Characters of the Virasoro algebra

For further applications it will be useful to know the so called characters of the representations of the Virasoro algebra. In a representation characterized by c and a highest weight h, these are generating functions for the number of linearly independent states at level n, hence corresponding to an eigenvalue $h+n$ for L_0. For a fixed $q = \exp 2i\pi\tau$ smaller than one in modulus ($\mathrm{Im}\,\tau > 0$)

the character appears as the trace in the given representation of
the operator q^{L_0}. In other words with $\dim(h + n)$ the dimension
of the space at level n we define

$$\chi_{(c,h)}(\tau) = \operatorname{Tr} q^{L_0 - c/24} = \sum_{n=0}^{\infty} \dim(h + n) q^{n+h-c/24} \qquad q = \exp 2i\pi\tau$$

(152)

The extra factor $q^{-c/24}$ is included in this definition for later
convenience. For general values c and h, when the highest weight
representation is such that the corresponding Verma module has
no singular vector, it is clear that

$$\chi_{(c,h)}(\tau) = \frac{q^{h-c/24}}{P(q)}$$

(153)

The factorized form of $P(q)$ shows that the series defining χ
converges for any $|q| < 1$. Moreover from equation (83), it is not
difficult to show that the leading term in the Hardy–Ramanujan
asymptotic expansion for $p(n)$, the number of partitions of n is

$$p(n) \underset{n \to \infty}{\sim} \frac{\exp \pi \sqrt{2n/3}}{4\sqrt{3}n} \left[1 + 0 \left(\frac{\ln n}{n^{1/4}} \right) \right]$$

(154)

The situation is more interesting for the degenerate representations, when the Verma module contains invariant subspaces. From
such diagrams as appear in figure 3, starting from an expression
such as given in (153) one must subtract similar terms. Furthermore at any stage such as shown on this figure, the sum of invariant subspaces is not a direct one. As a result starting from r and
s in the conformal grid the character formula for minimal representations obtained by Rocha–Caridi reads, with $\chi_{(c,h_{rs})} \equiv \chi_{r,s}$

$$\chi_{r,s}(\tau) = q^{-c/24} \frac{\sum_{k=-\infty}^{+\infty} q^{[(2kpp' + rp - sp')^2 - (p-p')^2]/4pp'} - (s \to -s)}{P(q)}$$

(155)

Here of course, as in (146), p and p' are coprime integers, $p > p'$,
r ranges between 1 and $p' - 1$, s between 1 and $p - 1$, with some
further restriction such as $rp - sp' > 0$ to avoid double counting.
Also $c = 1 - 6(p - p')^2/pp'$. Minimal unitary models are obtained
by setting $p' = m, p = m + 1$ with $m \geq 2$, integer.

i) In the minimal unitary case with $m = 2$, the irreducible representation is a trivial one-dimensional one. Setting $\chi \equiv 1$ in (155) leads to Euler's (pentagonal) identity

$$\prod_1^\infty (1 - q^n) = \sum_{n=-\infty}^{+\infty} (-1)^n q^{\frac{1}{2}n(3n+1)} \tag{156}$$

ii) For future use it is possible to present the expression (155) for the characters in a more compact form by introducing the following notation. Set $N = 2pp'$, an even integer. Instead of parametrizing the character by the couple r, s corresponding to $h_{r,s}$ in (147), use an integer modulo N, $\lambda \equiv rp - sp'$ mod$N$. Observe that if $w_0 \equiv r_0 p + s_0 p'$ mod$N$, with $r_0 p - s_0 p' = 1$ (according to the fact that p and p' are coprimes) we have $w_0^2 = 1$ mod$2N$ and $w_0 \lambda = rp + sp'$ mod$N$. Finally trade $P(q)$ for Dedekind's function

$$\eta(\tau) = q^{1/24} P(q) = q^{1/24} \prod_1^\infty (1 - q^n) \tag{157}$$

where we continue as above to use the notation $q = \exp 2i\pi\tau$. Then formula (155) can be rewritten as

$$\chi_\lambda(\tau) = \frac{\sum_{k=-\infty}^{+\infty} q^{(kN+\lambda)^2/2N} - q^{(kN+w_0\lambda)^2/2N}}{\eta(\tau)} \tag{158a}$$

showing that

$$\chi_\lambda(\tau) = \chi_{\lambda+N}(\tau) = \chi_{-\lambda}(\tau) = -\chi_{w_0\lambda}(\tau) \tag{158b}$$

These characters can be expressed in terms of elliptic functions. In appendix A, we recall some properties of elliptic θ-functions useful in the present context.

Many more nontrivial character formulas do exist, corresponding for instance to some of the irreducible invariant subspaces included in the reducible Verma modules. Also for $c = 1$ and $h = \frac{1}{4}\alpha^2$ (α integer) we have instead a linear chain of inclusions. Thus

$$\chi_\alpha(\tau) = \frac{q^{\alpha^2/4} - q^{(\alpha+2)^2/4}}{\eta(\tau)} \qquad \alpha \text{ integer.} \tag{159}$$

9.2   Examples

Let us now give some typical and elementary examples of conformal fields. We shall resume the general discussion in section 3.

9.2.1   Gaussian model

The simplest example, and the one which in a some sense is the building block of many of the explicit realizations, is the Gaussian, or massless, neutral free field model. The corresponding action reads, with $\varphi(\mathbf{x})$ real

$$S = \int \frac{\mathrm{d}^2 x}{2\pi} \frac{1}{2} \sum_{\mu=1,2} (\partial_\mu \varphi)^2 = \int \frac{\mathrm{d}^2 x}{2\pi} 2(\partial_{\bar{z}} \varphi)(\partial_z \varphi) \tag{160}$$

The normalization agrees with our definition of the energy-momentum tensor. With a sign fixed by equation (57), the latter takes the familiar form

$$T_{\mu\nu} = -\partial_\mu \varphi \partial_\nu \varphi + \tfrac{1}{2}\delta_{\mu\nu}(\partial\varphi)^2 \tag{161}$$

$$T \equiv T_{zz} = -(\partial_z \varphi)^2 \qquad \bar{T} \equiv T_{\bar{z}\bar{z}} = -(\partial_{\bar{z}}\varphi)^2 \qquad T_{z\bar{z}} = 0$$

The two-point function for φ involves an arbitrary scale R (infrared cutoff)

$$\langle \varphi(z_1, \bar{z}_1)\varphi(z_2, \bar{z}_2)\rangle = \tfrac{1}{2}\ln \frac{R^2}{z_{12}\bar{z}_{12}} \tag{162}$$

In the expression (161), a Wick ordering prescription has to be understood to give a definite meaning to correlation functions of T, as we discuss below. The logarithmic behaviour of the φ-correlations shows that it cannot be taken as a bona fide primary field (the tentative assignment of conformal weights is (0,0)). Rather, derivatives or exponentials of φ are candidates for observables. The corresponding correlations will have a more decent scaling behaviour, and should be such that the arbitrary scale R cancels out, as was discussed in chapter 4 of volume 1.

To obtain the central charge, we compute the $\langle TT \rangle$ correlation, omitting tadpole contributions i.e. self-contractions of φ's at the same point. This results in the expected analytic expression

$$\langle T(z_1)T(z_2)\rangle = 2\left(\partial_{z_1}\partial_{z_2}\langle \varphi(z_1,\bar{z}_1)\varphi(z_2,\bar{z}_2)\rangle\right)^2 = \frac{1}{2z_{12}^4} \tag{163}$$

with the complex conjugate for $\langle \bar{T}\bar{T}\rangle$. Consequently the central charge is unity

$$c = 1 \tag{164}$$

This was the rationale behind the normalization of the two-point function leading to the definition of the central charge.

Free fields allow us to understand how the anomaly computed for an infinitesimal conformal transformation gets promoted to the Schwarzian derivative for a finite one. The universal structure of the transformation law (50) implies that the calculation performed for $c = 1$ can be generalized by inserting any value for the central charge. The Wick prescription may be understood to stand for a short-distance subtraction procedure, after point splitting the arguments of T

$$T \rightarrow \lim_{z_1,z_2 \to z} -\partial_{z_1}\varphi \partial_{z_2}\varphi + \partial_{z_1}\partial_{z_2}\left(\frac{1}{2}\ln\frac{R}{z_{12}}\right)$$

to be inserted in correlation functions. In the above formula, only the leading short distance singularity of the correlation function does matter. Now, pretend that under a conformal map $\varphi(z,\bar{z}) = \tilde{\varphi}(u,\bar{u})$. In the new coordinate system with $u = f(z)$

$$T = \lim_{z_1,z_2 \to z} f'(z_1)f'(z_2)\left[-\partial_{u_1}\tilde{\varphi}\partial_{u_2}\tilde{\varphi} + \partial_{u_1}\partial_{u_2}\frac{1}{2}\ln\frac{\tilde{R}}{u_{12}}\right]$$
$$+ \frac{1}{2}\partial_{z_1}\partial_{z_2}\ln\frac{u_{12}}{z_{12}}$$

We have added and subtracted a term in such a way that the first part is recognized as $\tilde{T}(u)\,(\,\mathrm{d}u/\,\mathrm{d}z)^2$ in the coinciding point limit, while the second one yields the anomaly. Set

$$z_2 = z, \quad z_1 = z + \delta, \quad \frac{u_{12}}{z_{12}} = f' + \frac{1}{2}\delta f'' + \frac{1}{6}\delta^2 f''' + \cdots$$

$$\ln\frac{u_{12}}{z_{12}} = \ln f' + \frac{1}{2}\delta\frac{f''}{f'} + \delta^2\left(\frac{1}{6}\frac{f'''}{f'} - \frac{1}{8}\left(\frac{f''}{f'}\right)^2\right) + \cdots$$

and the anomaly is

$$\frac{1}{2}\left(\frac{\partial^2}{\partial z \partial\delta} - \frac{\partial^2}{\partial\delta^2}\right)\left[\ln f' + \frac{1}{2}\delta\frac{f''}{f'} + \delta^2\left(\frac{1}{6}\frac{f'''}{f'} - \frac{1}{8}\left(\frac{f''}{f'}\right)^2\right) + \cdots\right]$$
$$= \frac{1}{12}\left(\frac{f'''}{f'} - \frac{3}{2}\left(\frac{f''}{f'}\right)^2\right)$$

in agreement with equation (50) for $c = 1$.

Starting with the field φ we can define various families of operators, among them (Wick ordered) exponentials $e^{i\alpha\varphi}$. The quantity α is analogous to a charge, since correlation functions takes the form of Boltzmann weights involving classical Coulomb potentials (i.e. logarithmic in two dimensions) between pairs. Such correlations make sense only when the total charge vanishes, as we shall shortly explain. Wick ordering implies omission of self-contractions, and one finds

$$\left\langle \prod_j e^{i\alpha_j\varphi(z_j,\bar{z}_j)} \right\rangle_{\Sigma\alpha_j=0} = \prod_{j<k} \left(\frac{z_{jk}\bar{z}_{jk}}{R^2} \right)^{\frac{1}{2}\alpha_j\alpha_k} \tag{165}$$

If instead we were to keep self-contractions with an ultraviolet cutoff $\Lambda = a^{-1}$, the above expression would appear multiplied by a product of factors $\prod_j (a/R)^{\alpha_j^2/2}$, showing that Wick ordering amounts to an (infinite) multiplicative renormalization of $e^{i\alpha\varphi}$. However, this factor also depends on the infrared scale R. Choosing to normalize the operators at a given value R_0, we may study the remaining dependence on R of the form $R^{-u/2}$,

$$u = \sum_j \alpha_j^2 + \sum_{j\neq k} \alpha_j\alpha_k = (\Sigma\alpha_j)^2$$

Thus correlations are R-independent if and only if we have neutrality: $\Sigma\alpha_j = 0$. Should we violate this condition, correlations would vanish in the limit $R \to \infty$. This discussion also shows that the dimension of $e^{i\alpha\varphi}$ is $\frac{1}{2}\alpha^2$.

In particular

$$\left\langle e^{i\alpha\varphi(z_1,\bar{z}_1)} e^{-i\alpha\varphi(z_2,\bar{z}_2)} \right\rangle = \frac{1}{(z_{12}\bar{z}_{12})^{\frac{1}{2}\alpha^2}} \tag{166}$$

This means that $e^{i\alpha\varphi}$, or perhaps better the real forms $\cos\alpha\varphi$ and $\sin\alpha\varphi$, have conformal weights

$$h = \bar{h} = \tfrac{1}{4}\alpha^2 \tag{167}$$

i) When α is an integer, the associated character is given by equation (159) and the corresponding Verma module admits a singular vector at level $\alpha + 1$. This level is two if $\alpha = 1$, in which case $h = \bar{h} = \frac{1}{4}$ and the correlation (166) reduces to

$C_{12} = (z_{12}\bar{z}_{12})^{-\frac{1}{2}}$. This is in agreement with equation (87), where for $c = 1$, $h_+ = h_- = \frac{1}{4}$. Check that equation (95) is satisfied, as

$$\frac{\partial^2}{\partial z_1} C\left(z_{12}, \bar{z}_{12}\right) = \left(\frac{1}{4z_{12}^2} + \frac{1}{z_{12}}\frac{\partial}{\partial z_2}\right) C\left(z_{12}, \bar{z}_{12}\right)$$

ii) We have assumed that the field φ takes arbitrary real values. However, the Lagrangian is invariant under $\varphi \to \varphi + \text{cst}$ and $\varphi \to -\varphi$. This may be used to restrict to configurations, with the points φ and $\varphi + 2\pi\rho$ identified (ρ a constant) implying that φ takes its values on a circle (of radius ρ), as was the case for the XY-model. Nothing is changed in the correlations of vertex operators provided we assume that the charges α are of the form $\alpha = n/\rho$. We have also the possibility of including other operators corresponding to a frustration line, in the sense that, as $\mathbf{x}$ circles around one of the end points (a vortex), φ increases by an amount $2\pi m\rho$, with m an integer. These defects have been studied in the case of the XY-model. The correlation function of a pair $(m, -m)$ of vortices located at points 1 and 2 is a ratio of frustrated to unfrustrated partition functions. By shifting φ by φ_c, a multivalued harmonic solution (everywhere except at the sources), this ratio is equal to $\exp -S(\varphi_c)$. A similar interpretation can be given to the correlation of charges distinguished by a uniform harmonic function with δ-function sources at points 1 and 2. We can even generalize to a pair of composite $(n, m), (-n, -m)$ operators with

$$\varphi_c = \frac{in}{\rho}\text{Re}\left(\ln\frac{z - z_2}{z - z_1}\right) - \rho m \text{Im}\left(\ln\frac{z - z_2}{z - z_1}\right)$$

$$= \tfrac{1}{2}i\left(\frac{n}{\rho} + m\rho\right)\ln\left(\frac{z - z_2}{z - z_1}\right) - \tfrac{1}{2}i\left(\frac{n}{\rho} - m\rho\right)\ln\left(\frac{\bar{z} - \bar{z}_2}{\bar{z} - \bar{z}_1}\right)$$

$$\tag{168}$$

The occurrence of a factor i is such that in the shift $\varphi \to \varphi + \varphi_c$, the crossterm in the action yields $(in/\rho)[\varphi(1) - \varphi(2)]$. When one computes the corresponding classical action and performs the necessary subtractions, one finds

$$\exp(-S_{n,m}(1,2)) = \frac{1}{z_{12}^{2h_{n\,m}}\bar{z}_{12}^{2\bar{h}_{n,m}}}$$

$$\equiv \left\langle O_{n,m}(z_1, \bar{z}_1) O_{-n,-m}(z_2, \bar{z}_2) \right\rangle \tag{169}$$

with

$$h_{n,m} = \tfrac{1}{4}\left(\frac{n}{\rho} + m\rho\right)^2$$

$$\bar{h}_{n,m} = h_{n,-m} = \tfrac{1}{4}\left(\frac{n}{\rho} - m\rho\right)^2 \tag{170}$$

As a consequence, the generalized vertex operators $O_{n,m}$ have an integral spin equal to nm. The fact that the product of charge times vorticity (or magnetic charge) has to be an integer is reminiscent of the analogous property derived by Dirac for the quantum mechanical motion of a point charge in a three-dimensional monopole magnetic field. The operators $O_{n,m}$ have conformal weights depending on a continuous parameter ρ. They occur in the discussion of several models at $c = 1$, with a line of continuously varying exponents like the six vertex or Ashkin–Teller model (see section 3.6).

Similarly one can make use of the symmetry $\varphi \to -\varphi$ to restrict the values of the field to a so-called orbifold, here the quotient of the circle by a reflection with respect to a diameter.

9.2.2 Ising model

We want to cast in the present framework the critical Ising model presented in chapter 2 (volume 1). We recall that the action at criticality is effectively the one pertaining to a Majorana field. The action and two-point correlations read

$$S = \int \frac{d^2x}{2\pi}\psi\bar{\partial}\psi - \bar{\psi}\partial\bar{\psi} \tag{171}$$

$$\langle\psi(z_1)\psi(z_2)\rangle = 1/z_{12} \qquad \langle\bar{\psi}(\bar{z}_1)\bar{\psi}(\bar{z}_2)\rangle = 1/\bar{z}_{12}$$

These correlations are odd in the interchange $1 \leftrightarrow 2$, because of Fermi statistics, and one is analytic, the other antianalytic. Noether's theorem yields the energy–momentum tensor as

$$T = -\tfrac{1}{2}\psi\partial\psi \qquad\qquad \bar{T} = -\tfrac{1}{2}\bar{\psi}\bar{\partial}\bar{\psi} \tag{172}$$

with a normal ordering prescription as usual, so that

$$\langle T(z_1)\psi(z_2)\psi(z_3)\rangle = \frac{z_{23}}{2(z_{12})^2(z_{13})^2} \tag{173}$$

Equations (171) and (173) are consistent with the fact that ψ has weights $(\tfrac{1}{2}, 0)$, and $\bar{\psi}\ (0, \tfrac{1}{2})$. Thus ψ and $\bar{\psi}$ have half integral spin, as expected. The Ward identity is in agreement with these assignments, since it follows that

$$\langle T(z_1)\psi(z_2)\psi(z_3)\rangle = \left[\frac{1}{2z_{12}^2} + \frac{1}{2z_{13}^2} + \frac{1}{z_{12}}\partial_2 + \frac{1}{z_{13}}\partial_3\right]\langle\psi(z_2)\psi(z_3)\rangle$$

The central charge is obtained by computing

$$\langle T(z_1)T(z_2)\rangle = \tfrac{1}{4}\left\{\langle\psi(z_1)\partial\psi(z_2)\rangle\,\langle\partial\psi(z_1)\psi(z_2)\rangle\right.$$
$$\left. - \langle\psi(z_1)\psi(z_2)\rangle\,\langle\partial\psi(z_1)\partial\psi(z_2)\rangle\right\} \quad (174)$$
$$= \frac{1}{4z_{12}^4}$$

Consequently

$$c = \tfrac{1}{2} \quad (175)$$

It is natural at this point to expect that the model is built out from the minimal (unitary) representations of the Virasoro algebra for $c = \frac{1}{2}$, i.e. $m = 3$. We saw that the Kac table furnishes the weights 0, $\frac{1}{2}$, $\frac{1}{16}$, and we can tentatively identify the observables with integral spin, indeed zero, as being those with $h = \bar{h}$, through

	$(h, \bar{h})$	Δ	
identity	$(0, 0)$	0	I
energy density	$(\frac{1}{2}, \frac{1}{2})$	1	$\varepsilon = \bar{\psi}\psi$
spin	$(\frac{1}{16}, \frac{1}{16})$	$\frac{1}{8}$	σ

$$(176)$$

The energy density has a simple interpretation in terms of the Majorana field, while the spin σ has a nonlocal expression in $\psi, \bar{\psi}$. The critical exponent η is given by $\eta = 2\Delta_\sigma = \frac{1}{4}$ as expected. The logarithmic divergence of the integral over the energy–energy correlation $\langle\varepsilon\varepsilon\rangle$ implies $\alpha = 0$ (up to a logarithm) and $\nu = 1$. The other critical exponents are given by the standard scaling analysis as $\gamma = \frac{7}{4}$ and $\beta = \frac{1}{8}$. To summarize, once the value of the central charge $c = \frac{1}{2}$ is known, the minimal assumption gives the other exponents as well as a means to compute all critical correlations. Not much is gained in the present case, where integrability (and a long history) has provided us with the results beforehand. We can use this knowledge to check the formalism. But the way is now open to apply the same reasoning to other cases where much less is known, and the power of conformal field theory becomes obvious.

Since $\frac{1}{2}$ and $\frac{1}{16}$ are the two solutions $h_{\pm}$ of equations (87)–(90) implying the existence of singular vectors at level two, the corresponding correlation should satisfy a second order (partial) differential equation. According to chapter 2

$$\langle \varepsilon(1) \cdots \varepsilon(2n) \rangle = \left| \text{Pf} \frac{1}{z_{ij}} \right|^2 \tag{177}$$

$$\langle \sigma(1) \cdots \sigma(2n) \rangle^2 = \sum_{\substack{\alpha_i = \pm 1 \\ \Sigma \alpha_i = 0}} \prod_{i<j} \left(z_{ij} \bar{z}_{ij} \right)^{\frac{1}{4}\alpha_i \alpha_j} \tag{178}$$

Take the four point functions as an example. With

$$\langle \varepsilon(1)\varepsilon(2)\varepsilon(3)\varepsilon(4) \rangle = |E(z_1, z_2, z_3, z_4)|^2$$

$$E(z_1, z_2, z_3, z_4) = \frac{1}{z_{12} z_{34}} - \frac{1}{z_{13} z_{24}} + \frac{1}{z_{14} z_{23}}$$

One checks that

$$\left\{ \frac{3}{4} \frac{\partial^2}{\partial z_1^2} - \frac{1}{2} \left(\frac{1}{z_{12}^2} + \frac{1}{z_{13}^2} + \frac{1}{z_{14}^2} \right) - \left(\frac{1}{z_{12}} \partial_2 + \frac{1}{z_{13}} \partial_3 + \frac{1}{z_{14}} \partial_4 \right) \right\}$$
$$E(z_1, z_{,2} \, z_{,3} \, z_4) = 0$$

The case of the four spin correlation is more interesting, since it is not the modulus square of an analytic function, but rather a combination of the independent solutions of the corresponding equation. First rewrite explicitly the expression (178) for $n = 2$, with $r_{ij}^2 = z_{ij} \bar{z}_{ij}$, as

$$\langle \sigma(1)\sigma(2)\sigma(3)\sigma(4) \rangle^2$$

$$= \left(\frac{r_{14} r_{32}}{r_{12} r_{34} r_{13} r_{42}} \right)^{\frac{1}{2}} + \left(\frac{r_{13} r_{42}}{r_{12} r_{34} r_{14} r_{32}} \right)^{\frac{1}{2}} + \left(\frac{r_{12} r_{34}}{r_{13} r_{42} r_{14} r_{32}} \right)^{\frac{1}{2}}$$

The equation is

$$\left\{ \frac{4}{3} \frac{\partial}{\partial z_1^2} - \frac{1}{16} \left(\frac{1}{z_{12}^2} + \frac{1}{z_{13}^2} + \frac{1}{z_{14}^2} \right) - \left(\frac{1}{z_{12}} \partial_2 + \frac{1}{z_{13}} \partial_3 + \frac{1}{z_{14}} \partial_4 \right) \right\}$$
$$\langle \sigma(1)\sigma(2)\sigma(3)\sigma(4) \rangle = 0$$

The correlation is real, symmetric and positive. From the asymptotic factorization condition,

$$\lim_{R \to \infty} \langle \sigma(\mathbf{r}_1 + \mathbf{R})\sigma(\mathbf{r}_2 + \mathbf{R})\sigma(\mathbf{r}_3)\sigma(\mathbf{r}_4) \rangle = \langle \sigma(\mathbf{r}_1)\sigma(\mathbf{r}_2) \rangle \langle \sigma(\mathbf{r}_3)\sigma(\mathbf{r}_4) \rangle$$

its normalization follows from the one of the two-point function, here $\langle \sigma(1)\sigma(2) \rangle = \sqrt{2}/r_{12}^{1/4}$. Using global invariance, and the cross-ratio

$$x = \frac{z_{12}z_{34}}{z_{14}z_{32}}$$

with $x = 0, 1, \infty$ as z_1 coincides with z_2, z_3 or z_4, we write

$$\langle \sigma(1)\sigma(2)\sigma(3)\sigma(4)\rangle = \left(\frac{z_{14}z_{32}}{z_{12}z_{34}z_{13}z_{42}}\right)^{\frac{1}{8}} (c.c.)^{\frac{1}{8}} f(x,\bar{x})$$

The powers in the prefactors are in agreement with the weights of σ. The equation takes the hypergeometric form

$$\left\{x(1-x)\frac{\partial^2}{\partial x^2} + (\tfrac{1}{2}-x)\frac{\partial}{\partial x} + \tfrac{1}{16}\right\} f = 0$$

A general solution is a combination of $\sqrt{1\pm\sqrt{1-x}}$. We have the same equation for the $\bar{x}$ dependence. Consequently with $a_{\pm\pm}$ constants

$$f = \sum_{\pm\pm} a_{\pm\pm}\sqrt{1\pm\sqrt{1-x}}\sqrt{1\pm\sqrt{1-\bar{x}}}$$

Ignoring the overall multiplicative normalization, we are left with three constants to be determined by the symmetry (and reality) of the solution in any interchange $(z_i, \bar{z}_i) \leftrightarrow (z_j, \bar{z}_j)$. One also requires monodromy of the solution, namely that the correlation function be singled valued. This can be checked by following a path surrounding any of the singularities, $0, 1, \infty$. Analyzing these constraints gives $a_{++} = a_{--}$, $a_{+-} = a_{-+} = 0$. Restoring the normalization yields

$$\langle \sigma(1)\sigma(2)\sigma(3)\sigma(4)\rangle = \frac{1}{\sqrt{2}}\left|\frac{z_{14}z_{32}}{z_{12}z_{34}z_{13}z_{42}}\right|^{\frac{1}{4}} \times$$

$$\times \left\{\sqrt{1+\sqrt{1-x}}\sqrt{1+\sqrt{1-\bar{x}}} + \sqrt{1-\sqrt{1-x}}\sqrt{1-\sqrt{1-\bar{x}}}\right\}$$

Squaring this expression we see that it agrees with the expected result (178).

9.2.3 Three state Potts model

The Potts model was introduced in chapter 4 (volume 1). Here we specialize to the three state case. On a square lattice the partition function is

$$Z = \sum_{\varphi_x=0,\pm 2\pi/3} \exp \beta \sum_{(x,x')} \tfrac{2}{3}\left[\cos(\varphi_x - \varphi_{x'}) + \tfrac{1}{2}\right] \tag{179}$$

The "spin" $\sigma = e^{i\varphi}$ takes as values the cubic roots of the identity, and the exponent is such that the contribution of neighbouring spins is $\delta_{\sigma\sigma'}$. The model admits a global Z_3 symmetry, and obeys a duality relation with a fixed point

$$e^{\beta_c} = 1 + \sqrt{3} \tag{180}$$

More generally for a Q-state model $e^{\beta_c} = 1 + \sqrt{Q}$. The case $Q = 2$ is the Ising one (up to a rescaling of β). For $Q > 4$, it is known to have a first order transition while for $Q \leq 4$ the transition is continuous. Baxter (1980) has shown that for $Q = 3$, two critical exponents can be obtained

$$\alpha = \tfrac{1}{3} \qquad \beta = \tfrac{1}{9} \tag{181}$$

from the assumption that the three state Potts model is in the same universality class as the hard hexagon model which he solved.

The latter is a lattice gas model where "molecules" in the shape of hexagons have their centers on the nodes of a triangular lattice. Hexagons covering six elementary triangles are forbidden to overlap. One computes a generating function for the number of such configurations. The problem admits a Z_3 symmetry corresponding to the three sublattices on which the gas may "crystallize". Since both the hard hexagon and the three state Potts model admit a Z_3 symmetry and are "ferromagnetic" models with short-range interactions, it is most likely that they belong to the same universality class.

From the scaling relations $\nu d = 2 - \alpha$, and $\beta = \tfrac{1}{2}\nu(d - 2 + \eta)$ it follows that

$$\nu = \tfrac{5}{6} \qquad \eta = \tfrac{4}{15} \tag{182}$$

The real field $2\cos\varphi = \sigma + \bar{\sigma}$ should therefore have conformal weights $(\tfrac{1}{15}, \tfrac{1}{15})$. Similarly the thermal operator (coupled to $\beta - \beta_c$), denoted ε as before, has a dimension $\Delta_\varepsilon = \tfrac{1}{2}(d - \alpha/\nu) = \tfrac{4}{5}$. The corresponding conformal field theory should therefore contain the scalar fields

$$
\begin{aligned}
\sigma + \bar{\sigma} &\to A_{\frac{1}{15},\frac{1}{15}} \\
\varepsilon &\to A_{\frac{2}{5},\frac{2}{5}}
\end{aligned}
\tag{183}
$$

We now scan the Kac tables for the unitary minimal representations of the Virasoro algebra (Figure 4), to find that for

$$m = 5, \qquad c = \tfrac{4}{5} \qquad\qquad (184)$$

one finds indeed the two weights $\frac{1}{15}$ and $\frac{2}{5}$, with the second admitting a singular vector in the associated Verma module at level two again.

The table contains many more candidates as conformal weights, and the previous assignments are still tentative. It is however possible by an accurate numerical treatment, using the finite scaling method outlined in section 1.4, to confirm the value $c = \frac{4}{5}$. Further arguments will be presented in section 3.

Check the consistency of the above assignments using the three point function

$$\langle \varepsilon(z_1, \bar{z}_1) \cos \varphi(z_2, \bar{z}_2) \cos \varphi(z_3, \bar{z}_3) \rangle = \text{cst} \frac{1}{z_{12}{}^{h_\varepsilon} z_{13}{}^{h_\varepsilon} z_{23}{}^{2h_\sigma - h_\varepsilon}} \times (\text{c.c})$$

$$(185)$$

which should satisfy the differential equation

$$\left[\frac{3}{2(2h_\varepsilon + 1)} \partial_1^2 - \sum_{2,3} \left(\frac{h_\sigma}{(z_1 - z_k)^2} + \frac{1}{z_1 - z_k} \partial_k \right) \right]$$
$$\times \langle \varepsilon(1) \cos \varphi(2) \cos \varphi(3) \rangle = 0.$$

This requires that

$$h_\sigma = \frac{h_\varepsilon(1 - h_\varepsilon)}{2(2h_\varepsilon + 1)} \qquad\qquad (186)$$

and is indeed satisfied for $h_\varepsilon = \frac{2}{5}$, $h_\sigma = \frac{1}{15}$ (as well as $h_\varepsilon = \frac{1}{2}$, $h_\sigma = \frac{1}{16}$ in the Ising case).

It turns out that not all of the conformal weights occurring in the Kac table for $m = 5$ do appear for the Potts model. This will be discussed in the next section, when we look at requirements of a global nature. As we shall see, the correspondence between possible weights and scalar primary fields is established as follows

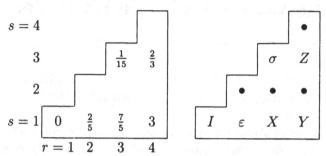

Only the rows with $s = 1, 3$ occur in the model. We list first the scalar fields including degeneracy (i.e. the number of independent such primary fields). For instance, the spin field is expected to have degeneracy two.

$$
\begin{array}{lll}
A_{0,0} & \text{identity} & I \\[4pt]
A_{\frac{2}{5},\frac{2}{5}} & \text{energy} & \varepsilon \\[4pt]
A_{\frac{7}{5},\frac{7}{5}} & & X \\[4pt]
A_{3,3} & & Y \\[4pt]
A_{\frac{1}{15},\frac{1}{15}} & \text{spin field} & \sigma \quad \text{2 such fields} \\[4pt]
A_{\frac{2}{3},\frac{2}{3}} & & Z \quad \text{2 such fields}
\end{array}
\tag{187}
$$

To this list should be added fields with nonzero angular momentum (respectively ± 3 and ± 1)

$$
\begin{array}{ll}
A_{0,3} & A_{3,0} \\[4pt]
A_{\frac{2}{5},\frac{7}{5}} & A_{\frac{7}{5},\frac{2}{5}}
\end{array}
\tag{188}
$$

allowing for a great diversity of observables.

It is possible to define an extension of the Q-state Potts model to continuous values of Q varying between 0 and 4. The central charge is given by the standard formula $c = 1 - 6/m(m + 1)$, with a relation

$$
Q = 4\cos^2 \frac{\pi}{m + 1}
\tag{189}
$$

The cases $m = 2, 3, 5$ correspond to percolation (although when $c = 0$ one uses an infinite sequence of nontrivial representations) Ising, and three state Potts model. The case $Q = 4$ appears as a limiting one with $m \to \infty$, $c \to 1$. See section 3.6.

Taking the value $c = \frac{4}{5}$ for granted in the case $Q = 3$, let us show a nontrivial application by computing, according to Dotsenko, the correlation function of four energy operators ε, given that it must satisfy a particular case of equation (95), namely

$$\left\{ \tfrac{5}{6}\partial_1^2 - \sum_{k=2,3,4} \left[\tfrac{2}{5}\frac{1}{z_{1k}^2} + \frac{1}{z_{1k}}\partial_k \right] \right\} \langle \varepsilon(1)\varepsilon(2)\varepsilon(3)\varepsilon(4) \rangle = 0 \qquad (190)$$

Set

$$\langle \varepsilon(1)\varepsilon(2)\varepsilon(3)\varepsilon(4) \rangle = \left(\frac{z_{13}z_{24}}{z_{12}z_{23}z_{34}z_{41}} \right)^{\frac{4}{5}} \times (\text{c.c})^{\frac{4}{5}} f(x, \bar{x})$$

$$x = \frac{z_{12}z_{34}}{z_{13}z_{24}} \qquad (191)$$

Substituting, we again obtain a hypergeometric equation

$$\left[x(1-x)\frac{\partial^2}{\partial x^2} + \tfrac{2}{5}(2x-1)\frac{\partial}{\partial x} - \tfrac{8}{25} \right] f(x, \bar{x}) = 0 \qquad (192)$$

Recall that the hypergeometric equation

$$\left[x(1-x)\frac{d^2}{dx^2} + (\gamma - (1+\alpha+\beta)x)\frac{d}{dx} - \alpha\beta \right] f(x) = 0 \qquad (193)$$

admits the general solution

$$f(x) = A\, F(\alpha, \beta; \gamma; x) + B\, x^{1-\gamma}(1-x)^{\gamma-\alpha-\beta} F(1-\alpha, 1-\beta; 2-\gamma; x) \qquad (194)$$

where F is the Gauss series

$$F(\alpha, \beta; \gamma; x) = 1 + \frac{\alpha\beta}{c}\frac{x}{1!} + \cdots + \frac{(\alpha)_n(\beta)_n}{(\gamma)_n}\frac{x^n}{n!} + \cdots$$

$$= \frac{\Gamma(\gamma)}{\Gamma(\alpha)\Gamma(\gamma-\alpha)} \int_0^1 dt\, t^{\alpha-1}(1-t)^{\gamma-\alpha-1}(1-xt)^{-\beta} \qquad (195)$$

using the notation $(\alpha)_n = \Gamma(\alpha+n)/\Gamma(\alpha) = \alpha(\alpha+1)...(\alpha+n-1)$.

Here $\alpha = -\tfrac{8}{5}$, $\beta = -\tfrac{1}{5}$, $\gamma = -\tfrac{2}{5}$. From the fact that $f(x, \bar{x})$ satisfies the same equation in $\bar{x}$, one derives

$$\langle \varepsilon(1)\varepsilon(2)\varepsilon(3)\varepsilon(4) \rangle = u_{11} \left| \frac{z_{13}z_{24}}{z_{12}z_{23}z_{34}z_{41}} \right|^{\frac{8}{5}} \left| F\left(-\tfrac{8}{5}, -\tfrac{1}{5}; -\tfrac{2}{5}; x\right) \right|^2$$

$$+ u_{12} \left(\frac{z_{13}z_{24}}{z_{12}z_{23}z_{34}z_{41}} \right)^{\frac{4}{5}} \frac{(\bar{z}_{13}\bar{z}_{24})^{-2}}{(\bar{z}_{12}\bar{z}_{23}\bar{z}_{34}\bar{z}_{41})^{-\frac{3}{5}}}$$

$$\times F\left(-\tfrac{8}{5},-\tfrac{1}{5};-\tfrac{2}{5};x\right) F\left(\tfrac{13}{5},\tfrac{6}{5};\tfrac{12}{5};\bar{x}\right)$$

$$+u_{21}\frac{(z_{13}z_{24})^{-2}}{(z_{12}z_{23}z_{34}z_{41})^{-\tfrac{3}{5}}}\left(\frac{\bar{z}_{13}\bar{z}_{24}}{\bar{z}_{12}\bar{z}_{23}\bar{z}_{34}\bar{z}_{41}}\right)^{\tfrac{4}{5}}$$

$$\times F\left(-\tfrac{8}{5},-\tfrac{1}{5};-\tfrac{2}{5};\bar{x}\right) F\left(\tfrac{13}{5},\tfrac{6}{5};\tfrac{12}{5};x\right)$$

$$+u_{22}\frac{\left|z_{13}z_{24}\right|^{-4}}{\left|z_{12}z_{23}z_{34}z_{41}\right|^{-\tfrac{6}{5}}}\left|F\left(\tfrac{13}{5},\tfrac{6}{5};\tfrac{12}{5};x\right)\right|^{2} \qquad (196)$$

Again we want to find the constants u_{ij} in such a way that the correlation be uniform. Set

$$\mathbf{F}=\begin{pmatrix} F_1(x) \\ F_2(x) \end{pmatrix} \equiv \begin{pmatrix} F(\alpha,\beta;\gamma;x) \\ x^{1-\gamma}(1-x)^{\gamma-\alpha-\beta}F(1-\alpha,1-\beta;2-\gamma;x) \end{pmatrix}$$

$$(197)$$

Under analytic continuation around a closed loop avoiding the singularities at $x=0$, 1, ∞, the vector $\mathbf{F}$ gets linearly transformed by a matrix with constant coefficients. These monodromy matrices are independent of continuous deformations of the paths avoiding the singularities. They are generated by two elementary loops around the points $x=0$ and $x=1$ (the loop around $x=\infty$ is equivalent to their product). Call g_0 and g_1 the respective actions on $\mathbf{F}$. One has obviously

$$^{g_0}\mathbf{F}=\begin{pmatrix} 1 & 0 \\ 0 & \omega_0 \end{pmatrix}\mathbf{F} \qquad \omega_0=\exp-2\mathrm{i}\pi\gamma \qquad (198)$$

To obtain g_1, it is convenient to transform $\mathbf{F}$ to an equivalent $\mathcal{F}$ adapted to find the regular and singular solutions at $x=1$, i.e.

$$\mathcal{F}=\begin{pmatrix} F(\alpha,\beta;\alpha+\beta+1-\gamma;1-x) \\ x^{1-\gamma}(1-x)^{\gamma-\alpha-\beta}F(1-\alpha,1-\beta;1+\gamma-\alpha-\beta;1-x) \end{pmatrix}$$

$$(199)$$

From the integral representation (195) one has

$$\mathbf{F}=\begin{pmatrix} A & B \\ \lambda A & \lambda'B \end{pmatrix}\mathcal{F}$$

$$A=\frac{\Gamma(\gamma)\Gamma(\gamma-\alpha-\beta)}{\Gamma(\gamma-\alpha)\Gamma(\gamma-\beta)} \qquad B=\frac{\Gamma(\gamma)\Gamma(\alpha+\beta-\gamma)}{\Gamma(\alpha)\Gamma(\beta)} \qquad (200)$$

$$\lambda=\frac{\Gamma(2-\gamma)\Gamma(\gamma-\alpha)\Gamma(\gamma-\beta)}{\Gamma(\gamma)\Gamma(1-\alpha)\Gamma(1-\beta)} \qquad \lambda'=\frac{\Gamma(2-\gamma)\Gamma(\alpha)\Gamma(\beta)}{\Gamma(\alpha+1-\gamma)\Gamma(\beta+1-\gamma)}$$

Since

$$^{g_1}\mathcal{F}=\begin{pmatrix} 1 & 0 \\ 0 & \omega_1 \end{pmatrix}\mathcal{F} \qquad \omega_1=\exp-2\mathrm{i}\pi(\gamma-\alpha-\beta) \qquad (201)$$

it follows that

$$^{g_1}\mathbf{F} = \frac{1}{\lambda' - \lambda} \begin{pmatrix} \lambda' - \lambda\omega_1 & \omega_1 - 1 \\ \lambda\lambda'(1 - \omega_1) & \lambda'\omega_1 - \lambda \end{pmatrix} \mathbf{F} \qquad (202)$$

The two transformations (198) and (202) become unitary for α, β, γ real and $\lambda\lambda' < 0$, which is verified in the present case, if we rescale the components of $\mathbf{F}$ as

$$\tilde{\mathbf{F}} = \begin{pmatrix} 1 & 0 \\ 0 & (-\lambda\lambda')^{-\frac{1}{2}} \end{pmatrix} \mathbf{F} \qquad (203)$$

leading to

$$^{g_0}\tilde{\mathbf{F}} = \begin{pmatrix} 1 & 0 \\ 0 & \omega_0 \end{pmatrix} \tilde{\mathbf{F}}$$

$$^{g_1}\tilde{\mathbf{F}} = \frac{1}{\lambda' - \lambda} \begin{pmatrix} \lambda' - \lambda\omega_1 & \sqrt{-\lambda\lambda'}(\omega_1 - 1) \\ \sqrt{-\lambda\lambda'}(\omega_1 - 1) & \lambda'\omega_1 - \lambda \end{pmatrix} \tilde{\mathbf{F}} \qquad (204)$$

As a result, the combination $|F_1|^2 - |F_2|^2/\lambda\lambda'$ is invariant under g_0 and g_1, hence under the full monodromy group. Correspondingly, up to an overall factor, the final expression for the correlation is

$$\langle \varepsilon(1)\varepsilon(2)\varepsilon(3)\varepsilon(4) \rangle$$

$$= \frac{\Gamma(\frac{6}{5})\Gamma(-\frac{1}{5})^2\Gamma(-\frac{8}{5})}{\Gamma(-\frac{2}{5})^2} \left| \frac{z_{13}z_{24}}{z_{12}z_{23}z_{34}z_{41}} \right|^{\frac{8}{5}} \left| F\left(-\frac{8}{5}, -\frac{1}{5}; -\frac{2}{5}; \frac{z_{12}z_{34}}{z_{13}z_{24}} \right) \right|^2$$

$$- \frac{\Gamma(\frac{13}{5})\Gamma(\frac{6}{5})^2\Gamma(-\frac{1}{5})}{\Gamma(\frac{12}{5})^2} \frac{|z_{13}z_{24}|^{-4}}{|z_{12}z_{23}z_{34}z_{41}|^{-\frac{6}{5}}} \left| F\left(\frac{13}{5}, \frac{6}{5}; \frac{12}{5}; \frac{z_{12}z_{34}}{z_{13}z_{24}} \right) \right|^2 \qquad (205)$$

Both coefficients are positive, since $\Gamma(-\frac{8}{5})$ as well as $-\Gamma(-\frac{1}{5})$ are positive. Moreover the full expression is symmetric in its arguments.

i) Check the symmetry of the correlation.

ii) Applying the same procedure in the case of correlations involving two spins and two energies, derive Dotsenko's expression

$$\langle \cos\varphi(1)\varepsilon(2)\cos\varphi(3)\varepsilon(4) \rangle$$

$$= \frac{\Gamma(-\frac{4}{5})\Gamma(\frac{3}{5})\Gamma(\frac{6}{5})\Gamma(-\frac{1}{5})}{\Gamma(\frac{2}{5})^2} \frac{|z_{13}|^{\frac{4}{3}}}{|z_{12}z_{23}z_{34}z_{41}|^{\frac{4}{5}}} \left| F\left(-\frac{4}{5}, \frac{3}{5}; \frac{2}{5}; \frac{z_{12}z_{34}}{z_{13}z_{24}} \right) \right|^2$$

$$- \frac{\Gamma(\frac{9}{5})\Gamma(\frac{2}{5})\Gamma(\frac{6}{5})\Gamma(-\frac{1}{5})}{\Gamma(\frac{8}{5})^2} \frac{|z_{13}|^{-\frac{16}{15}}|z_{24}|^{-\frac{12}{5}}}{|z_{12}z_{23}z_{34}z_{41}|^{-\frac{2}{5}}} \left| F\left(\frac{2}{5}, \frac{9}{5}; \frac{8}{5}; \frac{z_{12}z_{34}}{z_{13}z_{24}} \right) \right|^2 \qquad (206)$$

While it is true that the method is very powerful, the explicit expressions soon become very cumbersome. Alternative integral representations are discussed in the literature. Nevertheless it is comforting to see such nontrivial results emerging from a very general approach. They allow verification of the consistency of the theory, such as tested by the implementation of short-distance expansions.

Rather than carrying the investigation in this direction or presenting other models, we turn to the study of the same consistency conditions following ideas put forward by Cardy. A new idea will emerge from the study of finite size effects, namely modular invariance.

9.3   Finite size effects and modular invariance

In this section we present further developments and in particular, a classification of universality classes of minimal models.

9.3.1   Partition functions on a torus

In section 1.4, we have shown the relation between the central charge and finite size (or Casimir) effects in a periodic strip. We want to look at the matter in greater detail. As we recall, the transfer matrix per unit length on a periodic strip, with width L, reads

$$\mathcal{T} = \exp\left\{ -\frac{2\pi}{L} \left(L_0 + \bar{L}_0 - \tfrac{1}{12}c\right) \right\} \tag{207}$$

with the shift $\exp(\tfrac{1}{6}\pi c/L)$ being the above mentioned effect. The normalization is such that the free energy per unit area vanishes in the bulk. Hence, in a box of size L, M, with periodic conditions on both sides, the partition function reads

$$Z = \operatorname{Tr} \exp\left\{ -2\pi \frac{M}{L} \left(L_0 + \bar{L}_0 - \tfrac{1}{12}c\right) \right\} \tag{208}$$

A box with periodic boundary conditions has the topology of a torus, justifying the title of this section. Let us generalize slightly the construction by accompanying the "time" translation M by an additional "space" translation (figure 5). While in the

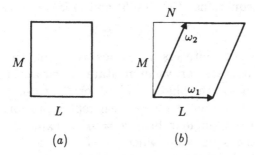

Fig. 5 (a) A rectangular periodic box. (b) More general periodic box; ω_1 and ω_2 are two generators of a corresponding lattice in complex notation, with $|\omega_1| = L$, $|\omega_2| = \sqrt{M^2 + N^2}$.

plane $L_0 + \bar{L}_0$ generates dilatations and $i(L_0 - \bar{L}_0)$ rotations, after a logarithmic map they become generators of time and space translations. Thus if we impose periodic boundary conditions according to the parallelogram of figure 5b, the partition function may be cast into the form

$$Z = \operatorname{Tr} \exp -2\pi \frac{M}{L} (L_0 + \bar{L}_0 - \tfrac{1}{12}c) + 2i\pi \frac{N}{L} (L_0 - \bar{L}_0)$$
$$= \operatorname{Tr} \left(q^{L_0 - c/24} \bar{q}^{\bar{L}_0 - c/24} \right) \tag{209}$$

where we have set

$$\omega_1 = L, \qquad \omega_2 = N + iM \qquad \tau = \omega_2/\omega_1$$
$$q = \exp 2i\pi\tau \qquad \bar{q} = \exp -2i\pi\bar{\tau} \tag{210}$$

The conventions are such that $\operatorname{Im} \tau > 0$, hence $|q| < 1$. The purpose of this choice of boundary conditions is to disentangle the dependence on the τ and $\bar{\tau}$ (or q and $\bar{q}$) variables. Suppose that we decompose the space of states into irreducible representations of the product of algebras $(V, \bar{V})$, each one pertaining to highest weights $(h, \bar{h})$. To these representations will be associated primary operators with the same weights. Let $N_{h,\bar{h}}$ be the number of times the representation $(h, \bar{h})$ occurs in this decomposition. If the number of distinct irreducible representations is finite (or denumerable), we can use the definition of conformal characters according to equations (152), (155) and (158) to write

$$Z = \sum N_{h,\bar{h}} \chi_{(c,h)}(\tau)\chi_{(c,\bar{h})}(\bar{\tau}) \tag{211}$$

Not only are the integers $N_{h,\bar{h}}$ positive, but the uniqueness of the ground state, or vacuum state, invariant under global transformations (in the plane $SL(2,C)$, on the torus translations), requires $N_{0,0} = 1$. Since we use real representations, $\chi_{c,\bar{h}}(\bar{\tau}) = \chi_{c,\bar{h}}(\tau)$. In a rectangular box, $q = \bar{q} = \exp(-2\pi M/L)$. Let us consider the strip limit where $M/L \to \infty$, $q \to 0$. If we assume that all primary operators have a positive dimension $\Delta = h + \bar{h} > 0$, we easily recover the limiting behaviour

$$Z \underset{M/L\to\infty}{\sim} \exp\tfrac{1}{6}\pi c M/L \tag{212}$$

as was claimed in section 1.4. Any state with $\Delta_n = h + \bar{h} + n$ will contribute an exponentially small correction, $\exp(-2\pi\Delta_n M/L)$. Thus the spectrum of the transfer matrix yields the set of all conformal dimensions in the theory, with the ratios of eigenvalues

$$\lambda_n/\lambda_0 = \exp(-2\pi\Delta_n/L) \tag{213}$$

such that $\ln(\lambda_n/\lambda_0)$ scales as L^{-1}. If however some operators carry a negative dimension, the leading behaviour (212) will be modified, as anticipated in section 1.4, and

$$Z \underset{M/L\to\infty}{\sim} \exp(\tfrac{1}{6}\pi c_{eff} M/L) \tag{214}$$

$$c_{eff} = c - 12\Delta_{inf} \qquad \Delta_{inf} < 0$$

Similarly, the succession of eigenvalues of the transfer matrix will be shifted by $-\Delta_{inf}$, as now

$$\lambda_n/\lambda_0 = \exp(-2\pi(\Delta_n - \Delta_{inf})/L) \tag{215}$$

This shows that one should be careful in interpreting numerical results.

A typical example of this phenomenon is the Lee–Yang edge singularity with $c = -\frac{22}{5}$, $\Delta_{inf} = -\frac{2}{5}$. One predicts (and indeed observes numerically) that $c_{eff} = \frac{2}{5}$. More generally in a minimal model (equations (146) and (147)) characterized by a pair (p,p') one has in the worst case

$$\Delta_{inf} = \frac{1-(p-p')^2}{2pp'} \qquad c_{eff} = 1 - \frac{6}{pp'} \geq 0 \tag{216}$$

The partition function in a periodic box has to satisfy global consistency constraints. From Euclidean invariance, space and time play a symmetric role. We can interchange them, so that in a rectangle of size L, M, the partition function assumes the two alternative expressions $Z = \mathrm{Tr}(T_L)^M = \mathrm{Tr}(T_M)^L$. None of these expressions exhibits this symmetry explicitly, and its implementation is a strong limitation on the choice of coefficients in the expansion (211). In more general terms, a torus can be considered as a unit cell (fundamental domain) when taking the quotient of the complex plane by a lattice of translations Λ, generated by two independent periods ω_1 and ω_2. Any two other periods ω_1' and ω_2' will also generate Λ, provided the relation between (ω_1, ω_2) and (ω_1', ω_2') is both ways linear with integral coefficients, i.e. invertible, and of unit determinant, in order to preserve the area as well as the orientation. This gives rise to the group $SL(2, \mathcal{Z})$. Moreover, from rotational and dilatation invariance, the modular ratio $\tau = \omega_2/\omega_1$ is the only significant variable. The above transformations act on τ as fractional linear transformations with integral coefficients and determinant unity, with matrices $\pm A$ are identified. This is the modular group $SL(2, \mathcal{Z})/Z_2$, (written as $PSL(2; \mathcal{Z})$) and the constraint of invariance is called modular invariance.

9.3.2 Kronecker's limit formula

Let us return to the Gaussian model to test modular invariance. The space of states is the Fock space of a free field. Eigenmodes correspond to a quantized momentum p_x taking integral values in units $2\pi/L$. Right movers have a positive p_x, left movers a negative one, the modes being associated with the propagation on a circle. We subtract the zero mode which would lead to an infinity by a multiplicative renormalization of Z. Each mode has an occupation number ranging from zero to infinity, and the central charge is $c = 1$. As a result we expect that the partition function is a product of Bose statistical factors of the form

$$Z_1 \propto (q\bar{q})^{-\frac{1}{24}} \prod_{n=1}^{\infty} \frac{1}{(1-q^n)(1-\bar{q}^n)} = \frac{1}{\eta(\tau)\eta(\bar{\tau})} \tag{217}$$

in agreement with equation (153) for $c = 1$, $h = 0$, using the definition (157) of Dedekind's function $\eta(\tau)$.

A direct proof of (217), with further generalizations, is known since the work of Kronecker in the 1880's in the context of number theory. The proof proceeds from a direct evaluation of the path integral and exhibits an unexpected factor, missing in (217) and essential for modular invariance.

Let k^i be the generators of the lattice dual to Λ,

$$\text{Re } k^i \bar\omega_j = \delta^i{}_j \qquad k^1 = -i\omega_2/A \qquad k^2 = i\omega_1/A \quad (218)$$

where $A = \text{Im}\,\omega_2\bar\omega_1$ is the area of the torus. The eigenvalues of the Laplacian are

$$E_{n_1,n_2} = (2\pi)^2 \left| n_1 k^1 + n_2 k^2 \right|^2 \tag{219}$$

with $n_{1,2}$ integers. The normalized zero mode eigenfunction on $\mathcal{T} = \mathcal{C}/\Lambda$ is $\varphi_0 = A^{-\frac{1}{2}}$. Omitting this zero mode, the functional integral expressing the partition function is formally

$$
\begin{aligned}
Z_1 &= \int \mathcal{D}\varphi \; A^{\frac{1}{2}} \delta\left(\int_{\mathcal{T}} \mathrm{d}^2 x \varphi\varphi_0 \right) \exp\left(-\tfrac{1}{2} \int_{\mathcal{T}} \mathrm{d}^2 x (\nabla\,\varphi)^2 \right) \\
&= A^{\frac{1}{2}} \prod_{n_1,n_2}' \; \frac{1}{E^{\frac{1}{2}}_{n_1,n_2}}
\end{aligned}
\tag{220}
$$

The purpose of the factor $A^{\frac{1}{2}}$ is to make the final result dimensionless. Note that the area A is modular invariant. In equation (220), the product over modes is ultraviolet divergent since E_{n_1,n_2} is unbounded for large n_1, n_2. The prime in the product stands for the omission of the contribution $n_1 = n_2 = 0$. A standard regularization (and renormalization) uses the so-called ζ-regularization. The name derives from the analytic continuation of Riemann's ζ function used in studying the distribution of primes. Introduce the analogous function

$$G(s) = \sum_{n_1,n_2}' \frac{1}{E^s_{n_1,n_2}} \tag{221}$$

absolutely convergent, hence analytic, for $\text{Re}\,s > 1$. As we shall see below, this function has a pole at $s = 1$, but admits an analytic continuation down to $s = 0$. Comparing the expressions (220) and

(221), one defines the renormalized Z_1 as

$$Z_1 = A^{\frac{1}{2}} \exp \tfrac{1}{2} \frac{dG}{ds}(0) \qquad (222)$$

As before, let τ stand for the ratio ω_2/ω_1 and $q = \exp 2i\pi\tau$. For $\mathrm{Re}\, s > 1$, we insert the definition of E_{n_1,n_2} in (221), resulting in

$$\left|\frac{2\pi\omega_1}{A}\right|^{2s} G(s) = \sum_{m,n}{}' \frac{1}{|m + n\tau|^{2s}} = 2\zeta(2s) + \sum_{n}{}' \left(\sum_{m} \frac{1}{|m + n\tau|^{2s}}\right)$$

$$(223)$$

This starting point of the calculation has explicit modular invariance, for if we substitute in $G(s)$, $\omega_1 = d\omega_1' + c\omega_2'$, $\omega_2 = b\omega_1' + a\omega_2'$, hence $\tau = (a\tau' + b)/(c\tau' + d)$, $ad - bc = 1$, we readily derive that $G(s)$ is invariant. Consequently, by analytic continuation, this will follow for Z_1.

Returning to equation (223), the first term involves Riemann's function $\zeta(s) = \sum_1^\infty m^{-s}$ with a simple pole at $s = 1$ and such that $2\zeta(0) = -1$, $2\zeta'(0) = -\ln 2\pi$. In the second term, the sum over m yields a periodic function of $n\tau$, of unit period. Hence it admits a Fourier series of the form

$$\sum_m \frac{1}{|m + n\tau|^{2s}} =$$

$$= \sum_p e^{2i\pi pn\mathrm{Re}\,\tau} \int_0^1 dy\, e^{-2i\pi py} \sum_m \frac{1}{[(m + y)^2 + n^2\mathrm{Im}\,\tau^2]^s}$$

$$= \sum_p \int_{-\infty}^{+\infty} dy\, e^{2i\pi p(n\mathrm{Re}\,\tau - y)} \frac{1}{[y^2 + n^2\mathrm{Im}\,\tau^2]^s}$$

$$= \frac{1}{\Gamma(s)} \sum_p \int_{-\infty}^{+\infty} dy\, e^{2i\pi p(n\mathrm{Re}\,\tau - y)} \int_0^\infty dt\, t^{s-1} e^{-t(y^2 + n^2\mathrm{Im}\,\tau^2)}$$

$$= \frac{\pi^{1/2}}{\Gamma(s)} \sum_p \int_0^\infty dt\, t^{s-3/2} e^{-\left[tn^2\mathrm{Im}\,\tau^2 + \pi^2(p^2/t) - 2i\pi pn\mathrm{Re}\,\tau\right]}$$

For $p = 0$ the integral reduces to $\Gamma(s - \tfrac{1}{2})\,|n\mathrm{Im}\,\tau|^{1-2s}$. We isolate this term, sum over $n \neq 0$, and use the functional equation for

Riemann's ζ function

$$2\pi^{-\frac{1}{2}s}\Gamma\left(\tfrac{1}{2}s\right)\zeta(s) = 2\pi^{-\frac{1}{2}(1-s)}\Gamma\left(\tfrac{1}{2}(1-s)\right)\zeta(1-s)$$

$$= -2\left(\frac{1}{s}+\frac{1}{1-s}\right)+\int_0^1\frac{dt}{t}\left(t^{-\frac{1}{2}s}+t^{-\frac{1}{2}(1-s)}\right)\sum_m{}'e^{-\pi m^2/t} \quad (224)$$

Changing t into $|\pi p/n\mathrm{Im}\,\tau|\,t$, yields

$$\Gamma(s)\left|\frac{2\pi\omega_1}{A}\right|^{2s}\left(\frac{\mathrm{Im}\,\tau}{\pi}\right)^{s-\frac{1}{2}}G(s) = 2\left(\frac{\mathrm{Im}\,\tau}{\pi}\right)^{s-\frac{1}{2}}\Gamma(s)\zeta(2s)$$

$$+2\left(\frac{\mathrm{Im}\,\tau}{\pi}\right)^{\frac{1}{2}-s}\Gamma(1-s)\zeta(2-2s)$$

$$+\sqrt{\pi}\sum_n{}'\sum_p{}'e^{2i\pi pn\mathrm{Re}\,\tau}\left|\frac{p}{n}\right|^{s-\frac{1}{2}}\int_0^\infty\frac{dt}{t}t^{s-\frac{1}{2}}e^{-\pi|pn|\mathrm{Im}\,\tau(t+t^{-1})}$$

$$(225)$$

The last double sum is an entire even function in $s-\frac{1}{2}$, hence the right-hand side is even in $s\to 1-s$, giving the required analytic continuation of $G(s)$. Since for $x>0$

$$\int_0^\infty\frac{dt}{t}t^{\pm\frac{1}{2}}\exp\left[-x(t+t^{-1})\right] = (\pi/x)^{\frac{1}{2}}e^{-2x}$$

and $\zeta(2)=\pi^2/6$, we find in the vicinity of $s=0$

$$G(s) = -1-2s\ln\left|\frac{A}{\omega_1}\right|+\tfrac{1}{3}s\pi\mathrm{Im}\,\tau$$

$$+s\sum_n{}'\sum_p{}'\frac{1}{|p|}e^{2i\pi pn\mathrm{Re}\,\tau-2\pi|pn|\mathrm{Im}\,\tau}+O(s^2)$$

Finally

$$G(0)=-1 \qquad G'(0)=-2\ln\left\{\left|\frac{A}{\omega_1}\right|\eta(q)\eta(\bar{q})\right\} \quad (226)$$

Noting that $|\omega_1|/A^{1/2}=1/(\mathrm{Im}\,\tau)^{1/2}$, we find the precise form of the partition function Z_1, from equation (222), as

$$Z_1 = \frac{1}{(\mathrm{Im}\,\tau)^{\frac{1}{2}}\eta(\tau)\eta(\bar{\tau})} \quad (227)$$

which differs from (217) by the prefactor $(\mathrm{Im}\,\tau)^{-\frac{1}{2}}$. The latter is essential to insure modular invariance, which was obvious at the start of the calculation, and its origin can be traced to the zero mode subtraction.

We deduce from equation (227) that $|\eta(\tau)|\,(\mathrm{Im}\,\tau)^{\frac{1}{4}}$ is modular invariant. To be precise, under the two generators $\tau \to \tau + 1$ and $\tau \to -\tau^{-1}$ of the modular group, the transformation law of $\eta(\tau)$ is

$$\eta(\tau + 1) = e^{2i\pi/24}\eta(\tau)$$
$$\eta(-1/\tau) = (\tau e^{-i\pi/2})^{\frac{1}{2}}\eta(\tau) \tag{228}$$

Under a general transformation $\tau \to \tau' = (a\tau + b)/(c\tau + d)$ we therefore have

$$\eta(\tau') = \varepsilon_A(c\tau + d)^{\frac{1}{2}}\eta(\tau) \tag{229}$$

where ε_A is a 24th root of unity. The number theoretic determination of these phases, not required here, is due to Jordan and Dedekind.

For the free field φ, the two-point function can easily be computed on the torus. It should satisfy

$$-\Delta\,\langle\varphi(\mathbf{x})\varphi(\mathbf{0})\rangle = 2\pi(\delta(\mathbf{x}) - 1/A) \tag{230}$$

The extra factor 2π on the r.h.s. is in agreement with the normalization at short distance (equation (162)), while the term $1/A$ represents the zero mode subtraction, making $-\Delta$ invertible in the orthogonal subspace. Both the correlation and the δ-function are understood to be doubly periodic. The solution to equation (230) is expressed in terms of one of the θ-functions (Appendix A) which we normalize as

$$F(z) = 2i\pi y^{-\frac{1}{2}}\frac{\theta_1(z, \tau)}{\theta_1'(0, \tau)}$$

$$= P(q)^{-2}(1 - y^{-1})\prod_{1}^{\infty}(1 - yq^n)(1 - y^{-1}q^n) \tag{231}$$

$$= P(q)^{-3}\sum_{-\infty}^{+\infty}(-y)^n q^{\frac{1}{2}n(n+1)}$$

where, as before, $q = \exp(2i\pi\tau)$ and $y = \exp(2i\pi z/\omega_1)$, and as usual $P(q) = \prod_{1}^{\infty}(1 - q^n)$. One verifies that

$$\Gamma_{12} = \exp\left[-2\,\langle\varphi(z_1, \bar{z}_1)\varphi(z_2, \bar{z}_2)\rangle\right]$$

$$= \exp\left(-2\pi\left[\frac{(\mathrm{Im}\,z_{12}/\omega_1)^2}{\mathrm{Im}\,\tau} + \mathrm{Im}\,z_{12}/\omega_1\right]\right)\left|\frac{\omega_1}{2\pi}F(z_{12})\right|^2 \tag{232}$$

One checks that Γ_{12} is doubly periodic, symmetric in the interchange $1 \leftrightarrow 2$, and behaves at short distance as

$$\Gamma_{12} \sim z_{12}\bar{z}_{12} \tag{233}$$

For later use, we look at Γ_{12} when z_{12} has rational coordinates in the ω_1, ω_2 basis. Let N, k and ℓ be integers, then

$$\exp -2 \left\langle \varphi(0)\varphi\left(\frac{k\omega_1 + \ell\omega_2}{N}\right)\right\rangle = \frac{A}{(2\pi)^2} Z_1(\tau,\bar{\tau})^2 \left|D_{\frac{k}{N},\frac{\ell}{N}}(\tau)\right|^2 \tag{234}$$

with

$$D_{k/N,\ell/N}(\tau) = q^{-\frac{1}{12}(6\ell(N-\ell)/N^2-1)}$$

$$\times \prod_{n=0}^{\infty} \left(1 - e^{2i\pi k/N}q^{n+\ell/N}\right)\left(1 - e^{-2i\pi k/N}q^{n+(N-\ell)/N}\right) \tag{235}$$

By construction $D_{k/N,\ell/N}(\tau)$ vanishes when both k and ℓ are multiples of N. Check that it satisfies the properties

$$\left|D_{k/N,\ell/N}(\tau)\right| = \left|D_{(N-k)/N,(N-\ell)/N}(\tau)\right|$$

$$= \left|D_{\ell/N,(N-k)/N}(-1/\tau)\right| \tag{236}$$

$$D_{k/N,\ell/N}(\tau+1) = e^{-\frac{1}{6}i\pi[6\ell(N-\ell)/N^2-1]}D_{(k+\ell)/N,\ell/N}(\tau) \tag{237}$$

$$D_{k/N,\ell/N+p}(\tau) = (-1)^p e^{-2i\pi kp/N}D_{k/N,\ell/N}(\tau)$$

$$D_{k/N+p,\ell/N}(\tau) = D_{k/N,\ell/N}(\tau) \tag{238}$$

The D-functions admit of the following alternative interpretation. Let us return to the computation of the free field partition function, but, instead of taking the field doubly periodic, let us require that, after a cycle around ω_1 or ω_2, it is multiplied by a phase (clearly the field is then complex). Precisely for λ, μ integers

$$\varphi(\mathbf{x} + \lambda\omega_1 + \mu\omega_2) = e^{2i\pi(k\mu-\ell\lambda)/N}\varphi(\mathbf{x}) \tag{239}$$

In complex notation, the proper modes correspond to eigenvalues of the form

$$-\left(n_2 + \frac{\ell}{N}\right)k^1 + \left(n_1 + \frac{k}{N}\right)k^2$$

$$= \frac{i}{A}\left\{\left(n_1 + \frac{k}{N}\right)\omega_1 + \left(n_2 + \frac{\ell}{N}\right)\omega_2\right\} \tag{240}$$

We do not require zero mode subtraction when k or ℓ (or both) are different from a multiple of N. Finally, instead of computing the square root of the inverse determinant, we simply compute the determinant. Therefore, with

$$G_{k/N,\ell/N}(s) = \left(\frac{A}{2\pi\,|\omega_1|}\right)^{2s} \sum_{m,n} \frac{1}{|m + n\tau + (k + \ell\tau)/N|^{2s}} \qquad (241)$$

we find after analytic continuation

$$\exp -\frac{d}{ds}\, G_{k/N,\ell/N}(s)\Big|_{s=0} = \left|D_{k/N,\ell/N}(\tau)\right|^2 \qquad (242)$$

Thus $D_{k/N,\ell/N}(\tau)$ appears as a (renormalized) determinant over "twisted" modes.

i) Verify equation (242).

ii) In the free field case, obtain the (constant) mean value of the energy-momentum on a torus. Performing a deformation $\delta x^\mu = \delta\varepsilon^{\mu\nu} x^\nu$, with $\delta\varepsilon$ an infinitesimal matrix, we have

$$\langle T_{\mu\nu}(\mathbf{x})\rangle\, \partial^\mu \delta x^\nu = \langle T_{\mu\nu}(\mathbf{x})\rangle\, \delta\varepsilon^{\mu\nu}$$
$$= \langle T(z)\rangle\left[\delta\varepsilon^{11} - \delta\varepsilon^{22} + i(\delta\varepsilon^{12} + \delta\varepsilon^{21})\right] + \text{c.c.}$$

with $\langle T(z)\rangle = \langle T\rangle$, independent of z,

$$\delta \ln Z_1 = -\frac{A}{2\pi}\,\langle T\rangle\left[\delta\varepsilon^{11} - \delta\varepsilon^{22} + i\left(\delta\varepsilon^{12} + \delta\varepsilon^{21}\right)\right] + \text{c.c.}$$

Inserting the expression (227) for Z_1, we obtain

$$\langle T\rangle = -\frac{\pi}{2A}\frac{\bar\omega_1}{\omega_1} - \frac{2\pi i}{\omega_1^2}\frac{\eta'(\tau)}{\eta(\tau)} = -\frac{\pi}{2A}\frac{\bar\omega_1}{\omega_1} + \frac{1}{2}\sum_{n,p}{}' \frac{1}{(n\omega_1 + p\omega_2)^2} \qquad (243)$$

where the last semiconvergent sum is understood as a double limit, summing first symmetrically on p, then on n.

9.3.3 Ising model

From the expression (171) for the action in the free Fermi case, and taking statistics into account, we derive that the partition function on a torus is the product of two Pfaffians $\mathrm{Pf}(\partial)\mathrm{Pf}(\bar\partial)$, with $\partial, \bar\partial$ the Cauchy Riemann operators. This is also the square root of the determinant of the Laplacian, $(\det -\Delta)^{1/2}$. However boundary conditions have to be chosen appropriately. According to the results of chapter 2 (volume 1), the fields ψ, $\bar\psi$ have to be antiperiodic along at least one of the generators ω_1, ω_2. Denote by 0 or $\frac{1}{2}$ the periodic or antiperiodic boundary conditions, in agreement with the notation of the previous subsection. We have just obtained the form of the $(\det -\Delta)$ as $|D_{k/2,\ell/2}(\tau)|^2$, with k, $\ell = 0, 1$. From chapter 2, it follows that we should have

$$c = \tfrac{1}{2} \qquad Z_{\frac{1}{2}} = \tfrac{1}{2}\left\{\left|D_{\frac{1}{2},\frac{1}{2}}(\tau)\right| + \left|D_{0,\frac{1}{2}}(\tau)\right| + \left|D_{\frac{1}{2},0}(\tau)\right|\right\}$$

(244)

while the term $\left|D_{0,0}\right|$ vanishes, reflecting the presence of a zero mode. This agrees with the results of Kaufmann, Ferdinand and Fisher. Moreover, should we insist in having one of the non vanishing terms in (244), then the other two follow from modular invariance. Indeed, according to equations (236) and (237), we obtain the action of $\tau \to \tau + 1$ and $\tau \to -\tau^{-1}$ as generating permutations on the three determinants

$$\left|D_{\frac{1}{2},\frac{1}{2}}\right| \qquad \left|D_{0,\frac{1}{2}}\right| \qquad \left|D_{\frac{1}{2},0}\right|$$

$$\tau \to \tau + 1 \qquad \left|D_{0,\frac{1}{2}}\right| \qquad \left|D_{\frac{1}{2},\frac{1}{2}}\right| \qquad \left|D_{\frac{1}{2},0}\right| \qquad (245)$$

$$\tau \to -\tau^{-1} \qquad \left|D_{\frac{1}{2},\frac{1}{2}}\right| \qquad \left|D_{\frac{1}{2},0}\right| \qquad \left|D_{0,\frac{1}{2}}\right|$$

This can be also understood by following the effect of the indicated change of basis (ω_1, ω_2) on the corresponding boundary conditions. As a result, the combination (244) is modular invariant. The various boundary conditions are called "spin structures". Three of those contribute to the partition function.

It is now possible to make contact with the expansion in characters as indicated in equation (211). The functions $D_{\frac{1}{2}k,\frac{1}{2}\ell}$ are naturally squares, so that defining

$$D_{\frac{1}{2},\frac{1}{2}}(\tau) = q^{-\frac{1}{24}} d_{\frac{1}{2},\frac{1}{2}}^2(\tau) \qquad D_{0,\frac{1}{2}}(\tau) = q^{-\frac{1}{24}} d_{0,\frac{1}{2}}^2(\tau)$$

$$D_{\frac{1}{2},0}(\tau) = 2q^{-\frac{1}{24}} d_{\frac{1}{2},0}^2(\tau)$$

(246)

we find

$$d_{\frac{1}{2},\frac{1}{2}}(\tau) = \prod_0^\infty (1 + q^{n+\frac{1}{2}}) \qquad d_{0,\frac{1}{2}}(\tau) = \prod_0^\infty (1 - q^{n+\frac{1}{2}})$$

$$d_{\frac{1}{2},0}(\tau) = q^{\frac{1}{16}} \prod_1^\infty (1 + q^n)$$

(247)

satisfying

$$d_{\frac{1}{2},\frac{1}{2}}(\tau)d_{0,\frac{1}{2}}(\tau)d_{\frac{1}{2},0}(\tau) = q^{\frac{1}{16}}$$
$$d^8_{\frac{1}{2},\frac{1}{2}}(\tau) = d^8_{0,\frac{1}{2}}(\tau) + 16d^8_{\frac{1}{2},0}(\tau) \tag{248}$$

One recognizes that the characters $\chi_{c,h}$, with $c = \frac{1}{2}$ and $h = 0, \frac{1}{2}$ or $\frac{1}{16}$, are expressed as

$$q^{\frac{1}{48}}\chi_{\frac{1}{2},0}(\tau) = \frac{1}{2}\left\{d_{\frac{1}{2},\frac{1}{2}}(\tau) + d_{0,\frac{1}{2}}(\tau)\right\}$$

$$= \frac{1}{2}\left\{\prod_0^\infty (1 + q^{n+\frac{1}{2}}) + \prod_0^\infty (1 - q^{n+\frac{1}{2}})\right\} \tag{249a}$$

$$= \frac{1}{P(q)}\sum_{k=-\infty}^{+\infty}\left(q^{[(24k+1)^2-1]/48} - q^{[(24k+7)^2-1]/48}\right) \underset{q\to 0}{\sim} 1$$

$$q^{\frac{1}{48}}\chi_{\frac{1}{2},\frac{1}{2}}(\tau) = \frac{1}{2}\left\{d_{\frac{1}{2},\frac{1}{2}}(\tau) - d_{0,\frac{1}{2}}(\tau)\right\}$$

$$= \frac{1}{2}\left\{\prod_0^\infty (1 + q^{n+\frac{1}{2}}) - \prod_0^\infty (1 - q^{n+\frac{1}{2}})\right\} \tag{249b}$$

$$= \frac{1}{P(q)}\sum_{k=-\infty}^{+\infty}\left(q^{[(24k+5)^2-1]/48} - q^{[(24k+11)^2-1]/48}\right) \underset{q\to 0}{\sim} q^{\frac{1}{2}}$$

$$q^{\frac{1}{48}}\chi_{\frac{1}{2},\frac{1}{16}}(\tau) = d_{\frac{1}{2},0}(\tau) = q^{\frac{1}{16}}\prod_1^\infty (1 + q^n) \tag{249c}$$

$$= \frac{1}{P(q)}\sum_{k=-\infty}^{+\infty}\left(q^{[(24k-2)^2-1]/48} - q^{[(24k+10)^2-1]/48}\right) \underset{q\to 0}{\sim} q^{\frac{1}{16}}$$

Consequently for the Ising model

$$Z_{\frac{1}{2}}(q,\bar{q}) = (q\bar{q})^{-\frac{1}{48}}\left\{\frac{1}{2}\left|d_{\frac{1}{2},\frac{1}{2}}(\tau)\right|^2 + \frac{1}{2}\left|d_{0,\frac{1}{2}}(\tau)\right|^2 + \left|d_{\frac{1}{2},0}(\tau)\right|\right\}$$

$$= \left|\chi_{\frac{1}{2},0}(\tau)\right|^2 + \left|\chi_{\frac{1}{2},\frac{1}{2}}(\tau)\right|^2 + \left|\chi_{\frac{1}{2},\frac{1}{16}}(\tau)\right|^2$$

$$\tag{250}$$

in agreement with the general discussion. This formula exhibits the expected operator content of the model, with three primary operators namely the identity, the energy density and the spin. The direct derivation confirms the value of the central charge, $c = \frac{1}{2}$, as well as the conformal weights $(0,0)$, $(\frac{1}{2},\frac{1}{2})$ and $(\frac{1}{16},\frac{1}{16})$ respectively. Modular invariance is also clear as we have seen above.

One can also compute various correlation on a torus. First let us consider fermionic fields. As we recall, in the plane

$$\langle \psi(z_1)\psi(z_2)\rangle_{\text{plane}} = \frac{1}{z_{12}} \qquad (251)$$

with the complex conjugate expression for the $\bar{\psi}$ fields. Assuming ψ of weights $(\frac{1}{2}, 0)$, it follows that on a periodic strip, with $z_{\text{plane}} = \exp(2i\pi/L)z_{\text{strip}}$

$$\langle \psi(z_1)\psi(z_2)\rangle_{\text{strip}(\frac{1}{2})} = \frac{\pi/L}{\sin(\pi z_{12}/L)} \qquad (252)$$

This is antisymmetric in the interchange $z_1 \leftrightarrow z_2$, and antiperiodic $\psi(z+L) = -\psi(z)$ (hence the index $\frac{1}{2}$). There exists however an other possibility, with a periodic field $\psi(z+L) = \psi(z)$ (index 0), namely

$$\langle \psi(z_1)\psi(z_2)\rangle_{\text{strip}(0)} = \frac{\pi/L}{\tan(\pi z_{12}/L)} \qquad (253)$$

On a torus, similar correlations satisfying

$$\bar{\partial}_1 \langle \psi(z_1)\psi(z_2)\rangle = \pi\delta^2(1,2) \qquad (254)$$

are distinguished by a double set of periodic or antiperiodic conditions forbidding the existence of zero modes. With $k, \ell = 0, 1$ not both 0, we write

$$\langle \psi(z_1)\psi(z_2)\rangle_{\frac{1}{2}k,\frac{1}{2}\ell} = \frac{2i\pi}{\omega_1} \frac{S_{\frac{1}{2}k,\frac{1}{2}\ell}(z_{12}, \tau)}{\sigma_{\frac{1}{2}k,\frac{1}{2}\ell}(\tau)} \qquad (255)$$

with the first index $\frac{1}{2}k$ referring to the behaviour in the direction ω_2, the second $\frac{1}{2}\ell$ in the direction ω_1. Using the notation $F(z)$ introduced in (231), and $y = \exp(2i\pi z/\omega_1)$, it is not difficult to check that the required expressions are

$$S_{\frac{1}{2},0}(z,\tau) = \frac{F(z - \frac{1}{2}\omega_1)}{F(z)} = \frac{y+1}{y-1} \prod_1^\infty \frac{(1+yq^n)(1+y^{-1}q^n)}{(1-yq^n)(1-y^{-1}q^n)}$$

$$S_{0,\frac{1}{2}}(z,\tau) = e^{-i\pi z/\omega_1} \frac{F(z - \frac{1}{2}\omega_2)}{F(z)}$$

$$= \frac{1}{y^{\frac{1}{2}} - y^{-\frac{1}{2}}} \prod_1^\infty \frac{(1 - yq^{n-\frac{1}{2}})(1 - y^{-1}q^{n-\frac{1}{2}})}{(1 - yq^n)(1 - y^{-1}q^n)}$$

$$S_{\frac{1}{2},\frac{1}{2}}(z,\tau) = e^{-i\pi z/\omega_1} \frac{F(z - \frac{1}{2}(\omega_1 + \omega_2))}{F(z)}$$

$$= \frac{1}{y^{\frac{1}{2}} - y^{-\frac{1}{2}}} \prod_{1}^{\infty} \frac{(1 + yq^{n-\frac{1}{2}})(1 + y^{-1}q^{n-\frac{1}{2}})}{(1 - yq^{n})(1 - y^{-1}q^{n})} \qquad (256)$$

The normalization factors are

$$\sigma_{\frac{1}{2},0}(\tau) = 2 \prod_{1}^{\infty} \left(\frac{1 + q^{n}}{1 - q^{n}} \right)^{2}$$

$$\sigma_{0,\frac{1}{2}}(\tau) = \prod_{1}^{\infty} \left(\frac{1 - q^{n-\frac{1}{2}}}{1 - q^{n}} \right)^{2} \qquad (257)$$

$$\sigma_{\frac{1}{2},\frac{1}{2}}(\tau) = \prod_{1}^{\infty} \left(\frac{1 + q^{n-\frac{1}{2}}}{1 - q^{n}} \right)^{2}$$

These expressions reduce correctly to (252) or (253) in the limit of a strip. The latter are referred to as Neveu–Schwarz and Ramond boundary conditions respectively in the context of string field theory. From equation (255) follows the expression of the energy-energy correlation on a torus, a weighted sum over three terms

$$\langle \varepsilon(z_{1}, \bar{z}_{1}) \varepsilon(z_{2}, \bar{z}_{2}) \rangle = \frac{1}{2} Z_{\frac{1}{2}}^{-1} \sum_{k,\ell}{}' \left| D_{\frac{1}{2}k,\frac{1}{2}\ell}(q) \right| \left| \langle \psi(z_{1}) \psi(z_{2}) \rangle_{\frac{1}{2}k,\frac{1}{2}\ell} \right|^{2}$$

$$(258)$$

The prime in the sum means that the $(0,0)$ sector is omitted. While it is true that $|D_{0,0}|$ vanishes, this sector may contribute to other quantities. For instance, we shall see in section 3.8 that $\langle \varepsilon(z, \bar{z}) \rangle$ is a nonvanishing constant, which arises only from this sector

$$\langle \varepsilon \rangle = \frac{\pi |\eta(\tau)|^{2}}{Z_{\frac{1}{2}}} \qquad (259)$$

Another example where this sector contributes is afforded by the spin–spin correlations (Di Francesco, Saleur and Zuber). Using the θ-functions notation (Appendix A),

$$\left| D_{\frac{1}{2},\frac{1}{2}}(\tau) \right| = |\theta_{3}(0,\tau)/\eta(\tau)|$$

$$\left| D_{0,\frac{1}{2}}(\tau) \right| = |\theta_{4}(0,\tau)/\eta(\tau)| \qquad (260a)$$

$$\left| D_{\frac{1}{2},0}(\tau) \right| = |\theta_{2}(0,\tau)/\eta(\tau)|$$

we have

$$Z_{\frac{1}{2}} = \frac{1}{2} \sum_{\nu=2}^{4} |\theta_{\nu}(0,\tau)/\eta(\tau)|$$

and, up to an overall factor,

$$
\langle \sigma(z_1,\bar{z}_1)\sigma(z_2,\bar{z}_2)\rangle = \frac{\displaystyle\sum_{\nu=1}^{4}\left|\theta_\nu(\tfrac{1}{2}z_{12},\tau)\right|\left|\frac{\theta_1'(0,\tau)}{\theta_1(z_{12},\tau)}\right|^{\frac{1}{4}}}{\displaystyle\sum_{\nu=2}^{4}\left|\theta_\nu(0,\tau)\right|}
$$

$$
\underset{z_{12}\to 0}{\sim}\frac{1}{(z_{12}\bar{z}_{12})^{\frac{1}{8}}} \tag{261}
$$

with $\theta_1(z,\tau)\underset{z\to 0}{\sim}\theta_1'(0,\tau)z$, $\theta_1'(0,\tau)=2\pi\eta(\tau)^3$.

The expression (261) behaves as expected at short distance. One can also convince oneself that the correlation is doubly periodic. The interesting point is that it is a sum of contributions arising from the four sectors in the form $\sum Z_\nu\langle\sigma\sigma\rangle_\nu / \sum Z_\nu$, where, in the numerator, the doubly periodic sector ($\nu=1$) gives a nonvanishing contribution in spite of the fact that $Z_\nu|_{\nu=1}=0$. Each of the partial quantities $Z_\nu\langle\sigma\sigma\rangle_\nu$ relative to a given spin structure satisfies a generalization of the Belavin, Polyakov, Zamolodchikov equations obtained by Eguchi and Ooguri in the form, valid for any operator A degenerate at level 2,

$$
\left\{\frac{3}{2(2h+1)}\partial_z^2 - 2\eta_1 z\partial_z\right.
$$

$$
-\sum_{i=1}^{n}\left(h_i\left[p(z-z_i)+2\eta_1\right]+\left[\zeta(z-z_i)+2\eta_1 z_i\right]\frac{\partial}{\partial z_i}\right)
$$

$$
\left.-2i\pi\frac{\partial}{\partial\tau}-2h\eta_1\right\}Z\langle A(z,\bar{z})A_1(z_1,\bar{z}_1)\cdots A_n(z_n,\bar{z}_n)\rangle=0 \tag{262}
$$

These equations involve Weierstrass's p and ζ functions with a double and simple pole at the origin as well as the η_1 constant

$$
\zeta(z,\tau)=\frac{\theta_1'(z,\tau)}{\theta_1(z,\tau)}+2\eta_1(\tau)z
$$

$$
p(z,\tau)=-\zeta'(z,\tau) \tag{263}
$$

$$
\eta_1(\tau)=-\tfrac{1}{6}\frac{\theta_1''(0,\tau)}{\theta_1'(0,\tau)}=-2i\pi\frac{\partial}{\partial\tau}\ln\eta(\tau)
$$

Beware that $\zeta(z,\tau)$ is not doubly periodic and the constant η_1 already appeared in equation (243). Using the same notations, the

$\varepsilon\varepsilon$ correlation (258) reads

$$\langle \varepsilon(z_1, \bar{z}_1)\varepsilon(z_2, \bar{z}_2)\rangle = \left| \frac{\theta_1'(0, \tau)}{\theta_1(z_{12}, \tau)} \right|^2 \frac{\sum_{\nu=2}^{4} |\theta_\nu(z_{12}, \tau)|^2 / |\theta_\nu(0, \tau)|}{\sum_{\nu=2}^{4} |\theta_\nu(0, \tau)|}$$

(264)

Higher order correlations on a torus can also be obtained in closed form.

9.3.4   The A–D–E classification of minimal models

We return to the discussion of section 3.1, in particular to equation (211), where the partition function is expanded as a sesquilinear form in Virasoro characters with integral nonnegative coefficients. Such an expansion is intimately related to a Hamiltonian formulation, with a preferred choice of time and space directions. It was claimed that the latter is in fact irrelevant, and that under a modular transformation, the partition function should be invariant, leading to constraints on the operator content of a given model. We found in the previous subsection, equation (250), a typical example involving such a sum over three terms with unit coefficients. In a sense, the Gaussian partition function (equation (227)) is also a limiting case of this situation. In this section, we report the results of imposing the constraint of modular invariance in the general case where the central charge assumes a rational value $c = 1 - 6(p - p')^2/pp'$, with the special, and important, unitary subcase such that p and p' are consecutive integers. Remarkably, a complete classification can be described in terms related to the Cartan–Killing classification of simple Lie algebras (to be specific, a subsequence called "simply laced"). This same A–D–E classification, where A stands for the Lie algebras of unitary groups, D for orthogonal groups in even dimension and E for the three exceptional Lie algebras E_6, E_7, E_8, occurs in several distinct instances, at first seemingly unrelated. An other famous case is the one of finite subgroups of three-dimensional rotations (up to conjugacy).

The fact that we can obtain a complete description of all minimal models is, to some extent, quite surprising and shows how deep is the present approach to critical models in two dimensions. Although all its implications have not yet been worked out, we know from the works on integrable models (Baxter, Andrews and Forrester, Pasquier) that one can construct lattice models which correspond to the predicted critical behaviour. On the other hand, the relation with integrable models is not yet completely clear. At any rate, it is of some importance to be able to obtain, even in a restricted range, a complete description of critical universality classes.

To carry out the previous program, we have to describe the behaviour of Virasoro characters under the action of the modular group. The latter is generated by the two basic transformations

$$
\begin{array}{lll}
T & \quad \tau \to \tau + 1 \\
S & \quad \tau \to -\tau^{-1}
\end{array}
\tag{265a}
$$

satisfying

$$
S^2 = (ST)^3
\tag{265b}
$$

In section 1.8, it was found convenient to label the representations by an integer λ defined modulo $N = 2pp'$, with a compact form of the characters given in equation (158). We also recall that under modular transformations, Dedekind's function, which appears in the denominator of the character formula, transforms as indicated in (228). Using Poisson's formula

$$
\sum_{n=-\infty}^{+\infty} f(n) = \sum_{p=-\infty}^{+\infty} \int \mathrm{d}x \; f(x) \; e^{2i\pi px}
\tag{266}
$$

it is readily found that

$$
T \qquad \chi_\lambda(\tau + 1) = e^{2i\pi\left(\lambda^2/2N - \frac{1}{24}\right)}\chi_\lambda(\tau)
\tag{267a}
$$

$$
S \qquad \chi_\lambda(-1/\tau) = \frac{1}{\sqrt{N}} \sum_{\lambda' \epsilon Z/NZ} e^{2i\pi\lambda\lambda'/N}\chi_{\lambda'}(\tau)
\tag{267b}
$$

In other words the action of T is diagonal, being, up to a phase, the multiplication by $\exp(2i\pi\lambda^2/2N)$, while the action of S is nothing but the finite Fourier transform over integers modulo N. The latter has a square equal to $\delta_{\lambda,-\lambda' \bmod N}$, but the equality $\chi_\lambda = \chi_{-\lambda}$ restores the property $S^2 = I$. Both transformations in (267) are compatible with the antisymmetry property $\chi_\lambda = -\chi_{\omega_0\lambda}$. We recall that ω_0 is given by $\omega_0 = r_0p + s_0p' \bmod.N$, with r_0, s_0 such that $r_0p - s_0p' = 1$, an identity expressing that the integers p and p' are coprime. We could of course use the symmetries of χ_λ to restrict the sum on the r.h.s. of (267b). This, however, would hide the relation between the Virasoro characters and an other set of similar functions arising in the study of the representations of the Kac–Moody algebra $A_1^{(1)}$, namely the algebra of local $SU(2)$ currents. An interesting class of highest weight representations of the corresponding infinite Lie algebra is also characterized by an anomaly (analogous to the central charge) called, in this context, the level, and denoted by k, an integer taking the values 0, 1,... These representations can be understood as a superposition of ordinary representations of $SU(2)$, the one with lowest spin being nondegenerate with a dimension $\lambda = 2j_{min} + 1$, and $2j_{min}$ ranging over integers from 0 to k. We present these matters in appendix C. The reason for introducing this subject here is the close relationship with the Virasoro characters and the quest for similar modular invariant partition functions.

Let us denote by $\chi_\lambda^{aff}(\tau)$ the $A_1^{(1)}$ characters, defined in appendix C. The previous Virasoro characters might be called by contrast, conformal ones. Specifically, for a level k representation of $A_1^{(1)}$, we set $N = 2(k+2)$, again an even integer, and the character reads

$$\chi_\lambda^{aff}(\tau) = \sum_{n=-\infty}^{+\infty} \frac{(Nn + \lambda)q^{(Nn+\lambda)^2/2N}}{\eta^3(\tau)} \tag{268}$$

We note the great similarity with the case of Virasoro characters (equation (158)), and observe that χ_λ^{aff} extends as a periodic (period N) odd function of λ

$$\chi_\lambda^{aff}(\tau) = \chi_{\lambda+N}^{aff}(\tau) = -\chi_{-\lambda}^{aff}(\tau) \tag{269}$$

In the present case the role of the involution $\lambda \to \omega_0 \lambda$ is simply played by the symmetry $\lambda \to -\lambda$. The similarity is further strengthened by the modular transformation properties, to be compared with (267)

$$\chi_\lambda^{aff}(\tau + 1) = e^{2i\pi\left(\lambda^2/2N - \frac{1}{8}\right)}\chi_\lambda^{aff}(\tau) \tag{270a}$$

$$\chi_\lambda^{aff}(-1/\tau) = \frac{-i}{\sqrt{N}} \sum_{\lambda' \in \mathcal{Z}/N\mathcal{Z}} e^{2i\pi\lambda\lambda'/N}\chi_{\lambda'}^{aff}(\tau) \tag{270b}$$

As suggested by Gepner and Witten, one can set forth a similar problem of classifying partition functions of the type (211) with affine characters replacing Virasoro characters, and coefficients restricted by analogous conditions, provided the sum over λ runs over a fundamental domain $1 \leq \lambda \leq \frac{1}{2}N - 1$ since $\chi_0^{aff} = \chi_{\frac{1}{2}N}^{aff} = 0$.

In both cases, the modular invariant solutions can be found in two steps. In the common first step, one simply looks for $N \times N$ matrices with arbitrary entries, call them $\mathcal{N}$, which commute with the two matrices

$$T_{\lambda,\lambda'} = e^{2i\pi\lambda^2/2N}\delta_{\lambda,\lambda'}$$
$$S_{\lambda,\lambda'} = \frac{1}{\sqrt{N}}e^{2i\pi\lambda\lambda'/N} \tag{271}$$

where λ and λ' as well as the labels of $\mathcal{N}$ run over integers modulo N.

It is found that such matrices $\mathcal{N}$ can be written as combinations of a linearly independent set defined as follows. Let δ be any positive divisor of $\frac{1}{2}N$, including unity, and α the greatest common divisor of δ and $\bar{\delta} = N/2\delta$, in compact notation $\alpha = (\delta, \bar{\delta})$. Clearly α^2 is a divisor of $\frac{1}{2}N$. Define the $N \times N$ matrix $(\Omega_\delta)_{\lambda,\lambda'}$ to have vanishing matrix elements unless α divides both λ and λ', in which case

$$(\Omega_\delta)_{\lambda,\lambda'} = \sum_{\xi \in Z/\alpha Z} \delta_{\lambda',\omega\lambda+\xi N/\alpha} \tag{272}$$

In this definition, the integer ω mod N/α^2 is obtained as follows. Since δ/α and $\bar{\delta}/\alpha$ are coprime, one can find two integers ρ and σ such that $\rho\bar{\delta}/\alpha - \sigma\delta/\alpha = 1$, then $\omega = \rho\bar{\delta}/\alpha + \sigma\delta/\alpha$ (mod N/α^2). One checks that $\omega^2 = 1$ (mod $2N/\alpha^2$), and this makes sense since in (272) it is required that $\lambda'/\alpha = \omega\lambda/\alpha$ (mod N/α^2).

It is not difficult to see that Ω_δ commutes with both T and S. It is slightly more involved to show, as did Gepner and Qiu, that any matrix $\mathcal{N}$ with this property is a linear combination of these Ω_δ's and that the latter are linearly independent. We remark that if $\delta = \frac{1}{2}N$, $\bar{\delta} = 1$, $\alpha = 1$, $\omega = 1$, and $\Omega_{\frac{1}{2}N} = I$.

The difficult task is to find the correct superposition of Ω_δ's such that $\sum \chi_\lambda^*(\tau)(\gamma_\delta\Omega_\delta)_{\lambda\lambda'}\chi_{\lambda'}(\tau)$, which is modular invariant, reduces, when summed over the fundamental (Virasoro or affine) characters, to a combination with nonnegative integral coefficients. The normalization in the affine case requires that the coefficient of $\chi_1^{*aff}(\tau)\chi_1^{aff}(\tau)$ be unity. In the Virasoro case, the coefficient of the invariant state, corresponding to $h = \bar{h} = 0$, should also be unity. Using the λ-notation, this corresponds to a unit coefficient for $\chi_{p-p'}^*(\tau)\chi_{p-p'}(\tau)$. The difficulty in both cases stems from an antisymmetry of the characters in $\lambda \to \omega_0 \lambda$ or $\lambda \to -\lambda$ respectively.

Let us first describe the solution for the affine partition functions, where we write the required combination as

$$Z_{affine} = \tfrac{1}{2} \sum_{\lambda, \lambda' \epsilon Z/NZ} \chi_\lambda^{aff*}(\tau) \left(\sum \gamma_\delta \Omega_\delta \right)_{\lambda \lambda'} \chi_{\lambda'}^{aff}(\tau) \qquad (273)$$

The coefficient $\tfrac{1}{2}$ takes into account the fact that, due to the antisymmetry $\chi_{-\lambda} = -\chi_\lambda$, each fundamental product $\chi_\lambda^* \chi_{\lambda'}$ is counted twice. Two infinite series and three exceptional solutions, fulfil all criteria. For conciseness we write, using as a label $n = \tfrac{1}{2}N$, the combination $\sum \gamma_\delta \Omega_\delta$ appearing in (273), and recall that Ω_n is the unit matrix

$$
\begin{array}{lll}
n \geq 2 & \Omega_n & (A_{n-1}) \\
n \text{ even } \geq 6 & \Omega_n + \Omega_2 & (D_{n/2+1}) \\
n = 12 & \Omega_{12} + \Omega_3 + \Omega_2 & (E_6) \qquad (274) \\
n = 18 & \Omega_{18} + \Omega_3 + \Omega_2 & (E_7) \\
n = 30 & \Omega_{30} + \Omega_5 + \Omega_3 + \Omega_2 & (E_8)
\end{array}
$$

As a consequence when n is odd the solution is unique. When n is even, there exist in general two distinct solutions, and three when $n = 12$, 18 or 30. The group theoretic meaning of this result is more clearly appreciated when we give the explicit form of (273)–(274) in table I.

To correlate this result with the A–D–E classification of simply laced, simple Lie algebras (appendix C), we note that the diagonal entries (0, 1, or 2) in the matrix (274), as shown in the table, are related to the degrees of the fundamental Casimir invariant polynomials pertaining to the corresponding Lie algebra. Recall that the enveloping algebra of a Lie algebra admits a subset of invariant polynomials. The latter can in turn be written as polynomials in a set of r of them, r being the rank of the Lie algebra. If we subtract 1 from the degrees of these fundamental invariant polynomials (the first one is always quadratic), then the above diagonal entries indicate how many polynomials of this degree are found in this set. For instance, corresponding to E_8, we find in the table that the Casimir invariants have degrees 2, 8, 12, 14, 18, 20, 24 and 30. These degrees are sometimes called Coxeter

Table I. Invariant partition functions for the Kac–Moody algebra $A_1^{(1)}$.

$n \geq 2$	A_{n-1}	$\displaystyle\sum_{\lambda=1}^{n-1} |\chi_\lambda|^2$
$n = 4\rho + 2$ $\rho \geq 1$	$D_{2\rho+2}$	$\displaystyle\sum_{\lambda_{odd}=1}^{2\rho-1} |\chi_\lambda + \chi_{4\rho+2-\lambda}|^2 + 2|\chi_{2\rho+1}|^2$
$n = 4\rho$ $\rho \geq 2$	$D_{2\rho+1}$	$\displaystyle\sum_{\lambda_{odd}}^{4\rho-1} |\chi_\lambda|^2 + |\chi_{2\rho}|^2 + \sum_{\substack{\lambda\ even=2}}^{2\rho-2} \left(\chi_\lambda^* \chi_{4\rho-\lambda} + \chi_{4\rho-\lambda}^* \chi_\lambda\right)$
$n = 12$	E_6	$|\chi_1 + \chi_7|^2 + |\chi_4 + \chi_8|^2 + |\chi_5 + \chi_{11}|^2$
$n = 18$	E_7	$|\chi_1 + \chi_{17}|^2 + |\chi_5 + \chi_{13}|^2 + |\chi_7 + \chi_{11}|^2 + |\chi_9|^2 + \left[(\chi_3 + \chi_{15})^* \chi_9 + \chi_9^* (\chi_3 + \chi_{15})\right]$
$n = 30$	E_8	$|\chi_1 + \chi_{11} + \chi_{19} + \chi_{29}|^2 + |\chi_7 + \chi_{13} + \chi_{17} + \chi_{23}|^2$

numbers. This correspondence is the reason of the labelling. Its origin is not at present fully understood.

It would seem that we have followed a side track in classifying these affine invariants, were it not for the fact that the solution of our original problem with Virasoro characters is simply related to the above one. Namely, recall that in this case $N = 2pp'$. Then, at the price of an irrelevant doubling of the original space, its turns out that we can think of Virasoro characters as carrying labels in a tensor space of dimension $2p' \times 2p$. Elements of the commutant of T and S can also be thought of as products $\Omega_\delta' \otimes \Omega_\delta$ and the corresponding integral positive partition functions are given by a pair of combinations chosen among those occurring in equation (274), with the role of n played successively by p' and p. Moreover, since p and p' are coprime, one of them at least is necessarily odd, the corresponding invariant being then of the A type. Consequently, and this is the main result of this section, one finds two infinite series and three exceptional subseries of minimal models consistent with the requirements of conformal and modular invariance. The partition functions are displayed in table II, where it is assumed for definiteness that p is odd. The particular unitary case will then mean that if m is odd $p \to m$, $p' = m + 1$ and if m is even $p \to m + 1$, $p' = m$. Thus, we find two infinite series and three pairs of exceptional models. The factor $\frac{1}{2}$ appearing in table II reflects the symmetry $\chi_{r,s} = \chi_{p'-r,p-s}$, when we return to the labelling in integers r, s ($\lambda = rp - sp'$) as used in the table, where each invariant is denoted by a pair of Lie algebras.

The analysis of each of the models occurring in the table is beyond the scope of the present discussion. We simply present a few remarks. First let us restrict our attention to the unitary case. In the $(A{-}A)$ series, the first candidate corresponds to $m = 3$, $p = 3$, $p' = 4$, thus in the present notation (A_3, A_2). This of course is the Ising model. In general, in this $A{-}A$ series (called the principal series), the partition function reads

$$Z = \sum_{1 \le s \le r \le m-1} \left| \chi_{r,s}(\tau) \right|^2 \tag{275}$$

and one obtains a generalization of the Ising model with only one scalar operator $A_{h,h}$ for each distinct entry in the Kac table. The next case, corresponding to $m = 4$, $c = \frac{7}{10}$, can be interpreted as the tricritical Ising model. In general, the unitary $A{-}A$ series

Table II. Invariant partition functions in terms of Virasoro characters.

$(A_{p'-1}, A_{p-1})$		$\dfrac{1}{2}\displaystyle\sum_{s=1}^{p-1}\sum_{r=1}^{p'-1}|\chi_{r,s}|^2$
$(D_{2\rho+2}, A_{p-1})$	$\begin{aligned}p' &= 4\rho+2 \\ \rho &\geq 1\end{aligned}$	$\dfrac{1}{2}\displaystyle\sum_{s=1}^{p-1}\left\{\sum_{r_{odd}=1}^{2\rho-1}|\chi_{r,s}+\chi_{4\rho+2-r,s}|^2 + 2|\chi_{2\rho+1,s}|^2\right\}$
$(D_{2\rho+1}, A_{p-1})$	$\begin{aligned}p' &= 4\rho \\ \rho &\geq 2\end{aligned}$	$\dfrac{1}{2}\displaystyle\sum_{s=1}^{p-1}\left\{\sum_{r_{odd}=1}^{4\rho-1}|\chi_{r,s}|^2 + |\chi_{2\rho,s}|^2 + \sum_{r_{even}=2}^{2\rho-2}(\chi_{r,s}^*\chi_{4\rho-r,s} + \chi_{4\rho-r,s}^*\chi_{r,s})\right\}$
(E_6, A_{p-1})	$p'=12$	$\dfrac{1}{2}\displaystyle\sum_{s=1}^{p-1}\{|\chi_{1,s}+\chi_{7,s}|^2 + |\chi_{4,s}+\chi_{8,s}|^2 + |\chi_{5,s}+\chi_{11,s}|^2\}$
(E_7, A_{p-1})	$p'=18$	$\dfrac{1}{2}\sum_{s=1}^{p-1}\{|\chi_{1,s}+\chi_{17,s}|^2 + |\chi_{5,s}+\chi_{13,s}|^2 + |\chi_{7,s}+\chi_{11,s}|^2 + |\chi_{9,s}|^2 + [(\chi_{3,s}+\chi_{15,s})^*\chi_{9,s} + \chi_{9,s}^*(\chi_{3,s}+\chi_{15,s})]\}$
(E_8, A_{p-1})	$p'=30$	$\dfrac{1}{2}\displaystyle\sum_{s=1}^{p-1}\{|\chi_{1,s}+\chi_{11,s}+\chi_{19,s}+\chi_{29,s}|^2 + |\chi_{7,s}+\chi_{13,s}+\chi_{17,s}+\chi_{23,s}|^2\}$

seems to describe the set of multicritical scalar models, of higher and higher degree (appendix A, chapter 5), with an effective Lagrangian of the form

$$\mathcal{L} = \tfrac{1}{2}(\partial\varphi)^2 + \varphi^{2k}$$

The case $k = 2$ is the Ising model again ($m = 3$), and a general k corresponds to the value $m = k + 1$. The central charge $c = 1-6/(k+1)(k+2)$ converges to the free field case ($c = 1$) as $k \to \infty$, in agreement with intuition. According to Zamolodchikov, the primary field with weights $h = \bar{h} = h_{2,2} = 3/4(k+1)(k+2)$ ($k \geq 2$) corresponds to the field φ. Higher powers : φ^n : have weights $h = \bar{h} = h_{n+1,n+1} = [(n+1)^2 - 1]/4(k+1)(k+2)$ for $n = 1, 2, \ldots,$ $k-1$, and $h = \bar{h} = h_{n-k+2,n-k+1} = [(n+3)^2 - 1]/4(k+1)(k+2)$ for $n = k, \ldots, 2k-2$. The field : φ^{2k-1} : is related to $\partial\bar{\partial}\varphi$ through the equation of motion. As will be discussed in the next subsection, all these models are naturally related to the spontaneous breaking of a Z_2 symmetry.

When $m = 5$ and $c = \frac{4}{5}$, we find two possibilities. According to the principal series, all primary operators are scalars, while a second possibility belongs to the (A, D) series (call it the complementary series), namely $p = 5$, $p' = 6$, thus (A_4, D_4). Using first the notation $\chi_{r,s}$, then the notation $\chi_{h_{rs}}$, the corresponding partition function reads

$$(A_4, D_4) \quad Z = \left|\chi_{1,1} + \chi_{4,1}\right|^2 + \left|\chi_{2,1} + \chi_{3,1}\right|^2 + 2\left|\chi_{3,3}\right|^2 + 2\left|\chi_{4,3}\right|^2$$

$$\equiv \left|\chi_0 + \chi_3\right|^2 + \left|\chi_{2/5} + \chi_{7/5}\right|^2 + 2\left|\chi_{1/15}\right|^2 + 2\left|\chi_{2/3}\right|^2$$

$$(276)$$

Assuming that the three state Potts model belongs to the value $\frac{4}{5}$ for the central charge, there are various ways to ascertain that (276) is the correct choice. One of them is to recognize that we need two operators with weights $\frac{1}{15}$, $\frac{1}{15}$, corresponding to the two $\sin\varphi$, and $\cos\varphi$ "spin operators". Another argument, to be described in the next subsection, shows that only (276) corresponds to the possibility of a Z_3 symmetry breaking, as should be the case. There is however an almost direct "microscopic" proof using a Coulomb gas description, which agrees with this choice, as further discussed in section 3.6. Assuming that equation (276) does correctly describe the content of the three state Potts model, we note the existence of operators with angular momentum ± 3,

± 1. Except for the principal A–A series, one always finds such tensorial fields.

Huse has observed that the A–A principal series was realized at the critical point of the restricted solid on solid models solved by Andrews, Baxter and Forrester. We have already mentioned that Pasquier has generalized their construction to include all other unitary cases listed in table II. Of course, the $i\varphi^3$ model or Lee Yang edge singularity (section 1.7) belongs also to the table. As we recall $c = -\frac{22}{5}$ and the partition function is in the (A, A) series

$$(A_4, A_1) \qquad Z = \left|\chi_{h=0}\right|^2 + \left|\chi_{h=-\frac{1}{5}}\right|^2 \qquad (277)$$

Proofs and comments on the results of this section will be found in the works quoted in the notes.

9.3.5   *Frustrations and discrete symmetries*

Assume that one of the theories described above is a critical limit of a lattice model, with nearest neighbour interactions between discrete variables $\sigma_{i,j}$ taking finitely many values. Assume furthermore that a finite group G operates globally on the variables in such a way that the action is invariant on an infinite lattice. When the latter is restricted to a finite range L, M, instead of considering periodic boundary conditions $\sigma_{L+1,j} = \sigma_{1,j}$ or $\sigma_{i,M+1} = \sigma_{i,1}$, as was implicit up to now, we can instead introduce twisted (or frustrated) boundary conditions in the form $\sigma_{L+1,j} = {}^{g_1}\sigma_{1,j}$, $\sigma_{i,M+1} = {}^{g_2}\sigma_{i,1}$, where $g_1, g_2 \in G$. Taking the continuous limit, we have therefore to envision frustrated partition functions Z_{g_1,g_2} which are now covariant under the modular group. To be specific, let us look for such circumstances where G is a cyclic group $G \equiv \mathcal{Z}/k\mathcal{Z}$, and the discrete variable $\sigma_{i,j}$ gives rise to one of the primary fields with conformal weights $(h, \bar{h})$. Furthermore let $\mathcal{Z}/k\mathcal{Z}$ act through phases, in such a way that the corresponding field $u_{h,\bar{h}}$ has to fulfil, with n_1 and n_2 integers,

$$u_{h,\bar{h}}(z + n_1\omega_1 + n_2\omega_2) = \exp\left[\frac{2i\pi}{k}\left(n_1 k_1 + n_2 k_2\right)\right] u_{h,\bar{h}}(z) \quad (278)$$

According to the definition of h and $\bar{h}$, we have

$$u_{h,\bar{h}}(z + n_1\omega_1) = \exp\left[2i\pi(h - \bar{h})n_1\right] u_{h,\bar{h}}(z) \qquad (279)$$

meaning that $k_1/k \equiv h - \bar{h}$ mod 1. Taking k_1 prime to k, and using the values of h and $\bar{h}$ in the Kac table for a minimal model, we conclude that k must divide $4pp'$ or $4m(m + 1)$ if we restrict ourselves to unitary models. This is a necessary condition for $\mathcal{Z}/k\mathcal{Z}$ to be a candidate symmetry group. Denote by Z_{k_1,k_2} the partition function corresponding to the boundary conditions (278). It is only invariant under a subgroup of the modular group which respect these conditions. In particular consider $Z_{k_1,0}$. If we set $\tau' = (a\tau + b)/(c\tau + d)$, it is seen that invariance of $Z_{k_1,0}$ requires $a = d = \pm 1$ mod k and $b = 0$ mod k, provided we assume k_1 and k coprime and a reality condition $Z_{k_1,0} = Z_{-k_1,0}$. This defines a subgroup, call it $\Gamma^0(k)$, of the full modular group Γ. The twisted boundary condition in the direction ω_1 does not prevent us from thinking of $Z_{k_1,0}$ as being the trace of the transfer matrix in a certain subspace of states which can be decomposed into irreducible parts according to the Virasoro algebras. Consequently, we must also find non-negative integers $N_{h,\bar{h}}$ such that

$$Z_{k_1,0} = \sum_{h,\bar{h}} N_{h,\bar{h}} \chi_h \bar{\chi}_{\bar{h}} \tag{280}$$

with the requirement that it be invariant under $\Gamma^0(k)$. If this is the case, acting with an arbitrary conformal transformation $\gamma \in \Gamma$ on the argument τ (and $\bar{\tau}$) of $Z_{k_1,0}$, must generate a function $Z_{k_1,0}(^\gamma\tau)$ invariant under the conjugate group $\gamma\Gamma^0(k)\gamma^{-1}$. The number of such functions is the number of (right) equivalence classes, $\gamma \sim \gamma'$ if $\gamma'^{-1}\gamma \in \Gamma^0(k)$, which is equal to

$$d_k = \tfrac{1}{2}k^2 \prod_{\substack{p \text{ prime} \\ p|k}} \left(1 - \frac{1}{p^2}\right) \qquad k \geq 3$$
$$d_2 = 3 \tag{281}$$

Here, the symbol $p \mid k$ means that p divides k. Each of the above functions is invariant under $\Gamma(k)$, the largest subgroup common to all $\gamma\Gamma^0(k)\gamma^{-1}$. This group, called the principal congruence subgroup (of level k) is an invariant subgroup of Γ. The finite quotient $\Gamma/\Gamma(k)$ is the modular group on integers modulo k. The d_k functions $Z_{k_1,0}(^\gamma\tau)$ correspond in fact to other boundary conditions $Z_{k'_1,k'_2} = Z_{-k'_1,-k'_2}$, each one being left invariant by a group $\gamma\Gamma^0(k)\gamma^{-1}/\Gamma(k)$ isomorphic to $\mathcal{Z}/k\mathcal{Z}$, the cyclic symmetry group. These other partition functions are again sesquilinear forms of the type (280), where, however, the coefficients are not restricted any longer to be positive integers. Of course, in this notation, $Z_{0,0}$ is the unfrustrated partition function derived in the previous section. The index of $\Gamma(k)$ in Γ, or the order of $\Gamma/\Gamma(k)$, is kd_k.

Let us apply this formalism to the Ising and three state Potts models, $m = 3$ and 5 in the unitary series. When $k \leq 4$, the group $\Gamma^0(k)$ is generated by the transformations $\tau \to \tau + k$ and $\tau \to \tau/\tau + 1$. For the Ising model, we expect a group of symmetry $\mathcal{Z}/2\mathcal{Z}$, i.e. $k = 2$. The most general form of the frustrated partition function $Z_{1,0}$ is, using the notation χ_h,

$$Z_{1,0} = (N_1 - N_2)\left[|\chi_0|^2 + \left|\chi_{\frac{1}{2}}\right|^2 + \left|\chi_{\frac{1}{16}}\right|^2\right] + N_2\left[\left|\chi_0 + \chi_{\frac{1}{2}}\right|^2 + \left|\chi_{\frac{1}{16}}\right|^2\right]$$

(282)

The first bracket is $Z_{0,0}$, the unfrustrated partition function, the second term involves the fermionic primary operators through $\chi_{\frac{1}{2}}\bar{\chi}_0$ and $\chi_0\bar{\chi}_{\frac{1}{2}}$. Acting with $\tau \to -\tau^{-1}$ generates $Z_{0,1}$ in the form

$$Z_{0,1} = (N_1 + N_2)\left[|\chi_0|^2 + \left|\chi_{\frac{1}{2}}\right|^2\right] + (N_1 - N_2)\left|\chi_{\frac{1}{16}}\right|^2 \qquad (283)$$

In the strip limit, the leading terms of $Z_{0,1}$ and $Z_{0,0}$ should be identical, as the effect of reversing the spins far apart should not affect the asymptotic leading behaviour. This fixes $N_1 + N_2 = 1$ and, assuming $N_2 \neq 0$ for a nontrivial $Z_{1,0}$, requires $N_1 = 0$, $N_2 = 1$.

As a byproduct, if $\tau = iM/L$, and in the limit $M/L \to \infty$, we expect that

$$Z_{1,0}/Z_{0,0} = \exp(-Mf) \qquad (284)$$

with f the interface free energy per unit length. In a self-dual model such as the present one, duality relates this interface energy with the behaviour of the spin-spin correlation function at criticality and yields for a finite system with width $L = |\omega_1|$, $f = \pi\eta/L$. Comparing with the above expressions, this produces correctly the value of the exponent $\eta = \frac{1}{4}$.

We can apply the same argument for the three state Potts model denoted (A_4, D_4) with $m = 5$ (see equation (276)). One can check that

$$Z_{1,0} = Z_{2,0} = (\chi_0 + \chi_3)^* \chi_{\frac{2}{3}} + (\chi_0 + \chi_3) \chi_{\frac{2}{3}}^*$$
$$+ \left(\chi_{\frac{2}{5}} + \chi_{\frac{7}{5}}\right)^* \chi_{\frac{1}{15}} + \left(\chi_{\frac{2}{5}} + \chi_{\frac{7}{5}}\right) \chi_{\frac{1}{15}}^* + \left|\chi_{\frac{2}{3}}\right|^2 + \left|\chi_{\frac{1}{15}}\right|^2 \qquad (285a)$$

is invariant under $\Gamma^0(3)$, in agreement with a $Z/3Z$ symmetry of the model. It exhibits the existence of a spin $\frac{1}{3}$ "pseudofermion". The ratio $Z_{1,0}/Z_{0,0}$ in the strip limit allows one to confirm the value $\eta = \frac{4}{15}$. On the other hand, the computation of $Z_{0,1}$ allows one to understand the behaviour of the primary operators with respect to the discrete symmetry groups. For instance for the Potts model,

$$Z_{0,1} = \left|\chi_0 + \chi_3\right|^2 + \left|\chi_{\frac{2}{5}} + \chi_{\frac{7}{5}}\right|^2 - \left|\chi_{\frac{2}{3}}\right|^2 - \left|\chi_{\frac{1}{15}}\right|^2 \qquad (285b)$$

In the generic case of the principal series of unitary (A, A) models, one finds a generalization of (282) in the form

$$Z_{1,0} = \frac{1}{2} \sum_{r=1}^{n-1} \sum_{s=1}^{m} \chi_{r,s} \chi_{r,m+1-s}^* \qquad (286)$$

also invariant under $\Gamma^0(2)$ indicating a $Z/2Z$ symmetry of the corresponding restricted solid on solid, or multicritical, model. The operator with lowest dimension in (286) yields the interface free energy

$$\begin{aligned} Lf &= 2\pi \, \min(h_{r,s} + h_{r,m+1-s}) \\ &= \pi \frac{(m-1)(m+3)}{4m(m+1)} \qquad m \text{ odd} \\ &= \pi \frac{(m-2)(m+2)}{4m(m+1)} \qquad m \text{ even} \end{aligned} \qquad (287)$$

The relation $Lf = \pi\eta$ seems to remain valid only up to $m = 4$ where it agrees with the value $\eta = \frac{3}{20}$ of the tricritical model.

Similarly, one can show that the two unitary theories with $m = 11$ or 12, corresponding to the Lie algebra E_6, have also a $Z/2Z$ symmetry. For instance, when $m = 11$

$$\begin{aligned} Z_{1,0} = \sum_{r_{odd}=1}^{9} &\Big\{ (\chi_{r,1} + \chi_{r,7}) (\chi_{r,5} + \chi_{r,11})^* \\ &+ (\chi_{r,1} + \chi_{r,7})^* (\chi_{r,5} + \chi_{r,11}) \\ &+ \left|\chi_{r,4} + \chi_{r,8}\right|^2 \Big\} \end{aligned} \qquad (288)$$

$$Z_{0,1} = \sum_{r_{odd}=1}^{9} \left| \chi_{r,1} + \chi_{r,7} \right|^2 + \left| \chi_{r,5} + \chi_{r,11} \right|^2 - \left| \chi_{r,4} + \chi_{r,8} \right|^2$$

By comparing with $Z_{0,0}$, we can figure out from this last expression which are the even or odd operators under the $Z/2Z$ group.

Find an argument to show that none of these minimal models admits a continuous symmetry, from the observation that they do not contain operators with dimensions $(1,0)$ and $(0,1)$ as candidates for the generating current (Friedan, Qiu, Shenker).

9.3.6 Nonminimal models

We have investigated cases with finitely many primary operators, a rational central charge smaller than unity, and conformal weights in the finite Kac table. There exists however a whole zoo of alternative theories which violate one or several of the above restrictions. We describe here some examples.

In section 2.1, when discussing the Gaussian model, we observed that instead of considering a field φ taking arbitrary real values, we could look at it as an angular variable with φ and $\varphi + 2\pi\rho$ identified. This led us to define (equations (169) and (170)) mixed operators $O_{n,m}(z, \bar{z})$ with conformal weights $h_{n,m} = \frac{1}{4}(n/\rho + \rho m)^2$, $\bar{h}_{n,m} = h_{n,-m}$, with n and m relative integers giving the intensity of a source of "electric" and "magnetic" charge located at the point $(z, \bar{z})$. We now investigate a similar situation on a torus. Since φ is an angular variable, it can wind around the unit circle a number of times as we describe a closed noncontractible path on the torus. These fields are such that

$$\varphi(z + k\omega_1 + k'\omega_2) - \varphi(z) = 2\pi\rho(km + k'm') \qquad (289)$$

with the variable $\bar{z}$ omitted. Let us consider the path integral in this sector. Fields fulfilling condition (289) may be written as a superposition $\varphi = \varphi_{per} + \varphi_{class}$ of a fluctuating periodic field φ_{per} and a specific solution to (289), φ_{class}, linear in $z, \bar{z}$, hence harmonic,

$$\varphi_{class}(z, \bar{z}) = 2\pi\rho \left\{ \frac{z}{\omega_1} \frac{m\bar{\tau} - m'}{\bar{\tau} - \tau} + \text{c.c.} \right\} \qquad (290)$$

The action is $S_{\text{periodic}} + \pi\rho^2 \, |m\tau - m'|^2 \, /\text{Im}\,\tau$ and the partition function in the m, m' sector reads

$$Z_{m',m}(\tau) = Z_1(\tau) \exp -\pi\rho^2 \frac{|m\tau - m'|^2}{\text{Im}\,\tau} \qquad (291a)$$

with $Z_1(\tau)$ the Gaussian partition function given by equation (227). From its definition, it is clear that under a modular transformation

$$Z_{m',m}\left(\frac{a\tau + b}{c\tau + d}\right) = Z_{am'+bm,cm'+dm}(\tau) \qquad (291b)$$

This shows that the modular group acts by linear transformations on the winding numbers m and m'. If we sum over all m, m' with a constant weight on the orbits of the group, we obtain a modular invariant. The simplest example is $Z_1(\tau) \equiv Z_{0,0}(\tau)$. As a second simple instance, we consider the other obvious invariant obtained by summing over all pairs m, m'. We append a prefactor ρ to normalize the large τ (small q) behaviour to obtain

$$Z(\rho, \tau) = \rho Z_1(\tau) \sum_{m',m} \exp\left(-\pi\rho^2 \frac{|m\tau - m'|^2}{\text{Im}\,\tau}\right) \qquad (292)$$

The Poisson formula applied to the m' summation yields, with $q = \exp 2i\pi\tau$,

$$Z(\rho, \tau) = \frac{1}{|\eta(\tau)|^2} \sum_{n,m} q^{\frac{1}{4}(n/\rho + m\rho)^2} \bar{q}^{\frac{1}{4}(n/\rho - m\rho)^2} \qquad (293)$$

where the weights $h_{n,m} = \frac{1}{4}(n/\rho + m\rho)^2$ and $\bar{h}_{n,m} = h_{n,-m}$ are recognized as those assigned to the operator $O_{n,m}$. This is called a Coulombic partition function, referring to logarithmic pair interactions between integral electric and magnetic charges. Comparing the expressions (292) and (293), we see that a factor $\rho/\text{Im}\,\tau^{\frac{1}{2}}$ has disappeared. For small q, the behaviour $(q\bar{q})^{-\frac{1}{24}}$ indicates that the central charge is unity. All dimensions $\Delta_{n,m} = \frac{1}{2}(n^2/\rho^2 + n^2\rho^2)$ are nonnegative and vary continuously with ρ, and angular momenta $S_{n,m} = nm$ are integers. For a generic ρ, $q^{h_{n,m}}/\eta(\tau)$ is a character of an irreducible representation ($c = 1, h_{n,m}$) of the Virasoro algebra. It is also clear that $Z(\rho, \tau) = Z(\rho^{-1}, \tau)$.

One application of this result is to the XY-model. Recall that the renormalization flow drives the two-point function to a critical one with exponent $\eta = \frac{1}{4}$, up to logarithmic corrections, indicating the presence of a marginal operator of dimension 2. Let us identify the spin operator of the XY-model with $O_{\pm 1,0}$ in such a way that $\eta = \frac{1}{4} = 2\Delta_{\pm 1,0}$. This suggests the choice $\rho = 2$. The corresponding magnetic or vortex operators $O_{0,\pm 1}$ then have dimension two, which signals that vortices become relevant. At any rate $Z(2,\tau) = Z(\frac{1}{2},\tau)$ is a consistent guess for an XY partition function at the end point of the critical line which should therefore have a central charge unity.

A further possibility is to consider φ as an angular variable of period $2\pi\rho$ with φ and $-\varphi$ identified. This is the orbifold model, and it can be understood as follows. The matrices with diagonal entries $e^{\pm i\varphi/\rho}$, $e^{\mp i\varphi/\rho}$ represent a conjugacy class within the group $SU(2)$. The corresponding partition function on a torus involves three more sectors where along any one or both generators the field changes sign. From our computation for the Ising model we deduce that in the orbifold case

$$Z_{\text{orb}} = \tfrac{1}{2} \left[Z(\rho;\tau) + 2 \left(|A(\tau)|^2 + |B(\tau)|^2 + |C(\tau)|^2 \right) \right] \quad (294a)$$

$$Z_{\text{Ising}} = \tfrac{1}{2} \left[\frac{1}{|A(\tau)|^2} + \frac{1}{|B(\tau)|^2} + \frac{1}{|C(\tau)|^2} \right] \quad (294b)$$

with

$$A(\tau) = \frac{q^{\frac{1}{48}}}{\prod_0^{\infty} \left(1 - q^{n+\frac{1}{2}} \right)} = \frac{1}{\eta(\tau)} \sum q^{\frac{1}{4}(2p+\frac{1}{2})^2}$$

$$B(\tau) = \frac{q^{\frac{1}{48}}}{\prod_0^{\infty} \left(1 + q^{n+\frac{1}{2}} \right)} = \frac{1}{\eta(\tau)} \sum (-)^p q^{\frac{1}{4}(2p+\frac{1}{2})^2} \quad (294c)$$

$$C(\tau) = \frac{1}{\sqrt{2}} \frac{q^{\frac{1}{48}}}{q^{\frac{1}{16}} \prod_1^{\infty}(1 + q^n)} = \frac{1}{\sqrt{2}\eta(\tau)} \sum (-1)^p q^{p^2}$$

Interestingly the combination appearing in (294a) simplifies to

$$Z_{\text{orb}}(\rho) = Z_{\text{orb}}(\rho^{-1}) = \tfrac{1}{2} [Z(\rho) + 2Z(2) - Z(1)] \quad (294d)$$

so that this describes a family of models with the same ρ-dependent dimensions as before, as well as some constant ones, and has been identified with the Ashkin–Teller critical line. For a description of this model of two spin systems $\sigma_i = \pm 1$, $\tau_i = \pm 1$

interacting via a four-spin term with

$$S = \sum_{(ij)} \beta_\sigma \sigma_i \sigma_j + \beta_\tau \tau_i \tau_j + \beta_{\sigma\tau} \sigma_i \sigma_j \tau_i \tau_j$$

we refer to the literature. One observes that $Z_{\text{orb}}(1) = Z(2)$ describe the same XY-model. One can picture the set of Coulombic ($O(2)$ invariant) and orbifold models at $c = 1$ as in figure 6 and point out on these two lines special points corresponding to various extra symmetries. Such is the case for $Z(1)$ which describes an $SU(2) \times SU(2)$ symmetric model, or $Z(\sqrt{2})$ which coincides with a Dirac spinor model, with

$$Z(\sqrt{2}) = \tfrac{1}{2} \left(\frac{1}{|A(\tau)|^4} + \frac{1}{|B(\tau)|^4} + \frac{1}{|C(\tau)|^4} \right) \qquad (295)$$

while $Z_{\text{orb}}(\sqrt{2})$ is Z_{Ising}^2. Other significant values are discussed in the references. There exist also three other interesting models constructed by Pasquier which do not lie on these two lines and correspond to exceptional affine simply laced Lie algebras

$$\hat{E}_6 \rightarrow \tfrac{1}{2} \left(2Z(3) + Z(2) - Z(1) \right)$$
$$\hat{E}_7 \rightarrow \tfrac{1}{2} \left(Z(4) + Z(3) + Z(2) - Z(1) \right) \qquad (296)$$
$$\hat{E}_8 \rightarrow \tfrac{1}{2} \left(Z(5) + Z(3) + Z(2) - Z(1) \right)$$

One can also use the previous formalism to generate partition functions for nonminimal models with central charge $c < 1$, describing the Q-state Potts model for $0 < Q < 4$ and the $O(n)$ model for continuous n varying between -2 and 2, the range where there exists a critical theory. These systems are defined for continuous n or Q through a high temperature expansion. It is possible to find a chain of plausible mappings between these models and integrable six vertex models or Coulomb gases with specific restrictions. The full derivations are outside the scope of this chapter and we shall content ourselves with a description of the results.

For ρ in the interval 1 to $\sqrt{2}$, set

$$n = -2\cos \pi \rho^2$$
$$e_0 = \pm (\rho^2 - 1) \mod 2 \qquad (297a)$$

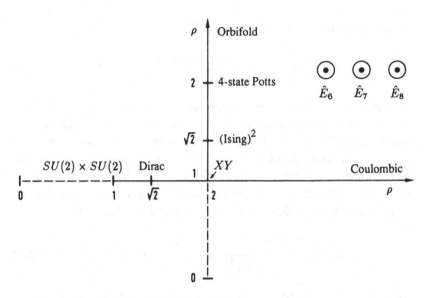

Fig. 6   Known $c = 1$ models. The horizontal line refers to $O(2)$ symmetric theories with a $2\pi\rho$-periodic field, the vertical one to orbifold models. The dotted parts are equivalent under $\rho \to \rho^{-1}$. The extra points depict three exceptional cases. The figure is taken from P. Ginsparg, "Curiosities at $c = 1$", Nucl. Phys. *B295 (FS 21)*(1988), 153-70.

in such away that n ranges between 2 and -2. The corresponding partition function is given by

$$Z(n; \tau) = \frac{\rho}{2} Z_1(\tau) \sum_{m', m} \exp -\frac{\pi\rho^2}{4} \frac{|m\tau - m'|^2}{\text{Im}\tau} \cos[\pi e_0(m', m)]$$

$$(297b)$$

In this expression (m', m) is the largest common positive divisor of m' and m. From our previous discussion, the most general modular invariant built out of a superposition of $Z_{m', m}$ would allow an arbitrary function of (m', m). There are two steps involved in deriving equations (297), the first is to show the origin of the relation between continuously varying n and ρ, the second to introduce the "defect charge" e_0. In the process of using a continuously variable n through a high temperature lattice expansion, one looses the property that Z is the limit of a trace of a transfer operator. In other words, by reexpanding $Z(n, \tau)$ as a series in fractional powers of q and $\bar{q}$, we have no guarantee that the coefficients are positive integers except in the case $\rho^2 = 1$, $\frac{4}{3}$, $\frac{3}{2}$ corresponding respectively to $n = 2, 1, 0$. For

$\rho^2 = 1$, $e_0 = 0$ mod 2, the cosine in (297*b*) reduces to unity, and $Z(n = 2, \tau)$ to $Z(\rho = \frac{1}{2}; \tau) = Z(\rho = 2, \tau)$, i.e. the *XY* partition function, as it should. Surprisingly when $n = 1$, the above expression agrees with the Ising partition function, while for $n = 0$, its value is unity. To give a hint on the role of e_0, we note that its purpose is to correct for weight factors attached to noncontractible loops on the torus.

In the strip limit, when $\tau \to +i\infty$, the leading term in the double sum (297*b*) corresponds to $m = 0$, leading to a behaviour, when summing over m',

$$Z(n; \tau) \sim (q\bar{q})^{(e_0^2/4\rho^2) - \frac{1}{24}}$$

for that determination of e_0 in the range $[0, 1]$, i.e. $e_0 = \rho^2 - 1$. Hence the central charge is given by

$$c = 1 - 6\frac{e_0^2}{\rho^2} = 1 - 6\frac{(\rho^2 - 1)^2}{\rho^2} \qquad (298a)$$

Let us parametrize as usual c in the form $1 - 6/m(m + 1)$. Collecting (297*a*) and (298*a*), we have

$$\rho^2 = \frac{m + 1}{m} \qquad n = 2\cos\frac{\pi}{m} \qquad e_0 = \frac{1}{m} \qquad c = 1 - \frac{6}{m(m + 1)}$$
$$(298b)$$

Let us now check that for $n = 1$, where $m = 3$, we do recover the known Ising result. Since $e_0 = \frac{1}{3}$, we split the sum in (297*b*) over values of (m', m) modulo 6, as

$$\sum_{m',m} = \sum_{(m',m)=0 \text{ mod } 6} + \frac{1}{2}\sum_{(m',m)=\pm 1 \text{ mod } 6}$$
$$- \frac{1}{2}\sum_{(m',m)=\pm 2 \text{ mod } 6} - \sum_{(m',m)=3 \text{ mod } 6}$$

Each piece can be related to a Coulombic partition function with a different value of ρ. For instance

$$\sum_{(m',m)=0 \text{ mod } 6} = \frac{1}{6}Z(3\rho, \tau)$$

taking into account the prefactor ρ in (297b). Similarly

$$\sum_{(m',m)=3 \text{ mod } 6} = \tfrac{1}{3}Z\left(\tfrac{3}{2}\rho,\tau\right) - \tfrac{1}{6}Z(3\rho,\tau)$$

$$\sum_{(m',m)=\pm2 \text{ mod } 6} = \tfrac{1}{2}Z(\rho,\tau) - \tfrac{1}{6}Z(3\rho,\tau)$$

$$\sum_{(m',m)=\pm1 \text{ mod } 6} = Z\left(\tfrac{1}{2}\rho,\tau\right) - \tfrac{1}{2}Z(\rho,\tau) - \tfrac{1}{3}Z\left(\tfrac{3}{2}\rho,\tau\right) + \tfrac{1}{6}Z(3\rho,\tau)$$

Putting everything together, we see that for $e_0 = \tfrac{1}{3}$ and $\rho = \sqrt{\tfrac{4}{3}}$, equation (297) reduces to

$$Z(n=1,\tau) = \tfrac{1}{2}\left[Z(\rho = 2\sqrt{3},\tau) - Z(2/\sqrt{3},\tau)\right] \qquad (299)$$

It is not obvious at this stage that we have reached the desired result. We leave it as an exercise for the reader to check that it is the case.

All unitary minimal partition functions of table II can be written as finite sums of Coulombic partition functions. It is left as a further nontrivial exercise to verify this point, as well as the fact that, for $n = 0$, $m = 2$, $c = 0$, $e_0 = \tfrac{1}{2}$, the partition function reduces to unity. As $n \to -2$, i.e. $m \to 1$, one finds $Z(n = -2,\tau) = 0$, but

$$\left.\frac{\partial}{\partial n}Z(n,\tau)\right|_{n\to-2} = \operatorname{Im}\tau\ \eta^2\bar{\eta}^2$$

a fermionic partition function equal to the inverse square of the Gaussian free field partition function. This is in agreement with the correspondence between $O(n)$ bosonic models with $n = -2$ and (complex) fermionic ones. One can also study the limit $n \to 0$ for the polymer problem.

Similar techniques can be applied to critical Potts models for continuous Q varying between 0 and 4. One sets

$$\rho^2 = \frac{4m}{m+1} \qquad Q = \left(2\cos\frac{\pi}{m+1}\right)^2$$

$$e_0 = \frac{2}{m+1} \qquad c = 1 - \frac{6}{m(m+1)} \qquad (300a)$$

so that everything is parametrized in terms of m. Then the partition function reads

$$Z(Q,\tau) = \tfrac{1}{2}\rho Z_1(\tau) \sum_{m',m} \exp\left\{-\tfrac{1}{4}\pi\rho^2 \frac{|m\tau - m'|^2}{\text{Im}\,\tau}\right\} \cos\left[\pi e_0(m,m')\right]$$

$$+ \tfrac{1}{2}(Q-1)\left[Z(\rho,\tau) - Z(\tfrac{1}{2}\rho,\tau)\right]$$

(300b)

One can look at particular cases, such as $Q = 1, 2, 3$, where, in the last case, one finds the 3-state Potts partition function given in equation (276). The limiting value for $m \to \infty$, $Q \to 4$ and $c \to 1$, is also of interest, and appears on figure 6. The first term in equation (300b) is analogous to the $O(n)$ partition function.

9.3.7 Correlations in a half plane

We have investigated in detail geometrically simple situations, the plane, the cylinder or the torus. In a sense, none of these exhibits genuine boundary effects. This is not true for a critical model in a half plane. Having in mind a discretization on a lattice as a regularization, we consider free boundary conditions. We assume that at criticality this amounts to requiring the vanishing of the magnetization at the boundary. More precisely, this is compatible with (i) the vanishing of correlations for large separations (ii) global conformal invariance in the upper half plane (i.e. under $SL(2,\mathcal{R})$), since the point at infinity belongs to the boundary.

The domain under consideration is the upper half plane $\text{Im}\,z > 0$. With an added point at infinity this is conformally equivalent to the interior of the unit disk, and is invariant under global real conformal transformations of $SL(2,\mathcal{R})/Z_2$, in the form of real Möbius transformations $z \to z' = (\alpha z + \beta)/(\gamma z + \delta)$, α, β, γ, δ real, $\alpha\delta - \beta\gamma = 1$. Thus the same transformation applies to the variable $\bar{z}$. In particular, this involves only real translations and dilatations. Correspondingly, even the two-point function becomes nontrivial. It has a different behaviour in the directions parallel or perpendicular to the boundary. We associate to each point in the upper half plane its symmetric image with respect to the real axis. Now let two points z_1 and z_2 be given in the upper half plane, together with their images, $\bar{z}_1$ and $\bar{z}_2$ (in the lower half plane). From those four complex numbers, we can form the real cross ratio u, invariant under $SL(2, \mathcal{R})$,

$$u = \frac{(z_1 - \bar{z}_1)(z_2 - \bar{z}_2)}{(z_1 - \bar{z}_2)(z_2 - \bar{z}_1)} \qquad\qquad 0 < u < 1 \qquad (301)$$

A real global conformal transformation which maps z_1, $\bar{z}_1$, z_2, $\bar{z}_2$ onto z'_1, $\bar{z}'_1$, z'_2, $\bar{z}'_2$ preserves the cross ratio, and has the form

$$z \to z' \qquad \frac{(z - \bar{z}_1)(z_2 - \bar{z}_2)}{(z - \bar{z}_2)(z_2 - \bar{z}_1)} = \frac{(z' - \bar{z}'_1)(z'_2 - \bar{z}'_2)}{(z' - \bar{z}'_2)(z'_2 - \bar{z}'_1)} \qquad (302)$$

If we look at a two-point correlation $G(1, 2)$ for the primary operator $A_{h,\bar{h}}$, global invariance implies

$$G(1, 2) \equiv \left\langle A_{h,\bar{h}}(z_1, \bar{z}_1) A_{h,\bar{h}}(z_2, \bar{z}_2) \right\rangle$$
$$= \left[\frac{(z'_1 - \bar{z}'_1)(z'_2 - \bar{z}'_2)}{(z_1 - \bar{z}_1)(z_2 - \bar{z}_2)} \right]^{h+h'} \left[\frac{(\bar{z}_1 - \bar{z}_2)(z'_1 - z'_2)}{(z_1 - z_2)(\bar{z}'_1 - z'_2)} \right]^{h-\bar{h}} G(1', 2')$$
$$(303)$$

The group $SL(2, \mathcal{R})$ depends on three real parameters. One can therefore impose on z'_1 and z'_2 three conditions, for instance to be pure imaginary and such that $z'_1 + z'_2 = 2\mathrm{i}$. Thus we may choose

$$z'_1 = \mathrm{i}(1 + \sqrt{1 - u}) \qquad\qquad z'_2 = \mathrm{i}(1 - \sqrt{1 - u}) \qquad (304)$$

For simplicity assume a scalar field, i.e. such that $h = \bar{h}$. We have then, with $z = x + \mathrm{i}y$,

$$G(1, 2) = g(u)/(y_1 y_2)^{2h} \qquad\qquad u = \frac{4 y_1 y_2}{(x_1 - x_2)^2 + (y_1 + y_2)^2}$$
$$g(u) = u^{2h} \left\langle A_{h,h} \left(\mathrm{i} \left[1 + \sqrt{1 - u} \right] \right) A_{h,h} \left(\mathrm{i} \left[1 - \sqrt{1 - u} \right] \right) \right\rangle$$
$$(305)$$

Thus global conformal invariance alone leaves us with an unknown function of a real variable $g(u)$.

In dimension larger than two, the subgroup of the global conformal group which leaves a half space invariant leads to a similar result, with y_1 and y_2 the distances to the boundary and $|x_1 - x_2|$ replaced by the length of the parallel component of the vector $\mathbf{r}_1 - \mathbf{r}_2$.

The two-point function in the present case is analogous to the four-point function in the plane with z_1, z_2, z_3, z_4 replaced by z_1, $\bar{z}_1$, z_2, $\bar{z}_2$. There is a similar doubling for higher correlations. To see this relation more precisely, we look at the energy–momentum tensor and require that in Cartesian coordinates, there be no energy–momentum flow accross the boundary. Using complex coordinates, this is equivalent to saying that on the boundary $T_{zz}(x) = T_{\bar{z}\bar{z}}(x)$. This enables one to extend $T(z) \equiv T_{zz}(z)$ analytically in the lower half plane through

$$\text{Im}\, z < 0 \qquad T(z) = \bar{T}(\bar{z}) \tag{306}$$

where obviously $\bar{z}$ is in the upper half plane, and as a result the right-hand side is well-defined. Thus we now use only the diagonal part of the product of two Virasoro algebras. Repeating the reasoning of section 1.3, one arrives at a formula similar to (40) (Cardy)

$$\langle T(z) A_1(z_1, \bar{z}_1) \ldots \rangle = \sum_p \left(\frac{h_p}{(z - z_p)^2} + \frac{\bar{h}_p}{(z - \bar{z}_p)^2} \right.$$

$$\left. + \frac{1}{(z - z_p)} \frac{\partial}{\partial z_p} + \frac{1}{(z - \bar{z}_p)} \frac{\partial}{\partial \bar{z}_p} \right) \langle A_1(z_1, \bar{z}_1) \ldots \rangle \tag{307}$$

One can derive analogous consequences as we did in the plane, when we have primary fields corresponding to degenerate representations, in which case we obtain differential equations for the correlations.

Let us illustrate this on our perennial example of the Ising model for the spin–spin correlation. Adapting the calculations done in section 2.2, one finds a hypergeometric equation admitting a solution through radicals, such that

$$\langle \sigma(z_1, \bar{z}_1) \sigma(z_2, \bar{z}_2) \rangle = \frac{1}{(y_1 y_2)^{\frac{1}{8}}} \frac{1}{\lambda^{\frac{1}{4}}} \left(a_+ \sqrt{1 + \lambda} + a_- \sqrt{1 - \lambda} \right)$$

$$\lambda = \sqrt{1 - u} = \sqrt{\frac{(x_1 - x_2)^2 + (y_1 - y_2)^2}{(x_1 - x_2)^2 + (y_1 + y_2)^2}} = \left| \frac{z_1 - z_2}{z_1 - \bar{z}_2} \right|$$

$$\tag{308}$$

Provided $a_+ + a_-$ is non zero, the short-distance behaviour for finite y_1 and y_2 is the expected one

$$\langle \sigma_1 \sigma_2 \rangle \underset{z_{12} \to 0}{\sim} \text{cst} \frac{(y_1 + y_2)^{\frac{1}{4}}}{(y_1 y_2)^{\frac{1}{8}}} \frac{1}{|z_{12}|^{\frac{1}{4}}} \tag{309}$$

If however, again for y_1 and y_2 finite, we let $|x_1 - x_2|$ tend to infinity, we have

$$\lambda \to 1 - 2y_1 y_2/(x_1 - x_2)^2$$

meaning that if $a_+ \neq 0$ the correlation tends to a constant. If we require that under such circumstances $\langle \sigma_1 \sigma_2 \rangle$ tends to zero, we are forced to the conclusion that $a_+ = 0$, which means that up to a multiplicative constant

$$\langle \sigma_1 \sigma_2 \rangle = \frac{\text{cst}}{(y_1 y_2)^{\frac{1}{8}}} \left(\lambda^{-\frac{1}{2}} - \lambda^{\frac{1}{2}} \right)^{\frac{1}{2}} \qquad (310)$$

As a consequence, when $|x_1 - x_2| \to \infty$ and y_1 and y_2 are finite, one finds

$$\langle \sigma_1 \sigma_2 \rangle \underset{|x_1-x_2|\to\infty}{\sim} \text{cst} \frac{(y_1 y_2)^{\frac{3}{8}}}{|x_1 - x_2|} \qquad (311a)$$

If the power of $|x_{12}|$ in the denominator is denoted $\eta_\parallel$, then

$$\eta_\parallel = 1 \qquad (311b)$$

a new nontrivial result from local conformal invariance.

Since the point at infinity is to be considered as part of the boundary, equation (311a) implies by global conformal invariance that the field vanishes on the boundary. For instance

$$\langle \sigma(iy_1)\sigma(iy_2) \rangle \underset{y_1 \to 0}{\sim} \text{cst} \frac{1}{y_2^{\frac{1}{4}}} \left(\frac{y_1}{y_2} \right)^{\frac{3}{8}} \qquad (312)$$

Thus the r.h.s. vanishes on the boundary, albeit with a fractional power (hence the function has an infinite derivative).

One can put the complete expression (310) in a suggestive form, which can be compared with the four-point function in the plane. Squaring $\langle \sigma_1 \sigma_2 \rangle$ and setting $z_{ij} = z_i - z_j$, $z_{i\bar{j}} = z_i - \bar{z}_j$ etc, one has

$$\langle \sigma_1 \sigma_2 \rangle^2 = \left(\frac{z_{1\bar{2}} z_{\bar{1}2}}{z_{1\bar{1}} z_{2\bar{2}} z_{12} z_{\bar{1}\bar{2}}} \right)^{\frac{1}{4}} - \left(\frac{z_{12} z_{\bar{1}\bar{2}}}{z_{1\bar{1}} z_{2\bar{2}} z_{1\bar{2}} z_{\bar{1}2}} \right)^{\frac{1}{4}} \qquad (313)$$

which shows the symmetry in the simultaneous exchange $1 \leftrightarrow 2$, $\bar{1} \leftrightarrow \bar{2}$ and the antisymmetry in $1 \leftrightarrow \bar{1}$ or $2 \leftrightarrow \bar{2}$, the latter being defined after analytic continuation.

Another application of this same geometric situation can be obtained through a conformal mapping as follows. The transformation $z \to t$

$$t = \frac{L}{i\pi} \ln z \qquad\qquad 0 < \mathrm{Arg} z < \pi \qquad (314)$$

maps the upper half z-plane onto a strip $0 < \mathrm{Re}\, t < \pi$ with free (as opposed to periodic) boundaries. If we now write $t = x + iy$, the above two-point correlation (squared) becomes

$$\langle \sigma_1 \sigma_2 \rangle^2 = \mathrm{cst} \frac{(\pi/L)^{\frac{1}{2}}}{(\sin(\pi x_1 L)\sin(\pi x_2/L))^{\frac{1}{4}}}$$

$$\times \left[\left| \frac{\sin(\pi/2L)(x_1 + x_2 + iy_1 - iy_2)}{\sin(\pi/2L)(x_1 - x_2 + iy_1 - iy_2)} \right|^{\frac{1}{2}} - (x_2 \to -x_2) \right] \qquad (315)$$

Note the translational invariance in the y direction, and the vanishing when x_1 or $x_2 = 0$ or L.

Using the mapping (314), we can also find, according to equation (50), the mean value $\langle T \rangle$ in the strip, assuming that it vanishes in the half plane. Thus for any central charge c

$$\langle T \rangle = \tfrac{1}{24} c\pi/L^2 \qquad (316)$$

Correspondingly, the free energy per unit length behaves as

$$F(L) = F_0 L + \tfrac{1}{24} c\pi/L \qquad (317)$$

The coefficient of the $1/L$ term is a quarter of the corresponding one on a periodic strip, where, instead of (316), we found $\langle T \rangle_{\mathrm{periodic}} = \tfrac{1}{6} c\pi/L^2$, suggesting that (316) follows by replacing $L \to 2L$ in this last expression. Observe nevertheless that the correlation functions are not recovered as easily, as testified by equation (315).

9.3.8   The vicinity of the critical point

The knowledge of critical properties should enable one to understand the whole critical domain, characterized by a finite correlation length, or in the particle language, a finite mass spectrum. Two directions are open for investigation, but neither of them seems to be fully elaborated. The first is to obtain a better understanding of the relationship between integrable models and their conformal invariant limit. There are many signs that these two subjects have much in common, judging from the mathematical background. While this looks perhaps more like a formal question, the second aspect, namely the study of massive continuous quantum field theories, is of primary importance. In this final section, we present some remarks on the subject.

Let us return to the computations performed in section 3.2 and examine the partition function of a complex massive scalar free field on a torus $\mathcal{T}$, satisfying boundary conditions as indicated in equation (239). Denote the corresponding partition function over complex fields as

$$D_{k/N,\ell/N}^{-2}(\mathcal{T},m) = \int \mathcal{D}(\varphi\bar\varphi)\exp - \int_{\mathcal{T}} \frac{\mathrm{d}^2x}{2\pi}\bar\varphi(-\Delta+m)\varphi$$

$$\varphi(\mathbf{x}+\lambda\boldsymbol{\omega}_1+\nu\boldsymbol{\omega}_2) = \exp 2\mathrm{i}\pi\frac{(k\mu-\ell\lambda)}{N}\varphi(\mathbf{x})$$

$$\tag{318}$$

Retracing the steps of section 3.2, we find after some laborious computations

$$D_{k/N,\ell/N}(\mathcal{T},m) = \exp -\pi\mathrm{Im}\,\tau\gamma_{\ell/N}$$

$$\left| \prod_{n=-\infty}^{+\infty}\left(1 - \mathrm{e}^{-2\mathrm{i}\pi\left[\frac{k}{N}+(n+\frac{\ell}{N})\mathrm{Re}\,\tau\right]-2\pi\mathrm{Im}\,\tau\sqrt{(n+\frac{\ell}{N})^2+t^2}}\right)\right| \tag{319}$$

where t stands for

$$t = \frac{m\,|\omega_1|}{2\pi} > 0 \tag{320}$$

With $A = |\omega_1|^2\,\mathrm{Im}\tau$ the area of the torus, $\psi(z) = \Gamma'(z)/\Gamma(z)$ the logarithmic derivative of Euler's Γ-function, and $\gamma = -\psi(1)$ Euler's constant, the quantities $\gamma_{\ell/N}$ are given by

$$\frac{\ell}{N} = 0, \quad \gamma_0 = \tfrac{1}{6} - t + t^2\ln\left[4\pi\mathrm{e}^{-\gamma}\left(\frac{\mathrm{Im}\tau}{A}\right)^{\frac{1}{2}}\right]$$

$$+ \tfrac{1}{2}t^4\int_0^1 \mathrm{d}\lambda(1-\lambda)\sum_{n=1}^{\infty}\frac{1}{(n^2+\lambda t^2)^{\frac{3}{2}}}$$

$$0 < \frac{\ell}{N} < 1, \quad \gamma_{\ell/N} = \tfrac{1}{6}\left[1 - \frac{6\ell(N-\ell)}{N^2}\right] \tag{321}$$

$$+ t^2\left\{\ln\left[4\pi\left(\frac{\mathrm{Im}\tau}{A}\right)^{\frac{1}{2}}\right] + \tfrac{1}{2}\psi(\ell/N) + \tfrac{1}{2}\psi(1-\ell/N)\right\}$$

$$+ \tfrac{1}{4}t^4\int_0^1 \mathrm{d}\lambda(1-\lambda)\sum_{n=-\infty}^{+\infty}\frac{1}{[(n+\ell/N)^2+\lambda t^2]^{\frac{3}{2}}}$$

We observe the occurrence of a logarithmic term which disappears in the difference $\gamma_{\ell/N} - \gamma_0$, a function of t only. Let us specialize the above calculation to a periodic scalar real field, in which case

$$Z_{\text{scalar}} = \frac{1}{D_{00}} = \exp \pi \text{Im} \, \tau \left\{ \tfrac{1}{6} - t + t^2 \ln \left[4\pi e^{-\gamma} \left(\text{Im} \, \tau / A \right)^{\frac{1}{2}} \right] \right.$$

$$+ \tfrac{1}{2} t^4 \int_0^1 \mathrm{d}\lambda (1 - \lambda) \sum_1^\infty (n^2 + \lambda t^2)^{-\frac{3}{2}} \right\}$$

$$\times \left\{ \prod_{n=-\infty}^{+\infty} \left(1 - \exp(2 i \pi n \text{Re} \, \tau - 2\pi \text{Im} \, \tau \sqrt{n^2 + t^2}) \right) \right\}^{-1} \quad (322)$$

The infinite product is real and positive. The derivation shows that Z_{scalar} is modular invariant, eventhough it is expressed in terms of variables adapted to a choice of basis. The quantity t measures in dimensionless units the departure from criticality. One can interpret the prefactor as $\exp \tfrac{1}{6} \pi \text{Im} \, \tau \, c(t, \text{Im} \, \tau)$, where $c(0, \text{Im} \, \tau) = 1$, as defining a "scale dependent central charge". The appearance of a $\ln(\text{Im} \, \tau / A)^{1/2}$ dependence in the t^2 coefficient reflects the necessity of an ultraviolet renormalization of the specific heat, the second derivative with respect to m. Finally, the denominator arises naturally if one uses a transfer matrix, by decomposing the fields in proper modes, with the discrete momenta $2\pi \, n / |\omega_1|$ associated to eigenfrequencies $[(2\pi n / |\omega_1|)^2 + m^2]^{1/2}$. By extracting the zero mode responsible for a potential divergence, we recover the critical result

$$\lim_{m \to 0} m \sqrt{A} Z(m) = \frac{1}{\sqrt{\text{Im} \tau} \, |\eta(\tau)|^2} \quad (323)$$

More interesting is the application of the previous formulas to obtain the partition function of the Ising model with m proportional to $T - T_c$. We know from chapter 2 that we have to sum over the four periodic or antiperiodic boundary conditions, with a sign appended, taken care of here by an analytic continuation in m which we take at first to be positive. Then

$$Z_{\text{Ising}} = \tfrac{1}{2} \sum_{\substack{\text{boundary} \\ \text{conditions}}} \int \mathcal{D}(\psi \bar{\psi}) \exp \int_T \frac{\mathrm{d}^2 x}{2\pi} \left\{ \psi \bar{\partial} \psi - \bar{\psi} \partial \bar{\psi} + m \bar{\psi} \psi \right\}$$

$$= \tfrac{1}{2} \left\{ D_{\frac{1}{2}, \frac{1}{2}}(m) + D_{0, \frac{1}{2}}(m) + D_{\frac{1}{2}, 0}(m) + D_{00}(m) \right\}$$

$$(324)$$

i.e.

$$
Z_{\text{Ising}} = \tfrac{1}{2} \exp \pi \text{Im}\tau \left\{ \tfrac{1}{12} - t^2 \ln \left[\pi e^{-\gamma} \left(\frac{\text{Im}\tau}{A} \right)^{\frac{1}{2}} \right] \right.
$$

$$
\left. - \tfrac{1}{2} t^4 \int_0^1 d\lambda (1 - \lambda) \sum_{n=1}^{\infty} \frac{1}{[(n - \tfrac{1}{2})^2 + \lambda t^2]^{\frac{3}{2}}} \right\}
$$

$$
\times \left[\sum_{\pm} \prod_{n=-\infty}^{+\infty} \left(1 \pm e^{2i\pi(n+\frac{1}{2})\text{Re}\tau - 2\pi\text{Im}\tau \sqrt{(n+\frac{1}{2})^2 + t^2}} \right) + \exp -\pi\text{Im}\tau \left\{ \tfrac{1}{4} - t \right. \right.
$$

$$
\left. + t^2 \ln 4 - \tfrac{1}{2} t^4 \int_0^1 d\lambda (1 - \lambda) \sum_{n=1}^{\infty} \left(\frac{1}{[(n - \tfrac{1}{2})^2 + \lambda t^2]^{\frac{3}{2}}} - \frac{1}{[n^2 + \lambda t^2]^{\frac{3}{2}}} \right) \right\}
$$

$$
\left. \times \sum_{\pm} \prod_{n=-\infty}^{+\infty} \left(1 \pm e^{2i\pi n \text{Re}\tau - 2\pi\text{Im}\tau \sqrt{n^2 + t^2}} \right) \right] \tag{325}
$$

Absolute value signs have been suppressed since all infinite products are real, and the expression reduces to $Z_{\frac{1}{2}}$ at criticality ($m = 0$). All terms sensitive to the sign of t are in the exponential factors of $D_{\frac{1}{2},0}$ and $D_{0,0}$ and their zero mode factors. Combining them we get a term $(\exp(\pi\text{Im}\,\tau t) \pm \exp(-\pi\text{Im}\,\tau t))$.

The relativistic spectrum of states is the expected one, very much as in the scalar case. Not only does the central charge become scale dependent, but so also do the dimensions of the various operators. For instance, the spin and disorder operator which were indistinguishable at criticality split into two operators with scale dependent dimensions related by duality, i.e. $t \leftrightarrow -t$. As was pointed out, the occurrence of a term in $t^2 \ln(\text{Im}\tau/A)^{\frac{1}{2}}$ in the central charge is a manifestation of the renormalization properties of the dimension one operator $\bar{\psi}\psi$. In a mass perturbation, an ultraviolet subtraction is required to second order, related to a logarithmic divergence in the specific heat.

The modular invariance of equation (325) is not explicit, but an expansion in powers of m exhibits this very clearly. Denote as above by $Z_{\frac{1}{2}}$ and Z_1 the critical partition functions of the Ising and Gaussian models respectively, and let $D_{ij} \equiv D_{ij}(0)$ (with i, $j = 0$, $\tfrac{1}{2}$, omitting the $(0,0)$ combination since $D_{00} = 0$). Then expanding (325) we get

$$
\ln Z_{\text{Ising}}(m) = \ln Z_{\frac{1}{2}} + \frac{mA^{\frac{1}{2}}}{2Z_1 Z_{\frac{1}{2}}} + m^2 A \left[\frac{1}{4\pi} \ln \left(\frac{Z_1 \sqrt{A} e^{\gamma}}{\pi} \right) \right.
$$
$$
\left. - \frac{1}{4\pi} \frac{\Sigma D_{ij} \ln D_{ij}}{Z_{\frac{1}{2}}} - \frac{1}{8(Z_1 Z_{\frac{1}{2}})^2} \right] + O((mA^{\frac{1}{2}})^3) \quad (326)
$$

The term linear in m corresponds to the nonvanishing expectation value of the energy operator in the spin sector. This effect was mentioned earlier. The arbitrary length scale involved in the term $m^2 \ln A^{\frac{1}{2}}$ appears as an additive renormalization of the specific heat as expected. Finally, the existence of odd as well as even terms in m shows that on a torus the maximum of the specific heat (replacing the divergence in the infinite plane) is not at $m = 0$ ($T = T_c$) but depends on the modular ratio. It is noticible that the specific heat contains an entropic contribution due to the various fermionic spin structures.

These exact results are specific to the Ising case, and only apply to one of the relevant couplings, namely to the mass. In general one can only treat perturbatively the vicinity of the critical point. However, such a perturbation theory is not trivial as it incorporates the knowledge of the critical quantities. As an indication of the procedure, let us look at a system on a torus, assuming one of the periods ω_1 to be real. The time evolution is dictated by a Hamiltonian

$$
H = H_0 + V
$$
$$
H_0 = \frac{2\pi}{\omega_1} (L_0 + \bar{L}_0 - \tfrac{1}{12}c) \qquad\qquad (327)
$$
$$
V = -\sum_i G_i \int_0^{\omega_1} \frac{du_1}{2\pi} \varphi_i(u_1, 0)
$$

The coordinate on the torus is written $u = u_1 + iu_2$, and the potential is a sum over local perturbations with couplings G to relevant fields of dimension smaller than two. For simplicity consider the case where the sum is limited to a unique term, with φ a primary scalar field of weights (h, h). Let us further restrict our attention to the computation of the running central charge, although the formalism is by no means limited to this. Let P

be the momentum operator, and $|0\rangle$ the ground state. Instead of computing the full partition function, consider the following matrix element, where T denotes time ordering according to u_2

$$Z_0 = \langle 0| \exp\{-\operatorname{Im} \omega_2 H + i P \operatorname{Re} \omega_2\} |0\rangle$$

$$= \exp \pi \operatorname{Im} \tau \tfrac{1}{6} c \, \langle 0| \, T \exp \frac{G}{2\pi} \int_0^{\operatorname{Im}\omega_2} du_2 \int_0^{\omega_1} du_1 \varphi(u_1, u_2) |0\rangle$$

$$(328)$$

It is convenient to return to the punctured plane using

$$x = \exp 2i\pi u/\omega_1 \qquad d^2 u = \left(\frac{\omega_1}{2\pi}\right)^2 \frac{d^2 x}{x\bar{x}}$$

$$d^2 u \; \varphi_{\text{torus}}(u, \bar{u}) = \left(\frac{\omega_1}{2\pi}\right)^{2-2h} \frac{d^2 x}{(x\bar{x})^{1-h}} \varphi_{\text{plane}}(x, \bar{x}) \qquad (329)$$

The integral is then in an annulus, $\rho = e^{-2\pi \operatorname{Im}\tau} \le |x| \le 1$, and ordering is in the radial direction. It is also natural to use a dimensionless coupling

$$g = G\left(\frac{|\omega_1|}{2\pi}\right)^{2-2h} \qquad (330)$$

In the previous discussion we had $G = m$, $h = \tfrac{1}{2}$, and t was used instead of g to recall that it played the role of a reduced temperature. On the other hand we can think of a magnetic perturbation with G the magnetic field and φ the spin field ($h = \tfrac{1}{16}$). Relevant perturbations will have $2 - 2h > 0$, in which case for fixed G, g grows as $|\omega_1| \to \infty$, and perturbation theory is dangerous unless $G \to 0$ to maintain a finite g. This is the most interesting case, as it may produce a flow to a different universality class. One can look at two typical possibilities. In the first, the conformal dimension $2h$ of φ is close to 2, and one may attempt a perturbation theory close to marginality, in the small parameter $2 - 2h$. Alternatively, one may perform a straightforward expansion in g. This expansion is *a priori* a safe one. For $2h$ a fractional number between 0 and 2, the ultraviolet behaviour is under control (except for a logarithmic divergence at second order when $2h = 1$, the case encountered above). The infrared singularities are suppressed by the finite geometry, and the series is likely to be convergent, as the physical quantities are analytic in g under those circumstances (this is exhibited in the previous example). We require only the correlations of the φ field at criticality since

$$Z_0(g, \text{Im}\tau) = e^{\frac{1}{6}\pi \text{Im}\,\tau c} \sum_{n=0}^{\infty} \left(\frac{g}{2\pi}\right)^n \int_{\rho \le |x_1| \le \cdots \le |x_n| \le 1}$$

$$\prod_1^n \frac{d^2x}{(x_1\bar{x}_1)^{1-h}} \, \langle 0 \,|\varphi(x_1, \bar{x}_1)\cdots\varphi(x_n, \bar{x}_n)|\, 0\rangle \quad (331)$$

If the Hamiltonian H has a ground state for $g \ne 0$, we presume that $\ln Z/\text{Im}\tau$ will have a finite limit as $\text{Im}\tau \to +\infty$. This justifies the definition of a running central charge according to

$$C(g) = 12 \lim_{\rho \to 0} \left\{ \frac{\ln Z_0(g, \rho)}{\ln 1/\rho} \right\} \quad (332)$$

If we assume that odd correlations vanish in the expansion (331), this will mean

$$C(g) = c + g^2 C_2 + g^4 C_4 + \cdots \quad (333)$$

with the zeroth order term being the usual central charge. Inserting the two-point correlation

$$\langle 0 \,|\varphi(x_1, \bar{x}_1)\varphi(x_2, \bar{x}_2)|\, 0\rangle = \frac{1}{|x_{12}|^{4h}} \quad (334)$$

which normalizes the coupling g, we derive that

$$C_2 = 6 \sum_{n=0}^{\infty} \frac{1}{n+h} \left(\frac{\Gamma(n+2h)}{n!\Gamma(2h)}\right)^2 = 6 \int_0^1 d\lambda \, \lambda^{h-1} F(2h, 2h; 1; \lambda)$$

$$(335)$$

meaningful for $2h < 1$. It is possible to recover in this way the running central charge of the Ising case for a thermal perturbation as computed previously (with due care paid to the logarithmic singularity in C_2) by performing similar calculations to all orders. One can also compute the operator dimensions along the flow. We leave it to the reader to pursue the matter further. It would of course be quite valuable to find a compact form for these expressions instead of power series.

Appendix 9.A   Jacobian θ-series and products

We collect here some formulas on elliptic θ-functions. We use a lattice of translations Λ generated by 1 and τ such that $\text{Im}\tau > 0$.

We have constantly set $q = \exp 2i\pi\tau$, $|q| < 1$. With z a variable in the complex plane, we also write $y = \exp 2i\pi z$. The four Jacobi θ-functions defined through

$$\theta_1(z,\tau) = -i\sum_{-\infty}^{+\infty}(-1)^n y^{n+\frac{1}{2}} q^{(n+\frac{1}{2})^2/2}$$

$$= -iy^{\frac{1}{2}}q^{\frac{1}{8}}\prod_1^{\infty}(1-q^n)\prod_0^{\infty}(1-yq^{n+1})(1-y^{-1}q^n)$$

$$\theta_2(z,\tau) = \sum_{-\infty}^{+\infty} y^{n+\frac{1}{2}} q^{(n+\frac{1}{2})^2/2}$$

$$= y^{\frac{1}{2}}q^{\frac{1}{8}}\prod_1^{\infty}(1-q^n)\prod_0^{\infty}(1+yq^{n+1})(1+y^{-1}q^n) \qquad (A.1)$$

$$\theta_3(z,\tau) = \sum_{-\infty}^{+\infty} y^n q^{\frac{1}{2}n^2}$$

$$= \prod_1^{\infty}(1-q^n)\prod_0^{\infty}(1+yq^{n+\frac{1}{2}})(1+y^{-1}q^{n+\frac{1}{2}})$$

$$\theta_4(z,\tau) = \sum_{-\infty}^{+\infty}(-1)^n y^n q^{\frac{1}{2}n^2}$$

$$= \prod_1^{\infty}(1-q^n)\prod_0^{\infty}(1-yq^{n+\frac{1}{2}})(1-y^{-1}q^{n+\frac{1}{2}})$$

are all entire functions of z. Their behaviour under translations in Λ is readily ascertained, as well as the location of their zeroes. The equality of the sums with the products is the content of Jacobi's triple product identity. The latter is a source of a large number of identities among which Euler's pentagonal identity

$$P(q) \equiv \prod_1^{\infty}(1-q^n) = \sum_{-\infty}^{+\infty}(-1)^k q^{\frac{1}{2}k(3k+1)} \qquad (A.2)$$

Other useful identities include (Euler)

$$\prod_1^{\infty}(1-q^{2n-1})(1+q^n) = 1 \qquad (A.3)$$

as well as Watson's quintuple product

$$\prod_1^\infty (1 - q^n)(1 - yq^n)(1 - y^{-1}q^{n-1})(1 - y^2 q^{2n-1})(1 - y^{-2}q^{2n-1})$$

$$= \sum_{-\infty}^{+\infty}(y^{3m} - y^{-3m-1})q^{\frac{1}{2}m(3m+1)}$$

$$(A.4)$$

Dedekind's function is

$$\eta(\tau) = q^{\frac{1}{24}} P(q) \qquad (A.5)$$

The function θ_1 vanishes at the origin. With prime meaning derivative with respect to z,

$$\frac{1}{2\pi}\theta_1'(0,\tau) = \tfrac{1}{2}\theta_2(0,\tau)\theta_3(0,\tau)\theta_4(0,\tau) = \eta^3(\tau) \qquad (A.6)$$

The quantities $\theta_i(0,\tau)$ $(2 \le i \le 4)$ are sometimes called the θ-constants. One has

$$\frac{\theta_2(0,\tau)}{\eta(\tau)} = 2\left[\frac{\eta(2\tau)}{\eta(\tau)}\right]^2 = 2\left[q^{\frac{1}{16}-\frac{1}{48}}\prod_1^\infty (1+q^n)\right]^2$$

$$\frac{\theta_3(0,\tau)}{\eta(\tau)} = \left[e^{\frac{-2i\pi}{48}}\frac{\eta\left(\frac{1}{2}(\tau+1)\right)}{\eta(\tau)}\right]^2 = \left[q^{-\frac{1}{48}}\prod_1^\infty (1+q^{n-\frac{1}{2}})\right]^2 \quad (A.7)$$

$$\frac{\theta_4(0,\tau)}{\eta(\tau)} = \left[\frac{\eta(\frac{1}{2}\tau)}{\eta(\tau)}\right]^2 = \left[q^{-\frac{1}{48}}\prod_1^\infty (1-q^{n-\frac{1}{2}})\right]^2$$

They satisfy

$$[\theta_3(0,\tau)]^4 = [\theta_2(0,\tau)]^4 + [\theta_4(0,\tau)]^4 \qquad (A.8a)$$

This is equivalent to

$$2\sum_{\substack{n_1,n_2,n_3,n_4 \\ \Sigma n_i \text{odd}}} q^{\left(n_1^2+n_2^2+n_3^2+n_4^2\right)}$$

$$= \sum_{n_1,n_2,n_3,n_4} q^{\left(n_1+\frac{1}{2}\right)^2+\left(n_2+\frac{1}{2}\right)^2+\left(n_3+\frac{1}{2}\right)^2+\left(n_4+\frac{1}{2}\right)^2} \quad (A.8b)$$

expressing that on a hypercubic four-dimensional lattice $\mathcal{L}$ the centers of hypercubes can be divided in two integral lattices $\mathcal{L}^\pm$ generated by the vectors $(\frac{1}{2}\varepsilon_1, \frac{1}{2}\varepsilon_2, \frac{1}{2}\varepsilon_3, \frac{1}{2}\varepsilon_4)$ with $\varepsilon_i = \pm 1$ and $\prod_i \varepsilon_i = \pm 1$. These two lattices are related by a symmetry with respect to any hyperplane perpendicular to an axis of coordinates

612   *9 Conformal Invariance*

and each one is isometric to the sublattice $\mathcal{L}_{odd} \subset \mathcal{L}$ ($\mathcal{L}_{odd}$ corresponding to an odd sum of coordinates).

At some point, the following Gauss sum is useful, with ω standing for the primitive Nth root of unity $\omega = e^{2i\pi/N}$

$$\frac{1}{\sqrt{N}} \sum_{n=0}^{N-1} \omega^{n^2} = \tfrac{1}{2}(1+i) + (-i)^{N-1}\tfrac{1}{2}(1-i) \qquad (A.9)$$

Equivalently, when $N \equiv 0, 1, 2, 3$ mod.4 the sum is $1+i$, 1, 0, i respectively.

The modular transformation properties follow from Poisson's formula. Set

$$\theta(\zeta,\tau) \equiv \theta_2(\zeta - \tfrac{1}{2},\tau) = \sum e^{2i\pi(n+\frac{1}{2})(\zeta-\frac{1}{2})}q^{\frac{1}{2}(n+\frac{1}{2})^2} \qquad (A.10)$$

Then, for a unimodular matrix $A = \pm \begin{pmatrix} a & b \\ c & d \end{pmatrix}$,

$$\eta\left(\frac{a\tau+b}{c\tau+d}\right) = \varepsilon_A(c\tau+d)^{\frac{1}{2}}\eta(\tau) \qquad (A.11)$$

$$\theta(\zeta + \mu\tau + \nu, \tau) = (-1)^{\mu+\nu} \exp -2i\pi \left(\tfrac{1}{2}\mu^2\tau + \mu\zeta\right)\theta(\zeta,\tau) \qquad (A.12)$$

$$\theta\left(\frac{\zeta}{c\tau+d}; \frac{a\tau+b}{c\tau+d}\right) = \varepsilon_A^3(c\tau+d)^{\frac{1}{2}} \exp\left(i\pi c\frac{\zeta^2}{c\tau+d}\right)\theta(\zeta,\tau) \ (A.13)$$

with ε_A a 24th root of unity.

The θ-functions enable one to construct doubly periodic meromorphic functions by ratios or derivatives. The prototype is the Weierstrass function with

$$p(z,\tau) = -\frac{\partial^2}{\partial z^2}\ln\theta_1(z,\tau)$$

$$= \frac{1}{z^2} + \sideset{}{'}\sum_{n,m}\left[\frac{1}{(z+n+m\tau)^2} - \frac{1}{(n+m\tau)^2}\right]$$

$$= \frac{1}{z^2} + 3z^2 \sideset{}{'}\sum_{n,m}\frac{1}{(n+m\tau)^4} + 5z^4 \sideset{}{'}\sum_{n,m}\frac{1}{(n+m\tau)^6} + \cdots$$

$$(A.14)$$

an even meromorphic function with a double pole at the origin (as well as at lattice translates) and two zeroes in the period parallelogram. It fulfils the nonlinear first order differential equation, with prime denoting derivative with respect to z, and τ being fixed

$$\tfrac{1}{4}p'^2 = p^3 - 15e_4 p - 35e_6 \qquad (A.15)$$

with

$$e_{2k}(\tau) = \sum_{n,m}' \frac{1}{(n+m\tau)^{2k}} \qquad (A.16)$$

The roots of the cubic polynomial on the r.h.s. in $(A.15)$ are, by symmetry, the values $p(\frac{1}{2})$, $p(\frac{1}{2}\tau)$, and $p\left(\frac{1}{2}(1+\tau)\right)$. Its discriminant, conventionally multiplied by 2^4, is related to the η-function through

$$2^4 \left[\left(p\left(\tfrac{1}{2}\right) - p\left(\tfrac{1}{2}\tau\right)\right) \left(p\left(\tfrac{1}{2}\right) - p\left(\tfrac{1}{2}(1+\tau)\right)\right) \left(p\left(\tfrac{1}{2}\tau\right) - p\left(\tfrac{1}{2}(1+\tau)\right)\right)\right]^2$$
$$= 2^4 3^5 5^2 \left(20e_4^3 - 49e_6^2\right) = (2\pi)^{12} \eta(\tau)^{24}$$
$$(A.17)$$

Appendix 9.B   Superconformal algebra

We have seen in various circumstances how commuting and anticommuting variables appear on equal footing. Here we describe a natural generalization of conformal transformations to the supercomplex plane parametrized by a pair of variables $\xi \equiv (z, \theta)$ and $\bar{\xi} \equiv (\bar{z}, \bar{\theta})$, with θ and $\bar{\theta}$ anticommuting.

An analytic superfield is a function of z and θ only, hence of the form

$$\phi(\xi) = \varphi_0(z) + \theta\varphi_1(z) \qquad (B.1)$$

A natural square root of the Cauchy–Riemann operator is defined as

$$D\phi(\xi) \equiv \left(\frac{\partial}{\partial\theta} + \theta\frac{\partial}{\partial z}\right)\phi(\xi) = \varphi_1(z) + \theta\frac{\partial}{\partial z}\varphi_0(z) \qquad (B.2)$$

with

$$D^2\phi(\xi) = \frac{\partial}{\partial z}\phi(\xi) \qquad (B.3)$$

Henceforth we write ∂ for $\partial/\partial z$. A supercontour integral means

$$\oint_C d\xi\,\omega(\xi) \equiv \oint_C dz \int d\theta\,[\omega_0(z) + \theta\omega_1(z)] = \oint_C dz\,\omega_1(z) \qquad (B.4)$$

while a definite integral

$$f(\xi_1, \xi_2) = \int_{\xi_2}^{\xi_1} d\xi\,\omega(\xi) \qquad (B.5)$$

is such that

$$D_1 f(\xi_1, \xi_2) = \omega(\xi_1) \qquad\qquad f(\xi_1, \xi_1) = 0 \qquad (B.6)$$

Thus

$$\int_{\xi_2}^{\xi_1} d\xi \equiv \theta_{12} = \theta_1 - \theta_2 \int_{\xi_2}^{\xi_1} d\xi \int_{\xi_2}^{\xi} d\xi' \equiv \xi_{12} = z_1 - z_2 - \theta_1\theta_2 \quad (B.7)$$

A superanalytic function admits a Taylor expansion

$$f(\xi_1) = \sum_0^\infty \frac{1}{n!} \xi_{12}^n \partial_2^n \left[1 + \theta_{12} D_2 \right] f(\xi_2) \qquad (B.8)$$

Since

$$\oint_{C_2} \frac{d\xi_1}{2i\pi} \xi_{12}^{-n-1} = 0 \qquad\qquad \oint_{C_2} \frac{d\xi_1}{2i\pi} \theta_{12} \xi_{12}^{-n-1} = \delta_{n,0} \qquad (B.9)$$

the residue theorems are

$$\oint_{C_2} \frac{d\xi_1}{2i\pi} f(\xi_1) \theta_{12} \xi_{12}^{-n-1} = \frac{1}{n!} \partial_2^n f(\xi_2)$$

$$\oint_{C_2} \frac{d\xi_1}{2i\pi} f(\xi_1) \xi_{12}^{-n-1} = \frac{1}{n!} \partial_2^n D_2 f(\xi_2) \qquad (B.10)$$

Superconformal transformations are such that the map $\xi \to \xi'$ satisfies

$$D = D(\theta') D' \qquad (B.11)$$

meaning that D transforms homogeneously. This is not the case of the general superanalytic transformations where

$$D = (D\theta') D' + [D(z') - \theta' D(\theta')] D'^2$$

Global superconformal transformations depend on three complex parameters (z_0, z_1, w_0) and two anticommuting ones θ_0, θ_1. They can be written

$$\xi \to \xi' \qquad z' = z_0 - \frac{w_0 - \theta\theta_0}{z - z_1 - \theta\theta_1} \qquad \theta' = \theta_0 + \frac{\theta - \theta_1}{z - z_1 - \theta\theta_1} \qquad (B.12)$$

This defines a supersymmetric extension of the group $SL(2, \mathcal{C})$. Local superconformal transformations satisfying $(B.11)$ corresponding to a supervector field $v(\xi)$ are generated by the super-energy–momentum tensor

$$T(\xi) = T_F(z) + \theta T_B(z) \qquad (B.13)$$

in the form

$$\delta_v \phi(\xi_2) = \oint_{C_2} d\xi_1 v(\xi_1) T(\xi_1) \phi(\xi_2) \qquad (B.14)$$

In formula $(B.13)$, $T_B(z)$ is the ordinary bosonic tensor of dimension 2 and $T_F(z)$ is its fermionic counterpart of dimension $\frac{3}{2}$, with the understanding that θ counts dimensionally as $z^{\frac{1}{2}}$. If ϕ is a primary superfield of dimension h, we have

$$\delta_v \phi = \left[v\partial + \tfrac{1}{2} D(v) D + h(\partial v) \right] \phi \qquad (B.15)$$

so that $(B.14)$ is equivalent to the short-distance expansion

$$T(\xi_1)\phi(\xi_2) \sim h \frac{\theta_{12}}{\xi_{12}^2} \phi(\xi_2) + \frac{1}{2\xi_{12}} D_2 \phi(\xi_2) + \frac{\theta_{12}}{\xi_{12}} \partial_2 \phi(\xi_2) + \cdots \quad (B.16)$$

The anomalous central charge appears in the corresponding formula for the product

$$T(\xi_1)T(\xi_2) \sim \tfrac{1}{6} \frac{c}{\xi_{12}^3} + \left[\tfrac{3}{2} \frac{\theta_{12}}{\xi_{12}^2} + \frac{1}{2\xi_{12}} D_2 + \frac{\theta_{12}}{\xi_{12}} \partial_2 \right] T(\xi_2) + \cdots \quad (B.17)$$

or

$$\delta_v T = \left[v\partial + \tfrac{1}{2} D(v) D + \tfrac{3}{2} \partial v \right] T + \tfrac{1}{12} c \partial^2 D v \qquad (B.18)$$

In finite form $(B.15)$ and $(B.18)$ read

$$\phi(\xi) = \phi'(\xi')(D\theta')^{2h}$$
$$T(\xi) = T'(\xi')(D\theta')^3 + \tfrac{1}{12} c \{\xi', \xi\} \qquad (B.19)$$

with the super-Schwarzian derivative written as

$$\{\xi', \xi\} = 2 \left(\frac{D^4 \theta'}{D\theta'} - 2 \frac{D^3 \theta'}{D\theta'} \frac{D^2 \theta'}{D\theta} \right) \qquad (B.20)$$

For a particular superconformal transformation

$$z' = f(z), \qquad \theta' = (\partial f(z)/\partial(z))^{\frac{1}{2}} \theta$$

$\{\xi', \xi\}$ reduces to $\theta \{z', z\}$ with $\{z', z\}$ the ordinary Schwarzian derivative. The Laurent expansion of T

$$T_F(z) + \theta T_B(z) = \sum \left(\frac{1}{2} \frac{G_n}{z^{n+\frac{3}{2}}} + \frac{\theta L_n}{z^{n+2}} \right) \qquad (B.21)$$

leads in the radial quantization scheme to the following set of commutation and anticommutation relations (with the notation

$$[A, B]_+ = AB + BA)$$

$$[L_m, L_n] = (m - n)L_{m+n} + \tfrac{1}{12}cm(m^2 - 1)\delta_{m+n,0}$$
$$[L_m, G_n] = \left(\tfrac{1}{2}m - n\right)G_{m+n} \qquad\qquad (B.22)$$
$$[G_m, G_n]_+ = 2L_{m+n} + \tfrac{1}{3}c\left(m^2 - \tfrac{1}{4}\right)\delta_{m+n,0}$$

In $(B.21)$ the sum over indices for the Virasoro generators runs on integers as usual. But for the G_n partners, there are two choices defining what are called the Neveu–Schwarz (N-S) sector, with $n \in \mathcal{Z}+\tfrac{1}{2}$ leading to single valued spinor fields (as well as T_F) in the z-plane, or the Ramond (R-) sector, with $n \in \mathcal{Z}$ and double valued fermionic fields (as well as T_F) in the z-plane. These properties are reversed in a periodic strip, according to $(B.19)$ where N-S fermionic fields are antiperiodic, R-fermionic fields periodic. Thus we have in fact two distinct superconformal algebras.

We have only discussed the ξ-dependence of fields. Of course similar properties hold with respect to $\bar{\xi}$. For short and with some abuse of language, we continue to omit this parallel dependence.

The operator L_0 acts as a Hamiltonian with the vacuum $|0\rangle$ being in the N-S sector, annihilated by the five generators L_0, $L_{\pm 1}$, $G_{\pm 1/2}$ of global superconformal transformations. The highest weight vectors $|h\rangle$ of the N-S sector are then in one-to-one correspondence with the superfields $\phi_h(\xi)$ through $|h\rangle = \phi_h(0)|0\rangle$.

The R-sector has the additional feature that G_0 acts as a supersymmetric charge. It commutes with L_0 and $G_0^2 = L_0 - \tfrac{1}{24}c$. Thus, in general, highest weight R-states come in pairs $|h^\pm\rangle$ with $|h^-\rangle = G_0|h^+\rangle$. One calls spin fields, those ordinary conformal fields $\theta^\pm(z)$ which intertwine the N-S and R sectors by generating the states $|h^\pm\rangle$ out of the vacuum, $\theta^\pm(0)|0\rangle = |h^\pm\rangle$. The condition for the existence of an unbroken supersymmetry generated by G_0 in the R-sector is that $G_0|h^+\rangle = 0$ (in other words the Verma module generated by $|h^-\rangle$decouples from the theory) and this is possible if and only if $h^+ = \tfrac{1}{24}c$.

Let F count the number of fermionic fields, then $\Gamma = (-1)^F$ is called the chirality operator. Paired R-states have opposite chirality as G_0 anticommutes with Γ. Spin fields of equal chirality are relatively local, while the opposite is true for spin fields of unequal chirality.

We restrict ourselves to the description of unitary representations of the superconformal algebras. If $c \geq \tfrac{3}{2}$ and $h \geq 0$ all representations afforded by Verma modules are equivalent to unitary ones. However when $c < \frac{3}{2}$ there only exists a discrete sequence of (degenerate) unitary representations. They are similar to the ones found in the unitary case for the ordinary conformal algebra. Let M denote an integer larger or equal to 2, then for this unitary series, in either the N-S or R-case the central charge is given by

$$c = \frac{3}{2}\left(1 - \frac{8}{M(M+2)}\right) \qquad (B.23)$$

while the weights belong to the finite table

$$h_{r,s} = \frac{[(M+2)r - Ms]^2 - 4}{8M(M+2)} + \tfrac{1}{32}(1 - (-1)^{r-s}) \quad (B.24)$$

$$1 \le r < M \qquad 1 \le s < M+2$$

Here $r - s$ is even for N-S representations and odd for R-representations. For $M = 2$ we have the trivial representation. When M is odd, $h = \frac{1}{24}c$ is not in the table, so that supersymmetry is broken in the R-sector, while for M even, $h_{\frac{1}{2}M,\frac{1}{2}M+1} = \frac{1}{24}c$ and supersymmetry can be preserved.

More generally one has formulas for the determinants of the contragredient forms as well as for characters. In particular for the unitary representations $(B.24)$, there are three types of characters. With the central charge understood, $q = e^{2i\pi\tau}$, and r and s labelling the weights, we denote them χ^{NS}, $\tilde{\chi}^{NS}$, χ^R

$$\chi_{r,s}^{NS}(\tau) = \mathrm{Tr}_{r,s}\, q^{L_0} \qquad\qquad r - s \equiv 0 \bmod 2$$

$$= \prod_{p=1}^{\infty}\left(\frac{1 + q^{p-\frac{1}{2}}}{1 - q^p}\right)\sum_{n=-\infty}^{+\infty}$$

$$\left\{q^{\left\{[2nM(M+2)+r(M+2)-sM]^2 - 4\right\}/8M(M+2)} - (s \leftrightarrow -s)\right\}$$

$$\tilde{\chi}_{r,s}^{NS}(\tau) = \mathrm{Tr}(-1)^F q^{L_0} \qquad\qquad r - s \equiv 0 \bmod 2$$

$$= \prod_{p=1}^{\infty}\left(\frac{1 - q^{p-\frac{1}{2}}}{1 - q^p}\right)\sum_{n=-\infty}^{+\infty}(-1)^{Mn}$$

$$\left\{q^{\left\{[2nM(M+2)+r(M+2)-sM]^2 - 4\right\}/8M(M+2)} - (-1)^{rs}(s \leftrightarrow -s)\right\}$$

$$\chi_{r,s}^{R}(\tau) = \mathrm{Tr}\, q^{L_0} \qquad\qquad r - s \equiv 1 \bmod 2$$

$$= q^{\frac{1}{16}}\prod_{p=1}^{\infty}\left(\frac{1 + q^p}{1 - q^p}\right)\sum_{n=-\infty}^{+\infty}$$

$$\left\{ q^{\left\{ [2nM(M+2)+r(M+2)-sM]^2-4 \right\}/8M(M+2)} - (s \leftrightarrow -s) \right\} \quad (B.25)$$

A classification of consistent minimal models (in particular unitary ones) can be achieved along the lines of section 3.4 using modular invariance on a torus. Let us just show here how the formalism relates to statistical physics by recording the interpretation of the first nontrivial model in the list $(B.23)$ corresponding to $M = 3$ (Friedan, Qiu and Shenker). It turns out to be unique, and to coincide with the case $m = 4$ for the ordinary conformal classification. In both cases $c = \frac{7}{10}$ and the possible weights in the Virasoro case were (figure 4) $h = 0$, $\frac{7}{16}$, $\frac{3}{2}$, $\frac{3}{80}$, $\frac{3}{5}$ and $\frac{1}{10}$, while, according to the stronger superconformal invariance, we have $h_{1,1} = 0$, $h_{2,2} = \frac{1}{10}$ in the N-S sector and $h_{1,2} = \frac{3}{80}$, $h_{2,1} = \frac{7}{16}$ in the R-sector.

This can be identified with the tricritical Ising model. Indexing by V the ordinary Virasoro representations, the Z_2 even sector is spanned by the two N-S representations

$$\begin{aligned} (0)_{\text{N-S}} &= (0)_V \oplus \left(\tfrac{3}{2}\right)_V \\ \left(\tfrac{1}{10}\right)_{\text{N-S}} &= \left(\tfrac{1}{10}\right)_V \oplus \left(\tfrac{3}{5}\right)_V \end{aligned} \quad (B.26)$$

while the Z_2 odd sector is the R-one, containing magnetic like operators, with

$$\begin{aligned} \left(\tfrac{3}{80}\right)_R &= \left(\tfrac{3}{80}\right)_V \\ \left(\tfrac{7}{16}\right)_R &= \left(\tfrac{7}{16}\right)_V \end{aligned} \quad (B.27)$$

Appending the usual prefactor $q^{-c/24}$ in the characters, one has

$$\chi_0^V \pm \chi_{\frac{3}{2}}^V = \left\{ \begin{matrix} \chi_0^{\text{N-S}} \\ \tilde{\chi}_0^{\text{N-S}} \end{matrix} = q^{-\frac{7}{240}} \prod_{n=0}^{\infty} \frac{(1 \pm q^{5n+\frac{5}{2}})(1 \pm q^{5n+\frac{3}{2}})(1 \pm q^{5n+\frac{7}{2}})}{(1 - q^{5n+3})(1 - q^{5n+4})} \right.$$

$$\chi_{\frac{1}{10}}^V \pm \chi_{\frac{3}{5}}^V = \left\{ \begin{matrix} \chi_{\frac{1}{10}}^{\text{N-S}} \\ \tilde{\chi}_{\frac{1}{10}}^{\text{N-S}} \end{matrix} = q^{-\frac{7}{240}+\frac{1}{10}} \prod_{n=0}^{\infty} \frac{(1 \pm q^{5n+\frac{1}{2}})(1 \pm q^{5n+\frac{1}{2}})(1 \pm q^{5n+\frac{9}{2}})}{(1 - q^{5n+1})(1 - q^{5n+4})} \right.$$

$$\chi_{\frac{3}{80}}^V = \chi_{\frac{3}{80}}^R = q^{-\frac{7}{240}+\frac{3}{80}} \prod_{n=0}^{\infty} \frac{(1 + q^{5n+2})(1 + q^{5n+3})(1 + q^{5n+5})}{(1 - q^{5n+1})(1 - q^{5n+4})}$$

$$\chi_{\frac{7}{16}}^V = \chi_{\frac{7}{16}}^R = q^{-\frac{7}{240}+\frac{7}{16}} \prod_{n=0}^{\infty} \frac{(1 + q^{5n+1})(1 + q^{5n+4})(1 + q^{5n+5})}{(1 - q^{5n+2})(1 - q^{5n+3})}$$

$$(B.28)$$

On a torus with modular ratio τ, we have the corresponding modular invariant partition function

$$Z(\tau,\bar\tau) = \left\{ \tfrac{1}{2} \left(\left|\chi_0^{N-S}(\tau)\right|^2 + \left|\tilde\chi_0^{N-S}(\tau)\right|^2 + \left|\chi_{\frac{1}{10}}^{N-S}(\tau)\right|^2 + \left|\tilde\chi_{\frac{1}{10}}^{N-S}(\tau)\right|^2 \right) \right.$$

$$\left. + \left|\chi_{\frac{3}{80}}^{R}(\tau)\right|^2 + \left|\chi_{\frac{7}{16}}^{R}(\tau)\right|^2 \right\}$$

$$= \left\{ \left|\chi_0^{V}(\tau)\right|^2 + \left|\chi_{\frac{3}{2}}^{V}(\tau)\right|^2 + \left|\chi_{\frac{1}{10}}^{V}(\tau)\right|^2 + \left|\chi_{\frac{3}{5}}^{V}(\tau)\right|^2 \right.$$

$$\left. \left|\chi_{\frac{3}{80}}^{V}(\tau)\right|^2 + \left|\chi_{\frac{7}{16}}^{V}(\tau)\right|^2 \right\} \qquad\qquad (B.29)$$

Appendix 9.C   Current algebra

Conformal field theory extends to systems with continuous symmetries. An example is afforded by generalized σ-models, with fields taking their values in a Lie group G, when the kinetic Lagrangian is supplemented by a topological term to implement a local $G_z \times G_{\bar z}$ invariance generated by a pair of conserved currents $J(z), \bar J(\bar z)(\bar\partial J = \partial\bar J = 0)$. The energy–momentum tensor is given by a quadratic expression in terms of the currents, with conformal weights $(1,0)$ and $(0,1)$ respectively. The coefficients of J and $\bar J$, in a Laurent expansion, understood as operators, generate a pair of commuting Kac–Moody (or affine) infinite Lie algebras. In the text we encountered as an auxiliary device the characters of the affine $SU(2)$ algebra and came accross the A–D–E classification of minimal models. All this justifies that we succinctly describe here some aspects of the interplay between group theory and conformal invariance.

C.1.   Simple Lie algebras

The reader is presumably familiar with the Cartan–Killing classification of simple Lie algebras over complex numbers and their finite dimensional representation. The following description can be complemented by the vast literature on the subject and is only intended to present some notation used in the text.

Simple Lie algebras come in four infinite families and five exceptional cases. The four families are realized as infinitesimal transformations corresponding to the classical complex matrix groups according to the table

$\ell \geq 1$ A_ℓ special linear group $SL(\ell + 1)$

$\ell \geq 2$ B_ℓ odd dimensional orthogonal group $SO(2\ell + 1)$

$\ell \geq 3$ C_ℓ symplectic group $Sp(2\ell)$

$\ell \geq 4$ D_ℓ even dimensional orthogonal group $SO(2\ell)$

The restriction on the rank ℓ, the maximal number of linearly independent commuting elements, which span a Cartan subalgebra, is to avoid isomorphisms in low dimensions, like $A_1 \sim B_1 \sim C_1$, $C_2 \sim B_2$, $D_2 \sim A_1 \times A_1$, $D_3 \sim A_3$. The five exceptional cases are less familiar; they are G_2, F_4, E_6, E_7, E_8, where the index gives the rank.

All Cartan, i.e. maximal abelian subalgebras are conjugate in the group. The elements in one of them, denoted K, can be simultaneously diagonalized in any representation, in particular in the adjoint representation. The set of these eigenvalues are elements in the dual space K^* and are called roots in the case of the adjoint representation. Let e_α (an element in the Lie algebra) be the common eigenvector, this means that for any k belonging to K one has

$$[k, e_\alpha] = \langle \alpha \,|k \rangle \, e_\alpha \qquad (C.1)$$

Simplicity of the Lie algebra implies that the roots span K^*. Moreover their exists an invariant symmetric nonsingular bilinear form on the algebra, denoted (,), i.e. such that

$$(a, b) = (b, a)$$
$$(a, b) = 0 \text{ for any } a \Rightarrow b = 0$$
$$([a, b] , c) = (a, [b, c]) \qquad (C.2)$$

Roots come in pairs $\{\alpha, -\alpha\}$, those being the only ones proportional to α. From Jacobi's identity, if two roots α and β are such that $\alpha + \beta \neq 0$ (i) if $\alpha + \beta$ is a root, $\left[e_\alpha, e_\beta\right]$ is proportional to $e_{\alpha+\beta}$, (ii) if $\alpha + \beta$ is not a root, $\left[e_\alpha, e_\beta\right] = 0$. If $\alpha + \beta = 0$, the elements

$$k_\alpha = [e_\alpha, e_{-\alpha}] \leftrightarrow (e_\alpha, e_{-\alpha}) \langle \alpha | \qquad (C.3)$$

span K (the correspondence means that $(k_\alpha, k) = (e_\alpha, e_{-\alpha}) \langle \alpha \, | k \rangle$) with the e_α's normalized in such a way that

$$(k_\alpha, k_\alpha) = 2 \, (e_\alpha, e_{-\alpha}) \neq 0 \qquad (C.4)$$

and the restriction of the bilinear form to K remaining nonsingular. The triplet $\{e_\alpha, e_{-\alpha}, k_\alpha\}$ generates an $SL(2)$ algebra

$$[k_\alpha, e_{\pm\alpha}] = \pm 2 e_{\pm\alpha} \qquad\qquad [e_\alpha, e_{-\alpha}] = k_\alpha \qquad (C.5)$$

with a familiar realization as two by two matrices

$$k_\alpha \to \begin{pmatrix} 1 & 0 \\ 0 & -1 \end{pmatrix} \qquad e_\alpha \to \begin{pmatrix} 0 & 1 \\ 0 & 0 \end{pmatrix} \qquad e_{-\alpha} \to \begin{pmatrix} 0 & 0 \\ 1 & 0 \end{pmatrix}$$
$$(C.6)$$

and an induced bilinear form

$$(a k_\alpha + b e_\alpha + c e_{-\alpha}, a k_\alpha + b e_\alpha + c e_{-\alpha}) = (k_\alpha, k_\alpha) \{a^2 + bc\}$$
$$= \tfrac{1}{2} (k_\alpha, k_\alpha) \, \mathrm{Tr} \begin{pmatrix} a & b \\ c & -a \end{pmatrix}^2$$
$$(C.7)$$

The element e_β is an eigenstate of k_α

$$\left[k_\alpha, e_\beta \right] = A_{\alpha,\beta} e_\beta$$
$$A_{\alpha,\beta} = \langle \beta \, | k_\alpha \rangle = \frac{2(k_\alpha, k_\beta)}{(k_\beta, k_\beta)} \qquad (C.8)$$

where $A_{\alpha,\beta}$ must be an integer as follows from studying the finite representation of the $SL_\alpha(2)$ algebra $(C.5)$ acting on the states $e_{\beta+s\alpha}$, s integer (k_α is the analog of twice the z component of angular momentum and has therefore integral eigenvalues). Since under duality it follows from $(C.3)$ and $(C.4)$ that

$$\tilde{k}_\alpha = \frac{2 k_\alpha}{(k_\alpha, k_\alpha)} \leftrightarrow \langle \alpha | \qquad (C.9)$$

it is natural to transfer the bilinear form on the dual root space K^* through

$$(\alpha, \beta) = \left(\tilde{k}_\alpha, \tilde{k}_\beta \right) \qquad (C.10)$$

in which case, we can also write

$$A_{\alpha,\beta} = \frac{2(\alpha, \beta)}{(\alpha, \alpha)} \qquad (C.11)$$

Assuming $(\alpha, \beta) \neq 0$, the elements $e_{\beta+s\alpha}$ span a representation of $SL_\alpha(2)$ which must contain an element with a k_α eigenvalue opposite to the one of e_β. This is for

$$s = -\frac{2(\alpha, \beta)}{(\alpha, \alpha)}$$

with the result that the set of roots contains with every root β the one reflected in the hyperplane perpendicular to α

$$w_\alpha(\beta) = \beta + s\alpha = \beta - \frac{2(\beta, \alpha)}{(\alpha, \alpha)}\alpha \qquad (C.12)$$

These reflections generate a point group called the Weyl group and relate the theory of simple Lie algebras with some aspects of crystallography. The Weyl group acts transitively on sets of basis roots $\{\alpha_1, \ldots, \alpha_\ell\}$, also called simple roots, in terms of which all roots can be written $\alpha = \sum n_i \alpha_i$ with all n_i's integers of the same sign (or zero). Accordingly the even number of roots is subdivided into positive or negative roots (denoted $\alpha > 0$ or $\alpha < 0$).

The 3ℓ elements $k_i \equiv k_{\alpha_i}$, $e_i^{(\pm)} \equiv e_{\pm\alpha_i}$, are generators for the Lie algebra and satisfy the following relations (Chevalley–Serre)

$$\left[k_i, k_j\right] = 0 \qquad \left[k_i, e_j^{(\pm)}\right] = \pm A_{ij} e_j^{(\pm)}$$

$$\left[e_i^{(+)}, e_j^{(-)}\right] = \delta_{ij} k_j \qquad (C.13)$$

$$\left(\mathrm{ade}_i^{(+)}\right)^{1-A_{ij}} e_j^{(+)} = \left(\mathrm{ade}_i^{(-)}\right)^{1-A_{ij}} e_j^{(-)} = 0 \qquad i \neq j$$

The $\ell \times \ell$ integral Cartan matrix

$$A_{ij} = \frac{2(\alpha_i, \alpha_j)}{(\alpha_i, \alpha_i)} \qquad (C.14)$$

therefore characterizes entirely the algebra up to simultaneous permutations of lines and columns. Its diagonal entries are equal to 2, and for $i \neq j$, $A_{ij} \neq 0$ implies $A_{ji} \neq 0$ (both being negative). The simply laced algebras A, D, E, are such that all roots are of equal length and as a consequence the Cartan matrix is symmetric.

The nonvanishing off-diagonal elements of A_{ij} are coded in a Coxeter–Dynkin diagram, where ℓ points are associated to the ℓ simple roots. Pairs of points $\{i, j\}$ are joined by $A_{ij} A_{ji} = 0, 1, 2, 3$ lines. If $A_{ij} = 0$ the corresponding roots are orthogonal and w_i and w_j commute. If the roots are not orthogonal $(w_i w_j)^{m_{ij}} = 1$, where corresponding to the possibilities of 1, 2 or 3 lines between

i and j, the integer m_{ij} equals 3, 4 or 6. The diagram is connected (otherwise the algebra would break in a direct sum of commuting subalgebras). The above description only characterizes the Weyl group, irrespective of the lengths of the roots. One shows that either the simple roots are of equal length, in which case there are no multiple lines in the diagram, or are subdivided into two groups, the long and short ones. Then if $A_{ij}A_{ji} = 1$, $(\alpha_i, \alpha_i) = (\alpha_j, \alpha_j)$ while if $(\alpha_i, \alpha_i) > (\alpha_j, \alpha_j)$ and $A_{ij} \neq 0$, then $A_{ji} = -1$ and A_{ij} equals -2 or -3. Thus it is sufficient to put an arrow on multiple lines pointing from the long to the short root. The possible sets of such diagrams are then depicted as follows

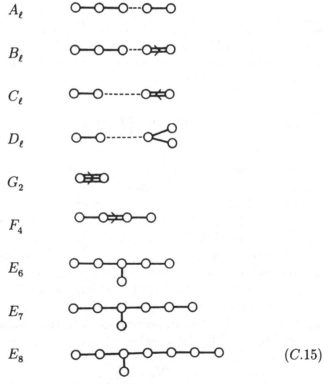

$$\begin{array}{ll} A_\ell & \\ B_\ell & \\ C_\ell & \\ D_\ell & \\ G_2 & \\ F_4 & \\ E_6 & \\ E_7 & \\ E_8 & \end{array} \qquad (C.15)$$

The roots $\{\alpha\}$ generate a lattice over integers, called the root lattice. Similarly the co-roots $2\alpha/(\alpha, \alpha)$ generate another lattice isomorphic up to scale with the preceding one, except in the case of B_ℓ and C_ℓ where this duality exchanges the root lattices. These two algebras have therefore the same Weyl group.

The root lattice is invariant under the semidirect product of the Weyl group and a discrete translation group. An equivalent description is that this inhomogeneous group is generated by $n+1$ reflections in the hyperplanes bounding a simplex (which is a fundamental domain). With the origin at one of the vertices, ℓ of the hyperplanes are orthogonal to ℓ simple roots and the $(\ell+1)$th hyperplane is orthogonal to a root $\boldsymbol{\alpha}_0$ expressed by an equation $\sum_0^\ell y_i \boldsymbol{\alpha}_i = 0, y_0 = 1$, with integral positive coefficients. These characteristic integers y_i obey several interesting relations. Their sum $h = \sum_0^\ell y_\ell$, called the Coxeter number, is related to the dimension of the Lie algebra, denoted dim, through

$$\dim = \ell(1+h) \qquad (C.16)$$

so that ℓh is even and equal to the number of roots. The notation h is standard and should not be confused with conformal weights. On the other hand, the product of y's is related to dimension of the Weyl group $|W|$ through

$$|W| = \ell! \left(\prod_0^\ell y_j \right) \det A_{ij} \qquad (C.17)$$

One defines analogous concepts for the co-roots and in particular a dual Coxeter number, denoted g.

The significance of the determinant of the Cartan matrix appears when one investigates the irreducible, finite-dimensional, representations of the Lie algebra. Those are associated to points of the lattice dual (over the integers $\mathcal{Z}$) of the co-root lattice, and are called weights. Thus for any weight $\mathbf{q}$ and any root $\boldsymbol{\alpha}$, $2(\mathbf{q}, \boldsymbol{\alpha})/(\boldsymbol{\alpha}, \boldsymbol{\alpha})$ is an integer. The weight lattice is also invariant under the Weyl group and contains as a sublattice the root lattice with an index equal to the determinant of the Cartan matrix. It is generated by ℓ fundamental weights $\mathbf{q}_i$, $1 \le i \le \ell$, satisfying

$$\frac{2(\mathbf{q}_i, \boldsymbol{\alpha}_j)}{(\boldsymbol{\alpha}_j, \boldsymbol{\alpha}_j)} = \delta_{ij} \qquad (C.18)$$

Dominant weights have nonnegative scalar products with the simple roots and any weight can be mapped on a dominant one by an element in the Weyl group. The sum $\mathbf{q}_m$ of fundamental weights satisfies

$$\mathbf{q}_m = \sum_1^\ell \mathbf{q}_i = \tfrac{1}{2} \sum_{\alpha>0} \alpha \qquad (C.19)$$

Finite-dimensional irreducible representations have a highest weight state (with a highest weight, a dominant one, denoted $\mathbf{q}$) annihilated by all $e_i^{(+)}$. The weights $\mathbf{q}_\mu$ associated to the eigenvalues of the generators of the Cartan subalgebra (i.e. the eigenvalue of k_α is $2(\alpha, \mathbf{q}_\mu)/(\alpha, \alpha)$) are of the form $\mathbf{q} - \Sigma \mu_i \alpha_i$, where the μ_i's are nonnegative integers. The weights of the adjoint representation are the roots, and the highest root is $\psi = -\alpha_0$, and is a long root.

The corresponding characters $\chi_\mathbf{q}$ are given by a formula due to Weyl, as a function of a ℓ-dimensional vector $\mathbf{x}$

$$\chi_\mathbf{q}(\mathbf{x}) = \sum_{\mathbf{q}'} \mathrm{mult}(\mathbf{q}') e^{2i\pi(\mathbf{q}',\mathbf{x})}$$

$$= \frac{\psi_{\mathbf{q}_m + \mathbf{q}}(\mathbf{x})}{\psi_{\mathbf{q}_m}(\mathbf{x})} \qquad (C.20)$$

$$\psi_\mathbf{q}(\mathbf{x}) = \sum_{w \in W} \varepsilon(w) e^{2i\pi(w(\mathbf{q}),\mathbf{x})}$$

Here $\mathrm{mult}(\mathbf{q}')$ is the multiplicity of the weight $\mathbf{q}'$ in the representation, and, in the sum over the Weyl group, $\varepsilon(w)$ is plus or minus one according to the parity of its expression in terms of generating reflections. The denominator admits a factorized form

$$\psi_{\mathbf{q}_m}(\mathbf{x}) = \prod_{\alpha>0} \left(e^{i\pi(\alpha,x)} - e^{-i\pi(\alpha,x)} \right) \qquad (C.21)$$

Taking the limit $\mathbf{x} \to 0$, yields the dimension of the representation

$$\dim_\mathbf{q} = \prod_{\alpha>0} \frac{(\alpha, \mathbf{q} + \mathbf{q}_m)}{(\alpha, \mathbf{q}_m)} \qquad (C.22)$$

The quadratic Casimir invariant (up to an arbitrary scale)

$$(\mathbf{q} + \mathbf{q}_m, \mathbf{q} + \mathbf{q}_m) - (\mathbf{q}_m, \mathbf{q}_m) = (\mathbf{q}, \mathbf{q} + 2\mathbf{q}_m) \qquad (C.23)$$

is insufficient to characterize the representation when $\ell > 1$. Among polynomials in the components of a weight, it is possible to single out those which are invariant under the Weyl group as being themselves expressible as polynomials in ℓ-fundamental homogeneous ones, which generalize the quadratic form $(\mathbf{q}, \mathbf{q})$.

The degrees of these polynomials are recorded in the last column of table III. The highest degree is the Coxeter number h, the product over all degrees is the order of the Weyl group, and twice their sum is $\dim(G) + \ell$, where $\dim(G)$ is the dimension of the Lie algebra. It will be recognized that, if one subtracts 1 from these degrees (obtaining what are called exponents), one recovers the nonzero diagonal entries (with their multiplicity) appearing in the A–D–E invariant partition functions discussed in the text. The table also includes a column for the dual Coxeter number g which is the value of the quadratic Casimir invariant $(C.23)$ in the adjoint representation when one normalizes the long roots to unity. Of course $h = g$ for simply laced groups. One has also the following formula (Freudenthal)

$$\dim(G) = \frac{24}{g} \frac{(\mathbf{q}_m, \mathbf{q}_m)}{(\boldsymbol{\alpha}_\ell, \boldsymbol{\alpha}_\ell)} \qquad (C.24)$$

where $(\boldsymbol{\alpha}_\ell, \boldsymbol{\alpha}_\ell)$ is the square length of a long root. The table includes a column for the dimension of the algebra, one for the order of the Weyl group, as well as one for the structure of the quotient of the weight to the root lattice considered as additive groups (the order being $\det A_{ij}$)

One could proceed to infinite Kac–Moody algebras by extending the previous description to degenerate Cartan matrices, or in a global fashion to loop groups (i.e. mappings from the circle into a Lie group) and their central extensions. We shall motivate this generalization by presenting the Wess–Zumino–Witten model. In parallel with the implementation of local scale invariance, the latter promotes a global symmetry to a local one in a tight conformally invariant structure.

C.2.   The Wess–Zumino–Witten model

The asymptotically free two-dimensional nonlinear σ-model is a massive field theory with a global continuous invariance group. In an attempt to find a nontrivial infrared fixed point, Witten was led in the case where the field variable u takes its values in a compact Lie group G, to add a topological term which suggests perturbatively the existence of such a critical theory. To simplify matters, we shall assume that the compact Lie group has a simple Lie algebra (which will then be extended over complex numbers).

Table III. Data for simple Lie algebras.

Algebra	dim	$|W|$	Weight/Root	g	degrees of invariants
A_ℓ $\ell \geq 1$	$\ell(\ell+2)$	$(\ell+1)$	$\mathbb{Z}/(\ell+1)\mathbb{Z}$	$\ell+1$	$2,3,\ldots,\ell+1$
B_ℓ $\ell \geq 2$	$\ell(2\ell+1)$	$2^\ell \ell$	$\mathbb{Z}/2\mathbb{Z}$	$2\ell-1$	$2,4,\ldots,2\ell$
C_ℓ $\ell \geq 3$	$\ell(2\ell+1)$	$2^\ell \ell$	$\mathbb{Z}/2\mathbb{Z}$	$\ell+1$	$2,4,\ldots,2\ell$
D_ℓ $\ell \geq 4$	$\ell(2\ell-1)$	$2^{\ell-1}\ell$	ℓ even: $\mathbb{Z}/2\mathbb{Z}\times\mathbb{Z}/2\mathbb{Z}$; ℓ odd: $\mathbb{Z}/4\mathbb{Z}$	$2(\ell-1)$	$2,4,\ldots,2(\ell-1),\ell$
G_2	14	12	id.	4	$2,6$
F_4	52	1 152	id.	9	$2,6,8,12$
E_6	78	51 840	$\mathbb{Z}/3\mathbb{Z}$	12	$2,5,6,8,9,12$
E_7	133	2 903 040	$\mathbb{Z}/2\mathbb{Z}$	18	$2,6,8,10,12,14,18$
E_8	248	696 729 600	id.	30	$2,8,12,14,18,20,24,30$

The topological term has a structure typical of an anomaly arising when studying fermions coupled to gauge fields. We shall not emphasize this aspect here, although part of the motivation for this bosonic construction is to find an equivalence with a fermionic one.

The starting point is the standard action

$$S^{(0)} = \frac{1}{2\lambda} \int \frac{d^2 x}{2\pi} \widetilde{\mathrm{Tr}} \partial_a u^{-1} \partial_a u \qquad \lambda > 0 \qquad (C.25)$$

To give a meaning to this expression, we assume that u takes its values in a faithful unitary representation of G and $\widetilde{\mathrm{Tr}}$ is the trace in this representation up to an appropriate factor (hence the tilde). Since

$$\widetilde{\mathrm{Tr}} \partial_a u^{-1} \partial_a u = -\widetilde{\mathrm{Tr}} (u^{-1} \partial_a u)(u^{-1} \partial_a u) \qquad (C.26)$$

and for u unitary, $u^{-1} \partial_a u$ is an antihermitian matrix in the representation of the Lie algebra, it is seen that $S^{(0)}$ is positive (the Boltzmann weight is understood as $\exp - S$). The same formula shows that all what is really needed is an invariant bilinear form on the Lie algebra, identified with the unique one (hence the restriction to a simple Lie algebra) again up to a factor. Thus

$$\widetilde{\mathrm{Tr}} (u^{-1} \partial_a u)(u^{-1} \partial_a u) \equiv (u^{-1} \partial_a u, u^{-1} \partial_a u) \qquad (C.27)$$

The convenient normalization, justified below, is that this bilinear form is such that the length of long roots is equal to one. This convention should be kept in mind when comparing with expressions appearing in the works of other authors.

In arbitrary coordinates or in curved space, with a metric g_{ab}, the Lagrangian in $S^{(0)}$ should be replaced by the conformal invariant expression $\sqrt{g} g^{ab} \widetilde{\mathrm{Tr}} (\partial_a u^{-1} \partial_b u)$. In any case, it is invariant under the global group $G \times G$, under $u \to u^{-1}$ as well as any reflection in the flat case.

To justify the structure of the added term $S^{(1)}$, one compactifies flat space to a sphere denoted here Σ_2 (due care has to be paid to a possible change of coordinate system). As a result, $x \to u(x)$ gives an image of Σ_2 into the group, which can be smoothly deformed to a point (the precise statement is that the second homotopy group $\pi_2(G)$ vanishes) so that we can extend $u(x)$ to a map $u(y)$ from the ball B (the interior of the sphere in three-space) to the group.

This enables one to write a second term

$$S^{(1)} = -\frac{ik}{24\pi} \int d^3y \varepsilon^{abc} \widetilde{\mathrm{Tr}}(u^{-1}\partial_a u)(u^{-1}\partial_b u)(u^{-1}\partial_c u)$$
$$\equiv -\frac{ik}{24\pi} \int_B (\omega_\wedge, \omega_\wedge\omega)$$
(C.28)

where ω is the one-form $\omega = u^{-1} du$, and in the second expression, the products are understood as exterior products on forms and interior products on Lie algebra elements (clearly $\omega_\wedge\omega = \frac{1}{2}[\omega_\wedge, \omega]$ belongs to the Lie algebra).

The extension from Σ_2 to B is not unique. By glueing two such extensions on Σ_2, one sees that the difference between two possible values of $S^{(1)}$ is given by a similar integral extending over a sphere Σ_3, giving an element in a class of the third homotopy group $\pi_3(G)$. For simple Lie groups, the latter admits a unique generator and the two values of $S^{(1)}$ differ at most by a multiple of the integral over this generator. The normalization of $\widetilde{\mathrm{Tr}}$ or $(\ ,\ )$ is chosen in such a way that this arbitrariness is an entire multiple of $2\pi i$, leaving $\exp{-S^{(1)}}$ invariant, and requiring k to be an integer. Thus we must have

$$\frac{k}{48\pi^2} \int_{\Sigma_3} (\omega_\wedge, \omega_\wedge\omega) \in \mathcal{Z} \qquad (C.29)$$

To see what is implied, consider an $SU(2)$ subgroup in G and represent faithfully u as $\exp i\psi\boldsymbol{\sigma}.\mathbf{n}$ where $\mathbf{n}$ is a unit vector on Σ_2 parametrized as $(\sin\theta\cos\varphi, \sin\theta\sin\varphi, \cos\theta)$, with $0 \leq \theta < \pi$, $0 \leq \varphi < 2\pi$, $\boldsymbol{\sigma}$ stands for the Pauli matrices and ψ ranges between 0 and π. This uses of course the fact that $SU(2)$ is topologically equivalent to Σ_3. A simple calculation yields, with $\omega = u^{-1} du$,

$$(\omega_\wedge, \omega_\wedge\omega) = 6(\sigma_3, \sigma_3)\sin^2\psi\, d\psi_\wedge \sin\theta\, d\theta_\wedge d\varphi$$

The homotopy generator is given by the identity map $\Sigma_3 \to SU(2)$. Since

$$\int_{\Sigma_3} \sin^2\psi\, d\psi_\wedge \sin\theta\, d\theta_\wedge d\varphi = 2\pi^2$$

we have in this case

$$\frac{k}{48\pi^2} \int_{\Sigma_3} (\omega_\wedge, \omega_\wedge\omega) = \tfrac{1}{4}k(\sigma_3, \sigma_3) \qquad (C.30)$$

The generator σ_3 is equivalent to k_α occurring in equations (C.5), (C.6). Therefore the quantization condition for the consistency of a path integral is that for any root α, $\frac{1}{4}k(k_\alpha, k_\alpha)$ be an integer. But $(\alpha, \alpha) = (\tilde{k}_\alpha, \tilde{k}_\alpha) = 4/(k_\alpha, k_\alpha)$. Thus we have for any root α,

$$k \in (\alpha, \alpha)\mathcal{Z} \qquad (C.31)$$

The most stringent condition is for long roots. We have agreed that the quadratic form is normalized in such a way that $(\alpha_\ell, \alpha_\ell) = 1$, in which case k has to be an integer. The same computation shows that k might be further restricted if G is not simply connected. For instance if G is $SO(3)$ instead of $SU(2)$, then k must be even.

We also see that, even though the three-form $(\omega_\wedge, \omega_\wedge\omega)$ is closed but not exact, one can parametrize it in order that it appears locally as a differential of a (nonuniform) two-form. For $SU(2)$ for instance, we could write

$$S^{(1)} = \frac{ik}{8\pi}(\sigma_3, \sigma_3)\int_{\Sigma_2} \varphi \sin^2\psi \, \mathrm{d}\psi_\wedge \sin\theta \, \mathrm{d}\theta$$

as a nonuniform integral over two-space with singularities at $\theta = 0$ and π.

We set the total action to be the sum

$$S = S^{(0)} + S^{(1)} \qquad (C.32)$$

and compute its variation to obtain the classical equations of motion

$$\delta S = \int \frac{\mathrm{d}^2 x}{2\pi}\left(u^{-1}\delta y, \partial_a\left[\frac{1}{\lambda}u^{-1}\partial_a u + \frac{1}{4}ki\varepsilon_{ab}u^{-1}\partial_b u\right]\right) \qquad (C.33)$$

which only involves a two-dimensional integral. Using complex coordinates $\{z, \bar{z}\}$, this yields

$$\left(\lambda^{-1} - \tfrac{1}{4}k\right)\bar{\partial}\left(u^{-1}\partial u\right) + \left(\lambda^{-1} + \tfrac{1}{4}k\right)\partial\left(u^{-1}\bar{\partial}u\right) = 0 \qquad (C.34)$$

When

$$\lambda = \frac{4}{|k|} \qquad (C.35)$$

we have either $\partial\left(u^{-1}\bar{\partial}u\right) = 0$ if k is positive, or $\bar{\partial}\left(u^{-1}\partial u\right) = 0$ if k is negative. Taking $k > 0$ for definiteness, a classical solution factorizes as

$$u_{\text{class}} = u_1(z)\bar{u}_2^{-1}(\bar{z}) \qquad (C.36)$$

With this value for λ, the action takes the form

$$S(u) = \frac{k}{8\pi} \left[\int d^2 x \tfrac{1}{2} \widetilde{\mathrm{Tr}}(\partial_a u^{-1} \partial_a u) \right.$$

$$\left. - \tfrac{1}{3}\mathrm{i} \int_B d^3 y \varepsilon^{abc} \widetilde{\mathrm{Tr}}(u^{-1}\partial_a u)(u^{-1}\partial_b u)(u^{-1}\partial_c u) \right] \quad (C.37a)$$

while

$$S(u + \delta u) = S(u) + k \int \frac{d^2 x}{2\pi} \left(u^{-1}\delta u, \partial(u^{-1}\bar{\partial} u) \right) \quad (C.37b)$$

Set $u_t = u e^{-tH}$ with H an element in the Lie algebra. Using the above equation, one can integrate $\dot{S}(u_t)$ from 0 to 1 to get the identity

$$S(uv^{-1}) = S(u) + S(v^{-1}) + k \int \frac{d^2 x}{2\pi} \left(v^{-1}\partial v, u^{-1}\bar{\partial} u \right) \quad (C.38)$$

Furthermore if we deform a solution $u_1(t, z)\bar{u}_2^{-1}(t, \bar{z})$ from $t = 0$ to 1 in such a way that at $t = 0$ it is the identity and at $t = 1$ it takes the value $(C.36)$, it is readily seen from equation $(C.37b)$ that the action remains constant and in fact equal to zero. Then equation $(C.38)$ shows that, for the particular choice $\lambda = 4/|k|$, the action is invariant under the gauge group $G_z \otimes G_{\bar{z}}$

$$u(z, \bar{z}) \rightarrow u_1(z)u(z, \bar{z})\bar{u}_2^{-1}(\bar{z}) \quad (C.39)$$

The question is now to find how this invariance reflects itself at the quantum level. Keeping λ and k as parameters, a one-loop computation shows that to leading order, the Callan–Symanzik β-function still vanishes at $\lambda = 4/|k|$, while it reveals a quantum anomaly (a Schwinger term) in the current commutators, the generators of the symmetry $(C.39)$. Rather than trying to derive from the path integral the precise nonperturbative answers, one can attempt to find a consistent set of correlations using the Ward identities of conformal invariance together with those of gauge invariance.

The conserved currents

$$\begin{array}{ll} J(z) = \tfrac{1}{2}ku\partial u^{-1} & \partial J = 0 \\ \bar{J}(\bar{z}) = \tfrac{1}{2}ku^{-1}\bar{\partial} u & \bar{\partial}\bar{J} = 0 \end{array} \quad (C.40)$$

take their values in the Lie algebra and the conservation laws which express analyticity for J (antianalyticity for $\bar{J}$) are just a restatement of the classical equations of motion (if we note that $\bar{\partial}J = -u(\partial\bar{J})u^{-1}$). As in the case of the energy–momentum tensor, these conservation laws are to be understood for insertions into correlation functions, up to contact terms.

Under an infinitesimal variation of the type $(C.39)$, with $\Omega(z)$ and $\bar{\Omega}(\bar{z})$ infinitesimal elements in the Lie algebra

$$u(z,\bar{z}) \rightarrow u(z,\bar{z}) - \Omega(z)u + u\bar{\Omega}(\bar{z})$$
$$u^{-1}(z,\bar{z}) \rightarrow u^{-1}(z,\bar{z}) - \bar{\Omega}(\bar{z})u^{-1} + u^{-1}\Omega(z) \qquad (C.41)$$

all this understood in a faithful matrix representation, the currents transform as

$$J \rightarrow J + [J,\Omega] + \tfrac{1}{2}k\partial\Omega = J + \delta_\Omega J$$
$$\bar{J} \rightarrow \bar{J} + [\bar{J},\bar{\Omega}] + \tfrac{1}{2}k\bar{\partial}\bar{\Omega} = \bar{J} + \delta_{\bar{\Omega}}\bar{J} \qquad (C.42)$$

If we take into account the fact that the currents are themselves the generators for infinitesimal gauge transformations, these equations are a concise statement equivalent to the existence of two infinite Kac–Moody algebras, very much as equation (47) encapsulated the Virasora algebras. It is understood up to here that J and $\bar{J}$ are classical fields appearing as arguments in correlation functions. In other words, their coefficients in an expansion over a Lie algebra basis do commute. If a field A has a local transformation law according to a representation of $G_z \otimes G_{\bar{z}}$, then equation $(C.26)$ implies that

$$\delta_\Omega A(z',\bar{z}') = \oint \frac{\mathrm{d}z}{2\mathrm{i}\pi}\,(\Omega(z),J(z))\,A(z',\bar{z}') \qquad (C.43)$$

where the contour encircles the point z', with a similar $\bar{z}$-formula for $\delta_\Omega A$. In deriving the normalization in $(C.43)$, it is to be remembered that $\mathrm{d}^2x = \mathrm{d}\bar{z} \wedge \mathrm{d}z/2\mathrm{i}$. This is entirely parallel to equation (41) for a coordinate transformation. We apply this formula to $J(z)$ instead of A, and expand J on a Lie algebra basis t_a, with

$$\Omega = \sum \omega^a t_a, \qquad J = \sum J^a t_a \quad \text{and} \quad (t_a, t_b) = \delta_{ab} \qquad (C.44)$$

We define the totally antisymmetric structure constants through

$$[t_a, t_b] = \mathrm{i}f_{abc}t_c \qquad (C.45)$$

Thus, in a unitary representation, the t's would be Hermitian and coincide in the case of $SU(2)$ with $\tfrac{1}{2}\sigma_a$. Equation $(C.43)$ translates into the short-distance expansion

$$J^a(z)J^b(z') = \tfrac{1}{2}k\frac{\delta_{ab}}{(z-z')^2} + \mathrm{i}\frac{f_{abc}}{(z-z')}J^c(z') + \cdots \qquad (C.46)$$

where the dots represent regular terms and use has been made of the total antisymmetry of the structure constants. The first term on the r.h.s. represents the quantum anomaly, analogous to the central charge for the Virasoro algebra. The expansion is in agreement with the commutative character of the coefficients $J^a(z)$ (i.e. it is invariant under a simultaneous exchange $(a,z) \Leftrightarrow (b,z')$). We omit for short the $\bar{z}$ counterpart. Formula $(C.46)$ is complemented by the standard ones giving the relation with the energy–momentum tensor and expressing that $J(z)$ has conformal weights $(1,0)$

$$T(z)T(z') = \frac{c}{2(z-z')^4} + \frac{2}{(z-z')^2}T(z')$$

$$+ \frac{1}{(z-z')}\partial T(z') + \cdots \qquad (C.47)$$

$$T(z)J^a(z') = \frac{1}{(z-z')^2}J^a(z') + \frac{1}{(z-z')}\partial J^a(z') + \cdots (C.48)$$

Using radial quantization, one can define corresponding operators $\hat{T}(z)$, $\hat{J}(z)$, acting on conformal fields hence also on a linear space of states. We expand these operators as Laurent series in powers of z and omit the hats on the operator coefficients for short

$$\hat{T}(z) = \sum_{-\infty}^{+\infty}\frac{L_n}{z^{n+2}} \qquad \hat{J}^a(z) = \sum_{-\infty}^{+\infty}\frac{J_n^a}{z^{n+1}} \qquad (C.49)$$

The short distance expansions are then equivalent to the set of comuutation rules

$$[L_n, L_m] = (n-m)L_{n+m} + \tfrac{1}{12}cn(n^2-1)\delta_{n+m,0} \qquad (C.50a)$$

$$[L_n, J_m^a] = -mJ_{n+m}^a \qquad (C.50b)$$

$$[J_n^a, J_m^b] = \mathrm{i}f_{abc}J_{n+m}^c + \tfrac{1}{2}kn\delta_{ab}\delta_{n+m,0} \qquad (C.50c)$$

The last equation defines the Kac–Moody algebra, the only possible central extension of the loop group algebra

$$[j_n^a, j_m^b] = \mathrm{i}f_{abc}j_{n+m}^c$$

Again $\bar{z}$ counterparts are understood with the same values for k and c. Unitarity of a representation requires

$$(J_n^a)^\dagger = J_{-n}^a \qquad (C.51)$$

A primary field, creating a highest weight state, will be characterized by a pair of conformal weights $(h, \bar{h})$ with respect to the Virasoro algebra, as well as a pair of finite-dimensional representations $(\mathcal{R}, \bar{\mathcal{R}})$ for the Lie algebras $\{J_0^a\}$, $\{\bar{J}_0^a\}$. Those can also be indexed by their dominant weight. Such a field evaluated at the origin (and with the barred counterparts omitted) when acting on the vacuum state generates a "ground state" $|h, \mathcal{R}\rangle$ annihilated by L_n and J_n^a for n positive.

From the specificity of the model, we expect that the energy-momentum tensor is expressible in terms of J and hence that the central charge is specified by the group G as well as the level k. Indeed at the classical level, the energy–momentum tensor is quadratic in the currents, corresponding to a structure suggested in the late sixties by Sugawara (in a fermionic context) and elaborated by many authors. Expressed in terms of operators this means that

$$\hat{T}(z) = \frac{1}{2x} \sum_a \; : \hat{J}^a(z)\hat{J}^a(z) : \qquad (C.52)$$

for an appropriate value of the coefficient x. Normal ordering requires that operators with an index $n < 0$ be pushed to the left of operators with an index $n \geq 0$. Obviously from the diagonal form of the scalar product on the Lie algebra, the expression $(C.52)$ is the only quadratic candidate for a group invariant (as $\hat{T}$ should be). We note the analogy with the scalar field case, where T was proportional to $(\partial\varphi)^2$ and is here proportional to its equivalent $\mathrm{Tr}\,\partial u^{-1}\partial u$.

The constant x occurring in $(C.52)$ is fixed by the consistency of the algebra. To see this, let us apply both sides of the equation to a highest weight state $|h, \mathcal{R}\rangle$ where h denotes the lowest value of L_0, and retain only the term in z^{-1}. This reads

$$\left(x L_{-1} - \sum_b J_{-1}^b J_0^b \right) |h, \mathcal{R}\rangle = 0 \qquad (C.53)$$

giving what might be termed a "null state" if we think of the module generated by $\{L_{-n}\} \{J_{-m}\} |h, \mathcal{R}\rangle$. Acting with L_1 and J_1^a

in turn, we find

$$\left(x\,[L_1, L_{-1}] - \sum_b [L_1, J_{-1}^b]\, J_0^b \right) |h, \mathcal{R}\rangle = 0$$

$$\left(x\,[J_1^a, L_{-1}] - \sum_b [J_1^a, J_{-1}^b]\, J_0^b \right) |h, \mathcal{R}\rangle = 0$$

$(C.54)$

To use the commutation rules $(C.50)$, we observe that acting in the irreducible representation $\mathcal{R}$ with dominant weight $\mathbf{q}_{\mathcal{R}}$, the operator $\sum_b J_0^b J_0^b$ yields the corresponding value of the Casimir invariant

$$C_{\mathcal{R}} = (\mathbf{q}_{\mathcal{R}}, \mathbf{q}_{\mathcal{R}} + 2\mathbf{q}_m) \qquad (C.55)$$

while from $(C.45)$, the similar quantity in the adjoint representation is

$$C_{\mathrm{adj}} \delta_{c_1, c_2} = \sum_{a,b} f_{abc_1} f_{abc_2}$$

$$= (-\alpha_0, -\alpha_0 + 2\mathbf{q}_m)\, \delta_{c_1, c_2}$$

$(C.56)$

It can be checked that C_{adj} is nothing but the dual Coxeter number (recall that $(\alpha_\ell, \alpha_\ell) = 1$). Returning to equations $(C.54)$, we find

$$2hx - C_{\mathcal{R}} = 0$$

$$2x - k - C_{\mathrm{adj}} \equiv 2x - k - g = 0$$

and therefore the normalization factor $2x$ is given by

$$2x = k + g \qquad (C.57)$$

while the conformal weight is related to the representation $\mathcal{R}$ through

$$h = \frac{C_{\mathcal{R}}}{k + g} \qquad (C.58)$$

For $SU(2)$, if we characterize the representation by the angular momentum j, knowing that g takes the value 2 (as follows from $(C.55)$ for $f_{abc} \to \varepsilon_{abc}$ or from $C_{\mathcal{R}} = j(j + 1)$ for $j = 1$), we have $2x = k + 2$ and $h = j(j + 1)/(k + 2)$.

The value of the central charge can be obtained by requiring that the vacuum state be invariant under the group G and therefore has weight $h = 0$. Thus computing

$$\langle 0 | L_2 L_{-2} | 0 \rangle = \langle 0 | [L_2, L_{-2}] | 0 \rangle = \tfrac{1}{2} c$$

$$= \frac{1}{4x^2} \left\langle 0 \left| \sum_{a,b} J_1^a J_1^a J_{-1}^b J_{-1}^b \right| 0 \right\rangle$$

and using the Kac–Moody commutation rules, one finds upon using the above value x that

$$c = \dim(G) \frac{k}{k+g} \qquad\qquad (C.59)$$

where $\dim(G)$ is the dimension of the Lie algebra of G. Thus for $SU(2)$, $c = 3k/(k+2)$. In general, for k integral, c ranges between the rank of the group (corresponding to $k = 1$ for simply laced groups) and its dimension as k goes to infinity.

By direct product, the construction extends to semisimple Lie algebras and can even include abelian factors (with corresponding value $g = 0$), where it coincides with the free field theory compactified on a circle.

There are various generalizations of the above construction. We shall limit ourselves to describing a remarkable observation made by Goddard, Kent and Olive. Suppose the simple Lie algebra of G contains a simple subalgebra corresponding to a subgroup H. If the level of G is k (the integer characterizing the anomaly in equation $(C.51c)$), when restricted to the subalgebra it might differ from k in the ratio of long roots, since, with $(\alpha_\ell, \alpha_\ell) = 1$ for G, the restriction of the quadratic form to the Lie algebra of H may induce a fractional value for the normalization of its long root. Assume that the first $\dim(H)$ indices refer to H, then

$$\hat{T}_G(z) = \frac{1}{k+g_G} \sum_1^{\dim(G)} : \hat{J}^a(z) J^a(z) :$$

$$\hat{T}_H(z) = \frac{1}{k+(\alpha_\ell, \alpha_\ell)_H g_H} \sum_1^{\dim(H)} : \hat{J}^a(z) \hat{J}^a(z) : \qquad (C.60)$$

Therefore the corresponding central charges are given by

$$c_G = \dim(G) \frac{k}{k+g_G} \qquad\qquad c_H = \dim(H) \frac{k'}{k'+g_H} \qquad (C.61)$$

where

$$k' = k \frac{(\alpha_\ell, \alpha_\ell)_G}{(\alpha_\ell, \alpha_\ell)_H} \geq k \qquad (C.62)$$

is greater or equal to k since $(\alpha_\ell, \alpha_\ell)_G/(\alpha_\ell, \alpha_\ell)_H$ is an integer. From the commutation rules $(C.50b)$, it follows that $L_n(G/H) \equiv L_n(G) - L_n(H)$ commutes with H currents and therefore also with $L_m(H)$. Thus the G-Virasoro algebra breaks into two commuting pieces

$$\hat{T}_G = \hat{T}_H + \hat{T}_{G/H} \qquad (C.63)$$

Since central charges are additive, it follows that for the coset space G/H

$$c_{G/H} = c_G - c_H = \dim G \frac{k}{k + g_G} - \dim H \frac{k'}{k' + g_H} \qquad (C.64)$$

Assume unitary representations. It follows that, if $c_{G/H} = 0$, the corresponding representation is trivial, and the representations of the combined Virasoro–Kac–Moody algebras for G and H are equivalent in spite of their very different appearance. This leads to nontrivial operator constructions. For instance if G is a simply laced Lie group, if $k = 1$, and if H is the maximal abelian torus (corresponding to the Cartan subalgebra), we have $c_G = c_H = \mathrm{rank}G = \ell$. Thus $c_{G/H} = 0$, and this leads to a presentation of the nonabelian theory in terms of ℓ periodic scalar fields and an associated construction for the remaining ("off-diagonal") currents in terms of exponentials of these scalar fields.

Another application is a construction of the series of minimal unitary representations of the Virasoro algebra. One possibility is to take the sequence

$$G = Sp(2(m-1)) \supset Sp(2(m-2)) \otimes Sp(2) = H$$

with the original level being $k = 1$ and the subgroup inheriting in this case a level $k' = k = 1$. We have

$$c_G = \frac{(m-1)(2m-1)}{m+1} = 2m - 5 + \frac{6}{m+1}$$

$$c_H = \frac{(m-2)(2m-3)}{m} + 1 = 2m - 6 + \frac{6}{m}$$

and therefore,

$$c_{Sp(2(m-1))_1/(Sp(2(m-2)\otimes Sp(2))_1} = 1 - \frac{6}{m(m+1)} \qquad (C.65a)$$

Another possibility is to use for G the group $SU(2) \times SU(2)$ and for H the diagonal subgroup $SU(2)$. In G the levels of the two factors can be taken at will. The choice $(m-2, 1)$ yields for H the level $m-1$ and for the central charge

$$\begin{aligned} c_{(SU(2)_{m-2} \times SU(2)_1)/SU(2)_{m-1}} &= \frac{3(m-2)}{m} + 1 - \frac{3(m-1)}{m+1} \\ &= 1 - \frac{6}{m(m+1)} \end{aligned} \qquad (C.65b)$$

i.e. once more the unitary-minimal series.

These observations can be used to complete the proof of the Friedan–Qiu–Shenker theorem that Virasoro representations with the above values for m integral larger or equal to 2 are the only unitary ones of central charge smaller than 1.

C.3. Representations and characters of Kac–Moody algebras

We can only sketch a few results from the representation theory of Kac–Moody algebras. We restrict ourselves to unitary representations and show first how one recovers in an algebraic setting the fact that k must be an integer (as opposed to the above topological derivation) and also derive the further restrictions on the representation $\mathcal{R}$ of G which lead to the so-called integrable representations of the Kac–Moody algebra.

Recall that for the Lie algebra of G in the adjoint representation the dominant weight is the positive (long) root

$$\psi = -\alpha_0 \qquad (C.66)$$

With this notation, let us indeed show that k is a nonnegative integer satisfying

$$k \geq \frac{2(\psi, \mathbf{q}_{\mathcal{R}})}{(\psi, \psi)} \geq 0 \qquad (C.67)$$

Although our expression is written with the convention that $(\psi, \psi) = 1$, we have reinstated an arbitrary normalization to remind us that $2(\psi, \mathbf{q}_{\mathcal{R}})/(\psi, \psi)$ is a nonnegative integer. To prove this, consider the $SU(2)_\alpha$ subalgebra in the Kac–Moody algebra, in the obvious generalization of the Chevalley–Serre basis, resulting from the commutation rules

$$[e_1^\alpha, e_{-1}^{-\alpha}] = k_0^\alpha - \tfrac{1}{2}k\,(e^\alpha, e^{-\alpha}) = k_0^\alpha - \frac{k}{(\alpha, \alpha)}$$

$$[k_0^\alpha, e_1^\alpha] = 2e_1^\alpha \qquad [k_0^\alpha, e_{-1}^{-\alpha}] = -2e_{-1}^{-\alpha}$$

This shows that the generators

$$\frac{k}{(\alpha, \alpha)} - k_0^\alpha, \ e_{-1}^{-\alpha}, \ e_1^\alpha$$

can play the role of

$$k_\alpha, e_\alpha, e_{-\alpha}$$

respectively, occurring as $SU(2)$ generators in equations $(C.5)$–$(C.6)$. The Kac–Moody representation can be decomposed into states with definite weights for the subalgebra $\{J_0^a\}$ i.e. satisfying

$$k_0^\alpha \,|\mathbf{q}\rangle = \frac{2(\alpha, \mathbf{q})}{(\alpha, \alpha)}\,|\mathbf{q}\rangle$$

We see that for any $\mathbf{q}$ in this decomposition $k - 2(\alpha, \mathbf{q})$ must be an integer. Applying this to the dominant root $\alpha = -\alpha_0 = \psi$ taking into account $(\psi, \psi) = 1$, we find that $k - 2(\psi, \mathbf{q})/(\psi, \psi)$ has to be integral and thus k is also an integer. Furthermore if the highest weight state corresponds to a representation $\mathcal{R}$ with dominant weight $\mathbf{q}_\mathcal{R}$, we must have

$$e_1^{-\alpha} \,|\mathbf{q}_\mathcal{R}\rangle = 0$$

For a unitary representation $(e_1^{-\alpha})^\dagger = e_{-1}^\alpha$. As a consequence

$$0 \leq \langle \mathbf{q}_\mathcal{R}|\, e_1^{-\alpha} e_{-1}^\alpha \,|\mathbf{q}_\mathcal{R}\rangle = \langle \mathbf{q}_\mathcal{R}|\, [e_1^{-\alpha}, e_{-1}^\alpha] \,|\mathbf{q}_\mathcal{R}\rangle$$

$$= \langle \mathbf{q}_\mathcal{R}|\, \left(\frac{k}{(\alpha, \alpha)} - k_0^\alpha\right) |\mathbf{q}_\mathcal{R}\rangle = \frac{k - 2(\alpha, \mathbf{q}_\mathcal{R})}{(\alpha, \alpha)} \langle \mathbf{q}_\mathcal{R}, \mathbf{q}_\mathcal{R}\rangle$$

Thus $k \geq 2(\alpha, \mathbf{q}_\mathcal{R})$, and the most stringent condition is for the dominant root ψ normalized to unity, leading to the restriction $(C.67)$ on the possible representations $\mathcal{R}$.

For instance for $SU(2)$, we have $(\psi, \mathbf{q}_R)/(\psi, \psi) = j$ the integral or half integral angular momentum, and it follows for the corresponding Kac–Moody algebra, denoted $A_1^{(1)}$, that k must be a nonnegative integer such that $k \geq 2j \geq 0$.

To conclude this appendix, we describe the Kac–Moody characters of the corresponding unitary representations on the set of states obtained by acting with a string of operators J^a_{-n} ($n > 0$) on the highest weight state $|q_{\mathcal{R}}\rangle$ and its partners giving a representation $\mathcal{R}$ of the underlying finite Lie group G.

Recall that we have an associated representation of the Virasoro algebra characterized by the values

$$c = \dim(G)\frac{k}{k+g} \qquad\qquad h = \frac{C_{\mathcal{R}}}{k+g} \qquad (C.68)$$

for the central charge and the conformal weight h, where $C_{\mathcal{R}} = (q_{\mathcal{R}}, q_{\mathcal{R}} + 2q_m)$. Of course the level k is a specified integer.

As in the finite-dimensional case, the character is a generating function for the multiplicities of the G weights, occurring at a degree n in a decomposition over states diagonalizing the Cartan subalgebra of $\{J^a_0\}$. Let these multiplicities be written $\text{mult}_n(q)$, and note that the corresponding eigenvalue of the grading operator L_0 is $h_n = h + n$.

With $\mathbf{x}$ an arbitrary ℓ-dimensional vector in root space and τ a complex number in the upper half plane, define the character as

$$\chi_{\mathcal{R},k}(\mathbf{x},\tau) = \sum_{n\geq 0}\sum_{\mathbf{q}} \text{mult}_n(\mathbf{q}) \exp 2i\pi\tau\left[(h+n)-\tfrac{1}{24}c\right]\exp 2i\pi(\mathbf{q},\mathbf{x})$$

$$(C.69)$$

where the sum over $\mathbf{q}$ is on the weight lattice of the finite Lie algebra of G, and we have included a factor $\exp(-\tfrac{1}{24}2i\pi\tau c)$, as in the text, in such a way that specialized to $\mathbf{x} = 0$, the above formula reads

$$\chi_{\mathcal{R},k}(\mathbf{0},\tau) = \text{Tr}_{\mathcal{R},k}\exp 2i\pi\tau\left(L_0 - \tfrac{1}{24}c\right) \qquad (C.70)$$

The Weyl–Kac expression for characters (generalizing equation $(C.20)$) yields χ as a ratio of two sums over the (finite) Weyl group of G, through

$$\chi_{\mathcal{R},k}(\mathbf{x},\tau) = \frac{\sum_{w\in W}\varepsilon(w)\theta_{w(q_{\mathcal{R}}+q_m),k+g}(\mathbf{x},\tau)}{\sum_{w\in W}\varepsilon(w)\theta_{w(q_m),g}(\mathbf{x},\tau)} \qquad (C.71)$$

where as above q_m is the sum of fundamental weights (equal to half the sum over positive roots) and the generalized θ-functions stand for

$$\theta_{\mathbf{q},k}(\mathbf{x},\tau) = \sum_{\beta\in\mathbf{q}+k\mathcal{L}}\exp\left\{\frac{2i\pi\tau}{k}\frac{(\beta,\beta)}{(\alpha_\ell,\alpha_\ell)}+2i\pi(\beta,\mathbf{x})\right\} \qquad (C.72)$$

Here $\mathcal{L}$ is the lattice generated on integers by the long roots and for simply laced groups corresponds to the full root lattice. We have again reinstated an arbitrary normalization for the invariant quadratic form. Of course when $k = 0$ and $\mathcal{R}$ is the identity, χ reduces to unity.

There exists a factorized form for the denominator in equation $(C.71)$, analogous to equation $(C.21)$, which leads to remarkable identities on θ-functions generalizing Jacobi's triple product identity (Appendix A). Inserting the definition $(C.72)$ into $(C.71)$ and dividing by a common factor precisely equal to the quantity $\psi_{\mathbf{q}_m}(\mathbf{x})$ occurring in Weyl's formula, we obtain more explicitly

$$\chi_{\mathcal{R},k}(\mathbf{x},\tau) = \frac{\displaystyle\sum_{\gamma \in \mathbf{q}_R + (k+g)\mathcal{L}} \exp\left[\frac{2i\pi\tau}{k+g} \cdot \frac{(\mathbf{q}_m + \gamma)^2}{\alpha_\ell^2}\right] \chi_\gamma(\mathbf{x})}{\displaystyle\sum_{\gamma \in g\mathcal{L}} \exp\left[\frac{2i\pi\tau}{g} \cdot \frac{(\mathbf{q}_m + \gamma)^2}{\alpha_\ell^2}\right] \chi_\gamma(\mathbf{x})} \qquad (C.73)$$

Specializing to $\mathbf{x} = 0$, with $\dim_{\mathbf{q}}$ the dimension of the $\mathcal{R}$-representation of dominant weight $\mathbf{q}$, we find

$$\chi_{\mathcal{R},k}(\mathbf{0},\tau) = \mathrm{Tr}_{\mathcal{R},k} \exp 2i\pi\tau \left(L_0 - \tfrac{1}{24}c\right)$$

$$= \frac{\displaystyle\sum_{\gamma \in \mathbf{q}_R + (k+g)\mathcal{L}} \exp\left[\frac{2i\pi\tau}{k+g} \cdot \frac{(\mathbf{q}_m + \gamma)^2}{\alpha_\ell^2}\right] \dim_\gamma}{\displaystyle\sum_{\gamma \in g\mathcal{L}} \exp\left[\frac{2i\pi\tau}{g} \cdot \frac{(\mathbf{q}_m + \gamma)^2}{\alpha_\ell^2}\right] \dim_\gamma} \qquad (C.74)$$

As a particular example, let us take the case where $G \equiv SU(2)$, with a unique pair of roots $\pm\alpha$, $\mathcal{L}$ is the root lattice $\alpha\mathcal{Z}$, the dual Coxeter number is $g = 2$, and $\mathbf{q}_m = \tfrac{1}{2}\alpha$. Then $\mathbf{q}_R = 2j(\tfrac{1}{2}\alpha)$, $\mathbf{q}_R + \mathbf{q}_m = (2j+1)(\tfrac{1}{2}\alpha)$ and $\gamma = p\alpha$ with p ranging over integers. Finally $\dim\mathbf{q}_R + (k+g)\gamma = 2j + 1 + 2(k+2)p$. The character reads

$$\chi_{j,k}(0,\tau) = \frac{\sum_{p\in\mathcal{Z}} \exp\left\{\frac{2i\pi\tau[2j+1+2(k+2)p]^2}{4(k+2)}\right\} [(2j+1) + 2(k+2)p]}{\sum_{p\in\mathcal{Z}} \exp\left\{\frac{2i\pi\tau[1+4p]^2}{8}\right\} [1 + 4p]}$$

$$(C.75)$$

Setting as in section (3.4)

$$N = 2(k+2) \qquad\qquad \lambda = 2j+1 \qquad (C.76)$$

this is seen to coincide with the expression quoted in equation (268) as

$$\chi_{j,k}(0,\tau) \equiv \chi_{\lambda,N}^{\text{aff}}(\tau) = \frac{\sum_{p\in Z} \exp\left[2i\pi\tau(\lambda + pN)^2/2N\right](Np+\lambda)}{\sum_{p\in Z} \exp\left[2i\pi\tau(1+4p)^2/8\right](4p+1)}$$

$$= \frac{\sum_{p\in Z} \exp\left[2i\pi\tau(\lambda + pN)^2/2N\right](Np+\lambda)}{\eta^3(\tau)}$$

$$(C.77)$$

upon using Jacobi's identity.

The Goddard–Kent–Olive construction of unitary representations of the minimal unitary series, translates into an identity among characters pertaining both to the affine Lie algebra of $SU(2)$ and to the Virasoro algebra. With $\varepsilon = 0$ or $\frac{1}{2}$, this reads

$$\chi_{j,k=m-2}^{\text{aff}}(\mathbf{x},\tau)\chi_{\varepsilon,k=1}^{\text{aff}}(\mathbf{x},\tau)$$

$$= \sum_{1\le 2j'+1\le m}' \chi_{c=1-6/m(m+1),h_{2j+1,2j'+1}}^{\text{Vir}}(\tau)\,\chi_{j',k=m-1}^{\text{aff}}(\mathbf{x},\tau) \quad (C.78)$$

where the prime means that the sum runs over those j' such that $j - j'$ is an integer if $\varepsilon = 0$ and a half integer if $\varepsilon = \frac{1}{2}$. Specializing to $\mathbf{x} = 0$, and using the (λ, N) notation as in $(C.76)$, this reads for $1 \le \lambda \le m - 1$ and $2\varepsilon = 0$ or 1

$$\chi_{\lambda,2m}^{\text{aff}}(\tau)\chi_{2\varepsilon+1,6}^{\text{aff}}(\tau)$$

$$= \sum_{\substack{1\le\lambda'\le m \\ \lambda-\lambda'\equiv 2\varepsilon\,\text{mod}\,2}} \chi_{\lambda(m+1)-\lambda'm,2m(m+1)}^{\text{Vir}}(\tau)\chi_{\lambda',2(m+1)}^{\text{aff}}(\tau) \quad (C.79)$$

Notes

An original reference on conformal invariance is A.M. Polyakov, *JETP Lett.* **12**, 381 (1970). The associated infinite Lie algebra in the two-dimensional case was introduced in the framework of dual models by M.A. Virasoro *Phys. Rev.* **D1**, 2933 (1970). Additional material on previous attempts at consistent conformally invariant theories can be found in the book by I.T. Todorov, M.C. Mintchev and V.B. Petkova, *Conformal Invariance in Quantum Field Theory*, Pubblicazione della classe di scienze della scuola normale superiore, Pisa (1978). The fundamental papers, at the origin of present applications, are those of A.A. Belavin, A.M. Polyakov, and A.B. Zamolodchikov, *J. Stat. Phys.* **34**, 763 (1984) and *Nucl. Phys.* **B241**, 333 (1984).

The collective work *Vertex Operators in Mathematics and Physics* edited by J. Lepowsky, S. Mandelstam and I.M. Singer, Springer Verlag, New York, (1985) contains many contributions, among them by D. Friedan, Z. Qiu and S. Shenker on unitary representations of the Virasoro and superconformal algebras, and their relevance to physics, and by A. Rocha Caridi on the corresponding characters.

The formula for the Kac determinant is announced in V.G. Kac, Springer Lecture Notes in Physics **94**, 441 (1979). We follow the derivation of B.L. Feigin and D.B. Fuchs in *Funct. Anal. and Appl.* **16**, 114 (1984), **17**, 241 (1983).

The articles by V.S. Dotsenko, *Nucl. Phys.* **B235** [FS 11] 54 (1984) and by the same author with V.A. Fateev, *Nucl. Phys.* **B240** [FS12] 312 (1984) and **B251** [FS 13] 691 (1985) discuss degenerate models and correlations.

The Ising partition function on a torus was derived in the work of Ferdinand and Fisher quoted in chapter 2. The relation between the central charge and the Casimir effect in a strip is discussed in H.W.J. Blöte, J.L. Cardy, M.P. Nightingale, *Phys. Rev. Lett.* **56**, 742 (1986) and I. Affleck, *Phys. Rev. Lett.* **56**, 746 (1986).

J.L. Cardy has written a review article in the Domb and Lebowitz series vol. **11** Academic Press, New York (1986). Some of his important contributions are *J. Phys.* **A17**, L385 (1984), *Nucl. Phys.* **B270** [FS 16] 186 (1986) on modular invariance, *Nucl. Phys.* **B240** [FS 12] 514 (1984) on the semi-infinite geometry, *Phys. Rev. Lett.* **54**, 1354 (1985) on the Lee–Yang edge singularity. The consequences of scaling in this case had previously been discussed by M.E. Fisher, *Phys. Rev. Lett.* **40**, 1610 (1978).

The interpretation of the principal series of minimal models in the unitary case by D.A. Huse, *Phys. Rev.* **B30**, 3908 (1984) is based on work by R.J. Baxter, *J. Phys.* **A13**, L61 (1980) and G.E. Andrews, R.J. Baxter and P.J. Forrester, *J. Stat. Phys.* **35**, 193 (1984). On this subject, see also A.B. Zamolodchikov, *Sov. J. Nucl. Phys.* **44**, 529 (1986).

Field theory on a torus is presented in a joint work with J.B. Zuber in *Nucl. Phys.* **B275** [FS 17] 580 (1986) while the *A–D–E* classification of minimal models stated as a conjecture in a work with A. Cappelli and J.B. Zuber, *Nucl. Phys.* **B280** [FS

18] 445 (1987) was proved by the same authors in *Comm. Math. Phys.* **113**, 1 (1987), and independently by A. Kato, *Mod. Phys. Lett.* **A2**, 585 (1987). An important relation with affine characters is discussed in D. Gepner and E. Witten, *Nucl. Phys.* **B278**, 493 (1986), D. Gepner, *Nucl. Phys.* **B287**, 111 (1987), and a piece of the puzzle proved by D. Gepner and Z. Qiu, *Nucl. Phys.* **B285** [FS 19] 423 (1987) as part of a larger discussion. Further work along the lines of integrable models is due to V. Pasquier, *Nucl. Phys.* **B285** [FS19], 162 (1987).

T. Eguchi and H. Ooguri have generalized the Belavin–Polyakov–Zamolodchikov equations to Riemann surfaces of arbitrary genus in *Nucl. Phys.* **B282**, 308 (1987). Explicit formulae for Ising correlation functions are found in P. Di Francesco, H. Saleur and J.-B. Zuber, *Nucl. Phys.* **B290** [FS20], 527 (1987).

Frustrations and symmetries are studied in G. von Gehlen, V. Rittenberg, *J. Phys.* **A19** L625 (1986), J.-B. Zuber, *Phys. Lett.* **B176**, 127 (1986) and J.L. Cardy, *Nucl. Phys.* **B275** [FS 17] 200 (1986).

The Ashkin–Teller (J. Ashkin and E. Teller, *Phys. Rev.* **64**, 178 (1943)), Potts, six and eight vertex models are presented in R.J. Baxter's book, *Exactly Solved Models in Statistical Mechanics* Academic Press, New York (1982). The interpretation of statistical models in terms of Coulomb gases is reviewed in B. Nienhuis, *J. Stat. Phys.* **34**, 731 (1984), with credit given to numerous contributors. In particular the combined charge-monopole operators and their correlations appear in the work of L.P. Kadanoff and A.C. Brown, *Ann. Phys.* (New York) **121**, 318 (1979). Their study in the conformal context is found in P. Di Francesco, H. Saleur and J.B. Zuber, *J. Stat. Phys.* **49**, 57 (1987), S.K. Yang, *Nucl. Phys.* **B285** [FS19] 183 (1987), S.K. Yang and H.B. Zheng, *Nucl. Phys.* **B285** [FS19] 410 (1987).

Deviations from criticality are discussed by A.B. Zamolodchikov *JETP Letters* **43**, 730 (1986) and in the work of one of the authors with H. Saleur, *J. Stat. Phys.* **48**, 449 (1987).

A reference to elliptic and modular functions, including Kronecker's limit formulas, is A. Weil *Elliptic Functions according to Eisenstein and Kronecker*, Springer Verlag, Berlin (1976).

The treatment of fermions in the context of conformal invariance originates in the work of P. Ramond, *Phys. Rev.* **D3**, 2415 (1971) and A. Neveu and J.H. Schwarz, *Nucl. Phys.* **B31**, 86 (1971).

For superconformal field theory and its applications to statistical physics, see the contribution of D. Friedan, Z. Qiu and S. Shenker in the book on *Vertex Operators in Mathematics and Physics* quoted above and their article in *Phys. Lett.* **151B**, 37 (1985) as well as M.A. Bershadsky, V.G. Knizhnik, M.G. Teitelman, *ibid.* 31 (1985), and H. Eichenherr, *ibid.* 26 (1985). P. Goddard, A. Kent and D. Olive, *Comm. Math. Phys.* **103**, 105 (1986) and A. Meurman and A. Rocha-Caridi, *Comm. Math. Phys.* **107**, 263 (1986) discuss the construction of characters. The classification of minimal superconformal models is investigated in D. Kastor, *Nucl. Phys.* **B280** [FS 18] 304 (1987), Y. Matsuo and S. Yahikozawa, *Phys. Lett.* **178B**, 211 (1986), and A. Cappelli, *Phys. Lett.* **185B**, 82 (1987).

The model of appendix C is described in A. Polyakov and P.B. Wiegmann, *Phys. Lett.* **B131**, 121 (1983), and E. Witten, *Comm. Math. Phys.* **92**, 455 (1984). The conformal structure is elaborated by V. Knizhnik and A.B. Zamolodchikov, *Nucl. Phys.* **B247**, 83 (1984), D. Gepner and E. Witten, *Nucl. Phys.* **B278**, 493 (1986), P. Goddard and D. Olive, *Int. J. of Mod. Phys.* **A1**, 303 (1983), where the last paper includes a review on Lie and Kac–Moody algebras. This subject is covered in many books among which J.P. Serre *Algèbres de Lie semi-simples complexes*, Benjamin, New York (1966) and J.E. Humphreys *Introduction to Lie Algebras and Representation Theory*, Springer Verlag, Heidelberg (1972). Kac–Moody algebras and their representation theory are described by V.G. Kac *Infinite Dimensional Lie Algebras*, 2nd edition, Cambridge University Press (1985), which contains numerous references to the mathematical literature. Finally it is worth recalling that a field theory in terms of currents was proposed by H. Sugawara, *Phys. Rev.* **170**, 1659 (1968) and developed by many authors, as part of current algebra, with results which anticipated many of the present ones, as testified in R. Dashen and Y. Frishman, *Phys. Rev.* **D11**, 2781 (1975).

Among important developments not covered here is the study of current correlations and their remarkable monodromy properties, as well as the relation between fermionic and bosonic constructions. The whole subject of two-dimensional conformal field theory has important applications to string field theory, an aspect briefly described in chapter 11.

10
DISORDERED SYSTEMS AND FERMIONIC METHODS

Real systems may involve various types of defects. It is therefore important to assess their relevance and to estimate their effect on the results established in pure cases. The emphasis here is on stability and robustness versus random perturbations. We present in the last section of this chapter the Harris criterion for estimating the effect of weak disorder on a critical system. At a more fundamental level, one can also look for new phenomena originating in the very existence of defects. For instance, the dynamics of dislocations in a crystal can explain the solid–liquid transition. Similarly, we were led to analyze the role of vortices when studying the XY-model. Random impurity potentials produce the localization of quantum wavefunctions (Anderson, 1958), which enables one to understand the transition between insulators and conductors. This same localization phenomenon appears in other contexts involving classical waves, such as optics or sound propagation. This subject has given rise to an intense activity. One cannot claim that it is fully understood at present, eventhough the weak disordered case seems under control, thanks to renormalization group arguments. The role of magnetic fields has opened a new area of research centered on the quantum Hall effect. Another domain which has stimulated vigorous and original developments is the one of magnetic systems with random and/or frustrated interactions, leading to the search for a spin glass phase, where magnetic moments can become frozen in random directions, with a plethora of metastable states, or valleys, in free energy. Applications of spin glass models range from neural systems, and optimization problems to the newly discovered high T_c superconductors. This list illustrates the diversity of the subject.

The archetype of systems with a Hamiltonian involving random parameters is the theory of random matrices elaborated by Wigner and Dyson in the context of nuclear spectra.

In this chapter, we offer simple illustrations of field theoretic methods, particularly the use of anticommuting variables. Our subjective choice of topics is mostly meant to invite the interested reader to pursue the study of this important branch of statistical physics.

10.1   One-dimensional models

10.1.1   Gaussian random potential

Consider a one-dimensional quantum Hamiltonian for a particle moving in a random potential with a Gaussian distribution. An ingenious method is known to obtain the spectrum (Halperin, 1965). The availability of an exact result enables one to test the heuristic replica trick, a method generally used in this context. We will also encounter in this formulation an elementary example of supersymmetry appearing when dealing with constrained systems, the Becchi-Rouet-Stora (BRS) supersymmetry, originally formulated in the quantization of gauge fields.

One is of course interested in various other properties, such as the structure of wavefunctions or the average of products of Green functions (to obtain transport coefficients). Unfortunately these are much more delicate questions, for which analytic answers are not known. We shall however be able to show that, quite generally, wavefunctions are localized in one dimension, when studying a discrete disordered system. This property holds true provided the random part of the Hamiltonian does not involve long-range correlations. This seems at first to confront us with a paradox. On the one hand, we shall obtain a continuous average spectrum, but localized wavefunctions indicate that they are normalizable, a property which under ordinary circumstances corresponds to a discrete spectrum. Here we have to face the unhappy fact that the classification of spectra involves not only a continuous and a discrete part, but also a singular one, the latter being infrequently encountered in unsophisticated applications. It is nonetheless most likely true that we encounter in disordered systems a typical situation where a singular part appears generically in the spectrum, and that its character is not accessible while computing an average value for the latter. It is also worthwhile to mention that, while we concentrate here on random Hamiltonians, similar properties would show up in the case where part of the operator, the potential say, would imply several incommensurate periods. This should be the case for instance for spectra pertaining to quasicrystals.

Let the Hamiltonian be

$$H = -\frac{\hbar^2}{2m}\frac{\partial^2}{\partial x^2} + V(x) \tag{1}$$

where the potential V is random, with zero mean, and no correlation between distinct points. This is an extreme case, but, as was said above, amenable to an exact treatment. We have to distinguish between quantum averages, which we denote here by round brackets, and averages over disorder, which we denote with usual brackets. The potential is described by a Gaussian process with a correlation written as

$$\langle V(x)V(y)\rangle = D\delta(x - y) \tag{2}$$

More precisely, all averages can be derived from Wick's theorem in the form

$$\left\langle \exp i \int dx g(x)V(x) \right\rangle = \exp -\tfrac{1}{2}D \int dx g^2(x) \tag{3}$$

assuming $g(x)$ square integrable.

We want to compute the average of the Green function at coinciding points. For complex z, introduce the quantity

$$G(z) = \left\langle \left(x \left| \frac{1}{z - H} \right| x \right) \right\rangle \tag{4}$$

It is x-independent due to invariance under translations (and reflection) after averaging over V. The imaginary part of $G(z)$, as z tends to the real axis, yields the average density of energy levels per unit length, as the total number of levels is proportional to the total length. To be rigorous, the system ought to be enclosed in a finite box of size L, and boundary conditions imposed on wavefunctions, for instance periodicity. Then $\rho(E)$, the average spectral density (per unit length), in the limit of an infinite box, is given by

$$\rho(E) = \text{Im}\frac{1}{\pi}G(E - \text{io}) = \langle (x \,|\delta(E - H)|\, x)\rangle \tag{5}$$

while the number of levels per unit length, with energies smaller that E, is

$$\mathcal{N}(E) = \int_{-\infty}^{E} dE' \rho(E') \tag{6}$$

Since the potential is unbounded, we expect a spectrum extending from minus to plus infinity. It is convenient to introduce dimensionless units for lengths and energies, according to

$$\lambda = \left(\frac{\hbar}{Dm^{\frac{1}{2}}}\right)^{\frac{1}{3}} \qquad \rho(E) = \left(\frac{m}{\hbar^2 D}\right)^{\frac{1}{3}} \bar{\rho}(\bar{E})$$

$$V = \lambda^{-2}\bar{V} \qquad H = \lambda^{-2}\bar{H} = \lambda^{-2}\left(-\tfrac{1}{2}\bar{\Delta} + \bar{V}\right) \tag{7}$$

$$x = \frac{\hbar}{m^{\frac{1}{2}}}\lambda\bar{x} \qquad \langle \bar{V}(\bar{x})\bar{V}(\bar{y})\rangle = \delta\left(\bar{x} - \bar{y}\right)$$

$$E = \lambda^{-2}\bar{E}$$

Henceforth we omit the bars, and note that in reduced units, D and $m/\hbar^2$ are effectively equal to unity.

At very high energy, the potential becomes negligible, so that the total number of states tends asymptotically to its classical limit (one state per cell $\Delta x \Delta p / 2\pi$)

$$\mathcal{N}(E) \underset{E \to \infty}{\sim} \int_{p^2 < 2E} \frac{dp}{2\pi} = \frac{(2E)^{\frac{1}{2}}}{\pi}$$

$$\rho(E) \underset{E \to \infty}{\sim} \frac{1}{\pi(2E)^{\frac{1}{2}}} \tag{8}$$

The behaviour for $E \to -\infty$ is less intuitive, and will result from the exact expression.

10.1.2   Fokker–Planck equation

With $V(x)$ chosen according to the Gaussian probability law, let $\psi(x)$ be the wavefunction corresponding to energy E, a solution of Schrödinger's equation. Of course $\psi(x)$ becomes a random functional of V. The logarithmic derivative f of ψ satisfies a first order Ricatti equation

$$f(x) = \psi'(x)/\psi(x)$$
$$f' + f^2 = 2(V - E) \tag{9}$$

An analogy with Brownian motion arises if we interpret the spatial coordinate x as playing the role of time.

We want to exploit the fact that the values of the potential at distinct points x are independent variables. Let $P(\xi, x)d\xi$ be the probability that $f(x)$ lies in the interval $\xi, \xi + d\xi$. Thus

$$P(\xi, x) = \langle \delta(f(x) - \xi) \rangle = \frac{1}{2\pi} \int d\alpha e^{-i\alpha\xi} \left\langle e^{i\alpha f(x)} \right\rangle \qquad (10)$$

Taking the x derivative of the last expectation value and using equation (9), we find

$$\frac{\partial}{\partial x} \left\langle e^{i\alpha f(x)} \right\rangle = i\alpha \left\langle f'(x)e^{i\alpha f(x)} \right\rangle$$

$$= -i\alpha \left\langle (f^2(x) + 2E - 2V(x))e^{i\alpha f(x)} \right\rangle$$

The characteristic property of the Gaussian law (3), with D now set equal to one, is

$$\langle V(x)\phi \rangle = \int dy \delta(x - y) \left\langle \frac{\delta\phi}{\delta V(y)} \right\rangle = \left\langle \frac{\delta\phi}{\delta V(x)} \right\rangle$$

Applying this in the above formula yields

$$\frac{\partial}{\partial x} \left\langle e^{i\alpha f(x)} \right\rangle = -i\alpha \left\langle \left(f^2(x) + 2E - 2i\alpha \frac{\delta f(x)}{\delta V(x)} \right) e^{i\alpha f(x)} \right\rangle \qquad (11)$$

Let us now pretend that we integrate the Ricatti equation forward in x, starting from the value of f at the point x_0, which we let recede to $-\infty$. Under such circumstances, $f(x)$ only depends on the values of V at points $y < x$. According to equation (9), the retarded function

$$\Gamma(x, y) = \frac{\delta f(x)}{\delta V(y)} \qquad\qquad y \le x \qquad (12)$$

vanishing for $y > x$, satisfies the equation

$$\left\{ \frac{\partial}{\partial x} + 2f(x) \right\} \Gamma(x, y) = 2\delta(x - y) \qquad (13)$$

with a solution given by

$$\Gamma(x, y) = 2\theta(x - y) \exp - \int_y^x dx' 2f(x') = 2\theta(x - y)\frac{\psi^2(y)}{\psi^2(x)} \qquad (14)$$

When $y \to x$, $\Gamma(x, y)$ seems to have the undefined limit

$$\Gamma(x, x) = 2\theta(0) \qquad (15)$$

which occurs frequently when dealing not too carefully with such problems. To handle this apparent difficulty, while maintaining the symmetry $x \to -x$, one can either discretize properly the model, or else replace in the $V - V$ correlation the singular δ-function by an approximation $u_\varepsilon(x - y)$ with a function u_ε symmetric, positive, strongly peaked at the origin, of unit integral and tending towards δ as ε tends to zero. This amounts to replacing the factor $\theta(x - y)$ in equation (14) by the integral of $u_\varepsilon(x')$ from $-\infty$ to $x - y$, and $\theta(0)$ by the same integral up to zero. Consequently, $\theta(0)$ is replaced by $\frac{1}{2}$, the average between $\theta(-0)$ and $\theta(+0)$, and therefore $\delta f(x)/\delta V(x) = \Gamma(x, x) = 1$. With this understood, we return to $P(\xi, x)$ and obtain, by integrating (11) with respect to α, the Fokker–Planck equation

$$\frac{\partial}{\partial x} P(\xi, x) = \frac{\partial}{\partial \xi} \left(\xi^2 + 2E + 2\frac{\partial}{\partial \xi} \right) P(\xi, x) \tag{16}$$

with ξ playing the role of configuration variable, and x the role of time. The structure of this equation entails the conservation of the total probability, $\int d\xi P(\xi, x)$, under mild assumptions on the behaviour of P for large ξ.

For large x, P tends to a stationary process, with

$$\lim_{x \to \infty} P(\xi, x) = p(\xi)$$

satisfying

$$\frac{\partial}{\partial \xi} \left(\xi^2 + 2E + 2\frac{\partial}{\partial \xi} \right) p(\xi) = 0 \tag{17}$$

Thus $p(\xi)$ describes an x-independent distribution of $\xi = \psi'/\psi$. Integrating equation (17) with respect to ξ, we have

$$\left(\xi^2 + 2E + 2\frac{\partial}{\partial \xi} \right) p(\xi) = p_0 \tag{18}$$

The constant p_0 is determined by the normalization condition

$$\int d\xi p(\xi) = 1 \tag{19}$$

which requires

$$\lim_{\xi \to \infty} \xi^2 p(\xi) = p_0 \tag{20}$$

We wish to obtain the relation between $p(\xi)$ and the spectral density. To do so, reinstate the box with x ranging from 0 to L, and, given a value ξ_0 for $f(0)$, require that $f(L) = \xi_L$. This leads to a discrete spectrum E_1, E_2, ... We let L tend to infinity and find

$$\rho(E) = \lim_{L \to \infty} \frac{1}{L} \sum_n \delta(E - E_n) = \lim_{L \to \infty} \frac{1}{L} \delta(f(L) - \xi_L) \left| \frac{\partial f}{\partial E}(L) \right| \tag{21}$$

We expect the limit on the r.h.s. to be finite. From equation (9)

$$\frac{\partial}{\partial x} \frac{\partial f}{\partial E} + 2f \frac{\partial f}{\partial E} = -2 \qquad \frac{\partial f}{\partial E}(x = 0) = 0$$

If we substitute for $\partial f / \partial E$ the function $g(x)$ defined through

$$\frac{\partial f}{\partial E} = g(x) \exp \left\{ -2 \int_0^x dy\, f(y) \right\} = g(x) \frac{\psi^2(0)}{\psi^2(x)}$$

the equation becomes

$$\frac{\psi^2(0)}{\psi^2(x)} g'(x) = -2$$

Thus

$$g(x) = -2 \int_0^x dy \frac{\psi^2(y)}{\psi^2(0)}$$

and

$$\frac{\partial f}{\partial E} = -2 \int_0^x dy \frac{\psi^2(y)}{\psi^2(x)} \tag{22}$$

We note that $\partial f / \partial E$ is of constant negative sign. Consequently, the quantity

$$\mathcal{P}(\xi, x) = \left\langle \left| \frac{\partial f}{\partial E} \right| \delta(f(x) - \xi) \right\rangle = - \left\langle \frac{\partial f}{\partial E} \delta(f(x) - \xi) \right\rangle \tag{23}$$

is related to the spectral density through

$$\rho(E) = \lim_{L \to \infty} \frac{1}{L} \mathcal{P}(\xi_L, L) \tag{24}$$

Proceeding as we did before for $P(\xi, x)$, one establishes that

$$\frac{\partial}{\partial x} \mathcal{P}(\xi, x) = \left\langle \left(2 + 2\xi \frac{\partial f}{\partial E} - \frac{\partial f}{\partial E} \frac{\partial}{\partial x} \right) \delta(f(x) - \xi) \right\rangle$$

$$= 2\mathcal{P}(\xi, x) + \left[(\xi^2 + 2E) \frac{\partial}{\partial \xi} + 2 \frac{\partial^2}{\partial \xi^2} \right] \mathcal{P}(\xi, x)$$

According to equation (24), $\mathcal{P}(\xi, x)$ behaves as $x\rho(E)$ for large positive x, hence the left hand side of this equation tends to $\rho(E)$

as $x \to +\infty$, independently of ξ. Let us multiply both sides by $p(-\xi)$ and integrate over ξ. Taking into account the normalization of $p(\xi)$, we get

$$\rho(E) = 2 \int d\xi \, p(\xi) p(-\xi)$$

$$+ \lim_{x \to +\infty} \int d\xi \, p(-\xi) \left[(\xi^2 + 2E) \frac{\partial}{\partial \xi} + 2 \frac{\partial^2}{\partial \xi^2} \right] \mathcal{P}(\xi, x)$$

Integrating the second term by parts, we find from (17) that it vanishes, and we obtain the required relation between the spectral density and the probability $p(\xi)$ in the form

$$\rho(E) = 2 \int d\xi p(\xi) p(-\xi) \qquad (25)$$

A more explicit formula follows from equation (18). Taking in the latter a derivative with respect to E, one has

$$\frac{\partial p_0}{\partial E} = 2 \frac{\partial}{\partial \xi} \frac{\partial p}{\partial E} + (\xi^2 + 2E) \frac{\partial p}{\partial E} + 2p$$

Premultiply by $p(\xi)$, integrate with respect to ξ, use equation (18) again and the normalization condition to find

$$\frac{\partial p_0}{\partial E} = \rho(E) + \int d\xi p(-\xi) \left\{ 2 \frac{\partial}{\partial \xi} + (\xi^2 + 2E) \right\} \frac{\partial p}{\partial E}$$

$$= \rho(E) + p_0 \int d\xi \frac{\partial p}{\partial E} = \rho(E)$$

Hence

$$p_0 = \mathcal{N}(E) \qquad (26)$$

Inserting this relation in equation (18), we see finally that $\mathcal{N}(E)$ is determined by the conditions

$$\mathcal{N}(E) = \left[2 \frac{\partial}{\partial \xi} + (\xi^2 + 2E) \right] p(\xi)$$

$$\int d\xi p(\xi) = 1 \qquad \lim_{\xi \to \infty} \xi^2 p(\xi) = \mathcal{N}(E) \qquad (27)$$

Equation (27) is nothing but the Fourier transform of Airy's equation. Taking into account the asymptotic condition, its solution reads

$$p(\xi) = \frac{\mathcal{N}(E)}{2} e^{-\frac{1}{6}\xi^3 - E\xi} \int_{-\infty}^{\xi} d\eta e^{\frac{1}{6}\eta^3 + E\eta} \qquad (28)$$

Normalization requires

$$\frac{1}{\mathcal{N}(E)} = \frac{1}{2}\int_{-\infty}^{+\infty} d\xi \int_{-\infty}^{\xi} d\eta\, e^{-\frac{1}{6}(\xi^3-\eta^3)-E(\xi-\eta)}$$

If we choose $\xi - \eta$ as an integration variable, we obtain the final expression as a single integral

$$\frac{1}{\mathcal{N}(E)} = \sqrt{\tfrac{1}{2}\pi}\int_{0}^{\infty} \frac{du}{\sqrt{u}}\exp-(\tfrac{1}{24}u^3 + Eu) \qquad (29)$$

When $E \to \infty$, u^3 can be neglected as compared to Eu, and we recover the semiclassical asymptotic form for $\mathcal{N}(E)$, as expected in (8). For $E \to -\infty$, the integral can be estimated using the saddle point method, with the result

$$\mathcal{N}(E) \underset{E\to-\infty}{\sim} \frac{\sqrt{-8E}}{2\pi}\exp-\tfrac{1}{12}(-8E)^{\frac{3}{2}} \qquad (30)$$

It is not too surprising to find that $\mathcal{N}(E)$ vanishes rapidly for $E \to -\infty$. The precise behaviour is somehow less expected. It is sometimes called a Lifshitz tail. The shape of $\mathcal{N}(E)$ is depicted on figure 1. Observe that $\mathcal{N}(E)^{-1}$ is an entire function with a series expansion

$$\mathcal{N}(E)^{-1} = \frac{\pi^{\frac{1}{2}}}{3^{\frac{5}{6}}}\sum_{0}^{\infty}\frac{(-2\cdot 3^{\frac{1}{3}}E)^n}{n!}\Gamma\left(\frac{2n+1}{6}\right) \qquad (31)$$

It is instructive to obtain the same Fokker–Planck equation (16), using a slightly different reasoning, valid for more general dynamical systems. We pursue the analogy between the coordinate x and a temporal variable. Let us start from equation (9), which we use in the form

$$V(x) = E + \tfrac{1}{2}\left(\frac{\partial f}{\partial x} + f^2\right)$$

To obtain the probability $P(\xi, x; \xi_0, x_0)$, we insert the above constraint in the distribution law for V, hence deriving the functional representation

$$P(\xi, x; \xi_0, x_0) = \int \mathcal{D}V\int \mathcal{D}(\xi)\prod_{x'}\delta\left(V - E - \tfrac{1}{2}\left[\frac{\partial \xi}{\partial x'} + \xi^2\right]\right)$$

$$\left|\frac{\mathcal{D}V}{\mathcal{D}\xi}\right|\exp-\tfrac{1}{2}\int_{x_0}^{x} dy\, V^2(y) \qquad (32)$$

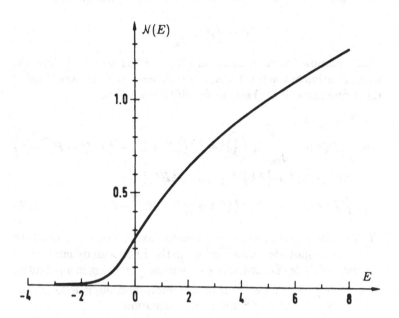

Fig. 1   The integrated density of states per unit length for a Gaussian random potential.

The integration over ξ satisfies the boundary conditions $\xi(x_0) = \xi_0$, $\xi(x) = \xi$. The Jacobian $J = |\mathcal{D}V/\mathcal{D}\xi|$ can be evaluated up to a multiplicative constant, assuming the mapping $\xi \leftrightarrow V$ to be one-to-one

$$J = J_0 \det \left(\frac{\mathrm{d}/\mathrm{d}x + 2\xi}{\mathrm{d}/\mathrm{d}x} \right) = J_0 \det \left(I + 2\frac{1}{\mathrm{d}/\mathrm{d}x}\xi \right)$$

where $(\mathrm{d}/\mathrm{d}x)^{-1}$ is understood as the free retarded Green function $\Gamma_0(x,y)$ satisfying $\partial/\partial x \Gamma_0(x,y) = \delta(x-y)$, i.e. $\Gamma_0(x,y) = \theta(x-y)$. Thus

$$J = J_0 \exp \mathrm{Tr}\ln \left(1 + \frac{1}{\mathrm{d}/\mathrm{d}x}2\xi \right) = J_0 \exp 2 \int_{x_0}^{x} \mathrm{d}y\Gamma_0(y,y)\xi(y)$$

Only the first term in the series expansion of the logarithm contributes in virtue of the retardation property of Γ_0. We have already settled the question of the ill-defined $\Gamma_0(y,y) = \theta(0)$ by arguing that it ought to be understood as $\frac{1}{2}$. Consequently

$$J = J_0 \exp \int_{x_0}^{x} \mathrm{d}y \xi \tag{33}$$

which we now insert in equation (32). Let us absorb J_0 into the normalization of the functional integral, and integrate over V using the δ-functions. If $\dot{\xi}(y)$ stands for $\mathrm{d}\xi/\mathrm{d}y$, we obtain

$$P(\xi, x; \xi_0, x_0)$$

$$= \int \mathcal{D}\xi \exp - \int_{x_0}^{x} \mathrm{d}y \left\{ \tfrac{1}{8}\dot{\xi}^2 + \tfrac{1}{2}\dot{\xi}\left(\tfrac{1}{2}\xi^2 + E\right) + \tfrac{1}{2}\left(\tfrac{1}{2}\xi^2 + E\right)^2 - \xi \right\}$$

$$= \exp - \left\{ \tfrac{1}{12}\xi^3 + \tfrac{1}{2}E\xi \right\} + \left\{ \tfrac{1}{12}\xi_0^3 + \tfrac{1}{2}E\xi_0 \right\}$$

$$\int \mathcal{D}\xi \exp - \int_{x_0}^{x} \mathrm{d}y \left\{ \tfrac{1}{8}\dot{\xi}^2 + \tfrac{1}{2}\left(\tfrac{1}{2}\xi^2 + E\right)^2 - \xi \right\} \tag{34}$$

The remaining path integral is a familiar expression for a quantum transition amplitude corresponding to the Euclidean dynamics of a fictitious particle of coordinate ξ with mass $\frac{1}{4}$ evolving in a potential $\frac{1}{2}\left(\frac{1}{2}\xi^2 + E\right)^2 - \xi$. As a result of this interpretation, it follows that P satisfies an analog of Schrödinger's equation

$$-\frac{\partial}{\partial x}P\left(\xi, x; \xi_0, x_0\right) = \mathrm{e}^{-\left\{\frac{1}{12}\xi^3 + \frac{1}{2}E\xi\right\}}$$

$$\left[-2\frac{\partial^2}{\partial \xi^2} + \tfrac{1}{2}\left(\tfrac{1}{2}\xi^2 + E\right)^2 - \xi \right] \mathrm{e}^{\left\{\frac{1}{12}\xi^3 + \frac{1}{2}E\xi\right\}} P\left(\xi, x; \xi_0, x_0\right) \tag{35}$$

Since

$$\mathrm{e}^{-\left(\frac{1}{12}\xi^3 + \frac{1}{2}E\xi\right)} \frac{\partial}{\partial \xi} \mathrm{e}^{\left(\frac{1}{12}\xi^3 + \frac{1}{2}E\xi\right)} = \frac{\partial}{\partial \xi} + \tfrac{1}{4}\xi^2 + \tfrac{1}{2}E$$

we can rewrite equation (35) as

$$\frac{\partial}{\partial \xi}\left[\xi^2 + 2E + 2\frac{\partial}{\partial \xi}\right] P\left(\xi, x; \xi_0, x_0\right) = \frac{\partial}{\partial x}P\left(\xi, x; \xi_0, x_0\right) \tag{36}$$

which is identical to the Fokker–Planck equation (16). We have also obtained in this way a path integral representation for the solution, as well as a natural connection between Euclidean quantum dynamics and the probabilistic theory of Brownian motion.

We can take advantage of this example to make a point concerning the treatment of the Jacobian. In the present case, we were fortunate in being able to compute it in closed form, due to its simple expression. However, this is the exception rather than the rule, and the variant treated below indicates a possibility of taking the calculations further, even without an analytic form for J. We want to replace J by a path integral over the exponential of an action. Let us assume again that the change of variable $V \leftrightarrow \xi$ is well-defined. This is the crucial assumption and (unfortunately) is not always true in more complex situations. If the hypothesis is fulfilled, the Jacobian J never vanishes, and the absolute value sign can be ignored, meaning that it can be taken as a determinant. It admits therefore a fermionic path integral representation over anticommuting variables ψ and $\bar{\psi}$

$$\det \frac{\delta V}{\delta \xi} = \int \mathcal{D}(\psi \bar{\psi}) \exp - \int_{x_0}^{x} dy_1 \int_{x_0}^{x} dy_2 \bar{\psi}(y_1) \frac{\delta V(y_1)}{\delta \xi(y_2)} \psi(y_2) \quad (37)$$

The added fermionic part in the action reads

$$\int_{x_0}^{x} dy_1 \int_{x_0}^{x} dy_2 \bar{\psi}(y_1) \frac{\delta V(y_1)}{\delta \xi(y_2)} \psi(y_2) = \int_{x_0}^{x} dy \bar{\psi}(y) \left(\frac{1}{2} \frac{\partial}{\partial y} + \xi \right) \psi(y)$$
$$(38)$$

Hence we obtain an effective Lagrangian

$$\mathcal{L}_{\text{eff}} = \frac{1}{2} \left(\frac{1}{2}(\dot{\xi} + \xi^2) + E \right)^2 + \bar{\psi} \left(\frac{1}{2} \dot{\psi} + \xi \psi \right) \quad (39a)$$

For a general variation of ξ, ψ and $\bar{\psi}$

$$\delta \mathcal{L}_{\text{eff}} = \left(\frac{1}{2}(\dot{\xi} + \xi^2) + E \right) \left(\frac{1}{2} \frac{d}{dx} + \xi \right) \delta \xi +$$
$$+ \delta \bar{\psi} \left(\frac{1}{2} \frac{d}{dx} + \xi \right) \psi + \bar{\psi} \left(\frac{1}{2} \frac{d}{dx} + \xi \right) \delta \psi + \bar{\psi} \psi \delta \xi \quad (40a)$$

This vanishes identically when

$$\delta \xi = \psi \delta \bar{a} \qquad \delta \psi = 0 \qquad \delta \bar{\psi} = \left(\frac{1}{2}(\dot{\xi} + \xi^2 + E \right) \delta \bar{a} \quad (41a)$$

where $\delta \bar{a}$ is an anticommuting variable, as if $\mathcal{L}_{\text{eff}}$ were "supersymmetric". This is an example of a BRS supersymmetry. It occurs whenever a constraint is introduced in a path integral, with an accompanying Jacobian, and the latter is represented using a Gaussian integral over anticommuting variables. Here the constraint is the dynamical equation (9). While the BRS symmetry originated in the context of the quantization of gauge fields to control the gauge fixing procedure, it is really of a general nature, independent of the geometric origin of the problem.

The structure of the variations (41) enable one to define a derivative operator s, such that for any functional of the fields

$$\delta \mathcal{A} = \int dx \frac{\delta \mathcal{A}}{\delta \bar{a}} \delta \bar{a} \equiv \int dx s \mathcal{A} \delta \bar{a} \quad (42a)$$

$$s\xi = \psi \qquad s\psi = 0 \qquad s\bar{\psi} = \left(\frac{1}{2}(\dot{\xi} + \xi^2) + E \right)$$

If one iterates s, which interchanges bosonic and fermionic fields, one has

$$s^2 \xi = 0 \qquad s^2 \psi = 0 \qquad s^2 \bar{\psi} = \left(\tfrac{1}{2} \frac{d}{dx} + \xi \right) \psi \qquad (43)$$

Thus s is nilpotent (of square equal to zero) provided ψ satisfies the equation of motion $\left(\tfrac{1}{2} d/dx + \xi \right) \psi = 0$. This is improperly called an on-shell condition, with reference to S-matrix concepts. In certain applications, it can be advantageous to define a genuine nilpotent operator, for instance when dealing with renormalization, where one studies counterterms satisfying $s\mathcal{A} = 0$. One knows *a priori* that an $\mathcal{A}$ of the form $s\mathcal{B}$ will solve this equation, and, under favorable circumstances, this is the most general solution. To ensure that $s^2 = 0$, the trick is to use a Fourier representation of the δ-functions which enforces the constraint. This amounts to adding a term in the Lagrangian with a Lagrange multiplier (field) α, initially pure imaginary, and to avoiding performing the integral over V. Thus the new effective Lagrangian reads

$$\mathcal{L}_{\text{eff}} = \tfrac{1}{2} V^2 + \alpha \left(\tfrac{1}{2} (\dot{\xi} + \xi^2) + E - V \right) + \bar{\psi} \left(\tfrac{1}{2} \frac{d}{dx} + \xi \right) \psi \qquad (39b)$$

and its variation

$$\begin{aligned}
\delta \mathcal{L}_{\text{eff}} = {} & (V - \alpha)\delta V + \delta\alpha \left(\tfrac{1}{2}(\dot{\xi} + \xi^2) + E - V \right) \\
& + \alpha \left(\tfrac{1}{2} \frac{d}{dx} \xi \right) \delta\xi + \delta\bar{\psi} \left(\tfrac{1}{2} \frac{d}{dx} + \xi \right) \psi \\
& + \bar{\psi} \left(\tfrac{1}{2} \frac{d}{dx} + \xi \right) \delta\psi + \delta\xi \bar{\psi}\psi \qquad (40b)
\end{aligned}$$

vanishes identically when

$$\delta\bar{\psi} = \alpha\delta\bar{a} \qquad \delta\xi = \psi\delta\bar{a} \qquad \delta\psi = 0 \qquad \delta\alpha = 0 \qquad \delta V = 0 \quad (41b)$$

The operator s becomes

$$s\bar{\psi} = \alpha \qquad s\xi = \psi \qquad s\psi = 0 \qquad s\alpha = 0 \qquad sV = 0 \qquad (42b)$$

and $s^2 = 0$. In this second form the definition of s is independent of the constraints. These are in fact coded in the effective Lagrangian. This hidden supersymmetry of the Fokker–Planck equation is only meaningful in the framework of a path integral representation.

10.1.3   *The replica trick*

The replica trick is a useful device when dealing with averages such as the one of the resolvant (4). Let us replace the complex argument z by $E - \mathrm{io}$ with E real and use the shorthand notation E, with an infinitesimal imaginary part implicitly understood. Also write $G(E)$ for $\left\langle \left(x \left| [H - E]^{-1} \right| x \right) \right\rangle$. If, instead of operators in an infinite dimensional space, we were dealing with finite matrices, the quantity to be averaged would be the ratio of a minor by the determinant of the matrix. It therefore involves the random potential both in the numerator and denominator and leads *a priori* to untractable calculations. One looks for a mean to avoid this difficulty.

First, we interpret $(H - E)^{-1}$ as the propagator of an auxiliary scalar field. Using translational invariance of the average, we write

$$
G(E) = \lim_{L \to \infty} \left\langle \frac{1}{L} \int_{-\frac{1}{2}L}^{\frac{1}{2}L} \mathrm{d}x \left(x \left| \frac{1}{H - E} \right| x \right) \right\rangle
$$

$$
= \lim_{L \to \infty} \frac{2}{\mathrm{i}L} \int_{-\frac{1}{2}L}^{\frac{1}{2}L} \mathrm{d}x
$$

$$
\times \left\langle \frac{\int \mathcal{D}\varphi\, \varphi^2(x) \exp \mathrm{i} \int\int_{-\frac{1}{2}L}^{\frac{1}{2}L} \mathrm{d}x\, \mathrm{d}y\, \varphi(x)\, (x\,|(H - E)|\,y)\, \varphi(y)}{\int \mathcal{D}\varphi \exp \mathrm{i} \int\int_{-\frac{1}{2}L}^{\frac{1}{2}L} \mathrm{d}x\, \mathrm{d}y\, \varphi(x)\, (x\,|(H - E)|\,y)\, \varphi(y)} \right\rangle
$$

$$
= \lim_{L \to \infty} \frac{2}{L} \frac{\partial}{\partial E} \left\langle \ln \left\{ \int \mathcal{D}\varphi \right. \right.
$$

$$
\left. \left. \exp \mathrm{i} \int\int_{-\frac{1}{2}L}^{\frac{1}{2}L} \mathrm{d}x\, \mathrm{d}y\, \varphi(x)\, (x\,|(H - E)|\,y)\, \varphi(y) \right\} \right\rangle \tag{44}
$$

The i in the exponential is chosen in agreement with the $E - \mathrm{io}$ prescription to ensure convergence of the path integral. It would seem that we have made little progress, since we now have to take the average of a logarithm. However, one notices that, with n a continuous variable,

$$
\frac{\partial}{\partial n} Z^n \xrightarrow[n \to 0]{} \ln Z \tag{45}
$$

For any integer n, Z^n is the partition function of n identical noninteracting replicas. They involve the same random potential linearly in the action, and this allows an explicit average over a Gaussian distribution. Provided an analytic continuation in n is meaningful, it will be possible to obtain the desired quantity (44). The method relies on the existence of this analytic continuation. It is therefore interesting to pursue the exercise in the present case under this assumption, to see whether it leads to an agreement with the previous result. The use of the replica trick in disordered problems has much in common with the technique used for polymer physics.

Let Φ denote an n-component field (of course n is thought at first as an integer). We have

$$G(E) = \lim_{L \to \infty} \frac{2}{L} \frac{\partial}{\partial E} \lim_{n \to 0} \frac{\partial}{\partial n}$$

$$\left\langle \int \mathcal{D}\Phi \exp i \int \int dx \, dy \Phi(x) \left(x \left| [H - E] \right| y \right) \Phi(y) \right\rangle \quad (46)$$

In effect the action reads

$$\int \int dx \, dy \; \Phi(x) \left(x \left| [H - E] \right| y \right) \Phi(y) = \int dx \mathcal{L}(\Phi) \quad (47)$$

with a Lagrangian

$$\mathcal{L}(\Phi) = \tfrac{1}{2}(\partial\Phi)^2 + (V - E)\Phi^2 \quad (48)$$

The average over V is readily obtained for any integer n since it simply requires a Gaussian integral to be performed. We assume that the result continues to make sense for arbitrary n. It reads

$$G(E) = \lim_{L \to \infty} \frac{2}{L} \frac{\partial}{\partial E} \lim_{n \to 0} \frac{\partial}{\partial n} \int \mathcal{D}\Phi$$

$$\exp i \int_{-\frac{1}{2}L}^{\frac{1}{2}L} dx \left[\tfrac{1}{2}(\partial\Phi)^2 - E\Phi^2 + \frac{i}{2}(\Phi^2)^2 \right] \quad (49)$$

In dimension higher than one, a similar expression would lead to a variant of φ^4 field theory. Luckily, in one dimension, we can turn this into a Schrödinger problem as follows. As $L \to \infty$ we expect the path integral to be dominated by a term of the form

$$\exp -\tfrac{1}{2}iLe(n) \quad (50)$$

Here $e(n)$ is the "ground state energy" of an associated Schrödinger equation, where the configuration variables Φ are the coordinates in a n-dimensional space, and once more x is interpreted as a temporal variable. Thus

$$e(n)\psi(\Phi) = \left\{ -\Delta_\Phi + 2E\Phi^2 - i(\Phi^2)^2 \right\} \psi(\Phi) \quad (51)$$

Since the potential term is complex, the notion of a ground state energy requires some qualification. If $e(n)$ admits an analytic continuation, we expect that

$$e(n) = ne_0 + n^2 e_1 + \cdots \tag{52}$$

$$G(E) = -\mathrm{i}\frac{\partial}{\partial E}e_0(E) \qquad \pi\rho(E) = \frac{\partial}{\partial E}\mathrm{Re}\,e_0(E) \tag{53}$$

Assuming that $e_0(E)$ vanishes for $E \to -\infty$, we obtain for the integrated density of states per unit length

$$\mathcal{N}(E) = \frac{1}{\pi}\mathrm{Re}\,e_0(E) \tag{54}$$

We require that the ground state be $O(n)$ invariant. The appearance of a symmetry among identical replicas is quite natural, and no symmetry breaking is expected in this one-dimensional context. Let us use instead of the vector argument Φ a scalar variable

$$q = \tfrac{1}{2}\Phi^2 \tag{55}$$

and rewrite equation (51), expanding every quantity in powers of n, thus exhibiting the continuation in this variable. In particular we write $\psi = \psi_0 + n\psi_1 + \cdots$, and

$$\left\{n\frac{\mathrm{d}}{\mathrm{d}q} + ne_0 + n^2 e_1 + \cdots\right\}\{\psi_0 + n\psi_1 + \cdots\}$$

$$= 2q\left\{-\frac{\mathrm{d}^2}{\mathrm{d}q^2} + 2E - 2\mathrm{i}q\right\}\{\psi_0 + n\psi_1 + \cdots\} \tag{56}$$

Hence

$$\left\{-\frac{\mathrm{d}^2}{\mathrm{d}q^2} + 2E - 2\mathrm{i}q\right\}\psi_0 = 0$$

$$\left\{e_0 + \frac{\mathrm{d}}{\mathrm{d}q}\right\}\psi_0 = 2q\left\{-\frac{\mathrm{d}^2}{\mathrm{d}q^2} + 2E - 2\mathrm{i}q\right\}\psi_1 \tag{57}$$

$$\cdots$$

We look for a solution $\psi_0(q)$ bounded at the origin, together with its derivative, and vanishing at infinity. It follows from the second equation that e_0 is minus the logarithmic derivative of ψ_0 at the origin

$$e_0 = -\left.\frac{d\psi_0/\,dq}{\psi_0}\right|_{q=0} \tag{58}$$

The first equation in (57) admits a continuation for $q < 0$, provided we define it as

$$\psi_0(-q) = \psi_0(q)^* \tag{59}$$

a condition compatible with the equation. We normalize the solution by requiring

$$\psi_0(0) = 1 \tag{60}$$

which insures continuity at the origin. Taking into account equations (58) and (54), the discontinuity of the derivative of ψ_0 at the origin produces an inhomogeneous additional term when extending the equation for ψ_0 in the range $-\infty < q < +\infty$

$$\left\{ -\frac{d^2}{dq^2} + 2E - 2iq \right\} \psi_0(q) = 2\pi \mathcal{N}(E)\delta(q) \tag{61}$$

We recognize that this is a Fourier transformed version of equation (27), if we set

$$p(\xi) = \int \frac{dq}{2\pi} e^{-iq\xi} \psi_0(q) \tag{62}$$

The condition $\psi_0(0) = 1$ is equivalent to the statement that the integral over $p(\xi)$ is normalized to unity. Equation (59) implies that $p(\xi)$ is real (and positive) and equation (61) translates into a real first order equation for $p(\xi)$

$$\left(\xi^2 + 2E + 2\frac{d}{d\xi} \right) p(\xi) = \mathcal{N}(E) \tag{63}$$

Thus, in the limit $n \to 0$, the variables $q = \frac{1}{2}\Phi^2$ and ξ are conjugate, and $\psi_0(q)$ and $p(\xi)$ are Fourier transforms of each other, a slightly surprising result for which we do not have a simple interpretation. We conclude that the use of the replica trick has been fully justified in the present example.

i) Compute $\text{Im} e_0(E)$ and show that

$$
\begin{aligned}
e_0(E) &= \pi \mathcal{N}(E) - \frac{1}{2}i \frac{\mathcal{N}'(E)}{\mathcal{N}(E)} \\
G(E) &= -i\frac{d}{dE} e_0(E) = \frac{1}{2}\left(\frac{\rho(E)^2}{\mathcal{N}(E)^2} - \frac{\rho'(E)}{\mathcal{N}(E)} \right) - i\pi\rho(E)
\end{aligned}
\tag{64}
$$

The solution $\psi_0(q)$ of equation (57) in the range $0 < q < \infty$, normalized to $\psi_0(0) = 1$, is a stationary point of the effective action

$$S(\eta) = \int_0^\infty dq \left\{ \left(\frac{d\eta}{dq} \right)^2 + 2(E - iq)\eta^2 \right\} \tag{65}$$

If we substitute ψ_0 and integrate the kinetic term by parts, we find

$$S(\psi_0) = -\psi_0(0) \frac{d\psi_0(0)}{dq}$$

$$+ \int_0^\infty dq \; \psi_0(q) \left[-\frac{d^2}{dq^2} + 2(E - iq) \right] \psi_0(q) = e_0(E) \tag{66}$$

The derivative $de_0(E)/dE$ is given by the partial derivative of $S(\eta)$ with respect to E evaluated for $\eta = \psi_0$

$$\frac{de_0(E)}{dE} = 2 \int_0^\infty dq \psi^2(q)$$

$$= 2 \int_0^\infty dq \int_{-\infty}^{+\infty} d\xi_1 \, d\xi_2 e^{iq(\xi_1 + \xi_2)} p(\xi_1) p(\xi_2)$$

$$= 2i \int_{-\infty}^{+\infty} d\xi_1 \, d\xi_2 \frac{p(\xi_1) p(\xi_2)}{\xi_1 + \xi_2 + i\varepsilon} \tag{67}$$

Taking the real part leads to

$$\rho(E) = \operatorname{Re} \frac{1}{\pi} \frac{de_0(E)}{dE} = 2 \int_{-\infty}^{+\infty} d\xi p(\xi) p(-\xi) \tag{68}$$

which reproduces formula (25). Let us now use the virial theorem in the form

$$0 = \int_0^\infty dq \frac{d\psi_0(q)}{dq} \left[-\frac{d^2}{dq^2} + 2(E - iq) \right] \psi_0(q)$$

$$= -E\psi_0^2(0) + \tfrac{1}{2} \left(\frac{d\psi_0}{dq}(0) \right)^2 - 2i \int_0^\infty dq \; q\psi_0 \frac{d\psi_0}{dq}$$

where the values at $q = 0$ are understood in the limit $q \to +0$. But $\psi_0(0)^2 = 1$, $(d\psi_0(0)/dq)^2 = e_0^2$, and an integration by part shows that the last integral is $-\tfrac{1}{2} de_0(E)/dE$ according to (67). Thus

$$i\frac{de_0}{dE} + e_0^2 = 2E \tag{69}$$

which is the required relation between the real part of $e_0(E)$, i.e. $\pi \mathcal{N}(E)$, and its imaginary part. Taking the imaginary part of both sides yields

$$\frac{d\mathcal{N}}{dE} + 2\mathcal{N} \operatorname{Im} e_0 = 0$$

which agrees with equation (64).

ii) Express $\mathcal{N}(E)$ with the help of the two independent solutions $Ai(x)$ and $Bi(x)$ of Airy's equation

$$\frac{\mathrm{d}^2 u}{\mathrm{d}x^2} = xu \tag{70}$$

and show that

$$\mathcal{N}(E) = \frac{2^{\frac{1}{3}}}{\pi^2 \left\{ Ai(-2^{\frac{1}{3}}E)^2 + Bi(-2^{\frac{1}{3}}E)^2 \right\}} \tag{71}$$

iii) Show that the second coefficient $e_1(E)$ in the expansion (52) for $e(n)$ is given by

$$e_1(E) = -\int_0^\infty \mathrm{d}q \psi_0(q) \frac{1}{2q} \left[e_0 + \frac{\mathrm{d}}{\mathrm{d}q} \right] \psi_0(q) \tag{72}$$

and that its value allows an estimate of the fluctuations of the spectrum around its mean value $\rho(E)$. More precisely in a box of size L, with L tending to infinity, let $\nu_L(E)\,\mathrm{d}E$ be the total number of energy levels in the range E, $E + \mathrm{d}E$. We have

$$\lim_{L\to\infty} \frac{1}{L} \langle \nu_L(E) \rangle = \rho(E)$$

$$\lim_{L\to\infty} \frac{1}{L} \{ \langle \nu_L(E_1)\nu_L(E_2) \rangle - \langle \nu_L(E_1) \rangle \langle \nu_L(E_2) \rangle \} = \sigma(E_1, E_2) \tag{73}$$

with

$$e_0(E) = \lim_{\varepsilon \to +0} \frac{1}{i} \int \mathrm{d}E' \rho(E') \ln(E' - E + i\varepsilon)$$

$$e_1(E) = \lim_{\varepsilon_i \to +0} \frac{1}{2}i \int \int \mathrm{d}E_1' \, \mathrm{d}E_2' \sigma(E_1', E_2')$$
$$\ln(E_1 - E + i\varepsilon_1) \ln(E_2 - E + i\varepsilon_2) \tag{74}$$

where we take the principal branch of the logarithm in the complex plane with a cut along the negative real axis. In particular for $E \to +\infty$

$$e_1 \sim \frac{-i}{4(2E)} + \frac{9}{16(2E)^{\frac{5}{2}}} + \cdots \tag{75}$$

10.1.4   *Random one-dimensional lattice*

As a second example, we investigate the Laplacian on a random one-dimensional lattice. Such lattices have been introduced in any dimension to restore continuous symmetries on average, while keeping an ultraviolet cutoff. We present a detailed discussion in chapter 11. For our purpose, the random lattice is only used as an elementary illustration of one-dimensional disordered discrete systems. The model is a variant of random harmonic chains studied by Dyson (1953).

Let us consider a set of N points drawn at random independently and with uniform probability in an interval L. In the limit where both N and L tend to infinity in a fixed ratio $\delta = 1/a = N/L$, δ being the density, the successive intervals ℓ between the ordered points become independent variables with a common Poisson distribution

$$p(\ell)\, \mathrm{d}\ell = \frac{\mathrm{d}\ell}{a} \mathrm{e}^{-\ell/a} \qquad (76)$$

Henceforth we use the microscopic scale a as a unit of length, such that effectively $\delta^{-1} = a = 1$. Our points generate a lattice and we take as a definition of the Laplacian of a scalar field φ defined on the ordered vertices

$$(\Delta\varphi)_n = \frac{\varphi_{n+1} - \varphi_n}{\ell_n} - \frac{\varphi_n - \varphi_{n-1}}{\ell_{n-1}} \qquad (77)$$

We study the spectrum of eigenvalues $\Omega \geq 0$ such that

$$(\Delta + \Omega)\varphi = 0 \qquad (78)$$

Let $\rho(\Omega)\, \mathrm{d}\Omega$ denote the spectral density per unit length, i.e. per site in the present units. It satisfies the normalization condition

$$\int_0^\infty \mathrm{d}\Omega \rho(\Omega) = 1 \qquad (79)$$

which states that, for N sites and given boundary conditions, we have N eigenvalues. This reflects the existence of a short-distance cutoff, and implies that the spectral density decreases fast enough for large Ω. In contradistinction to the case of a regular lattice, the spectrum is here unbounded. We shall see that the high Ω tail results from strongly localized states around (unprobable) regions with small separations between lattice points.

We can transform the problem into one with a random potential on a regular lattice. Define

$$Q_n = \frac{\varphi_{n+1} - \varphi_n}{\ell_n} \qquad (80)$$

with the obvious meaning that Q_n corresponds to the gradient of the field φ. These quantities satisfy the equation

$$Q_{n+1} - 2Q_n + Q_{n-1} = -\Omega \ell_n Q_n \qquad (81)$$

which is a Schrödinger equation on a regular lattice with potential term $-\Omega \ell_n$, at zero effective energy, where Ω plays the role of coupling constant. Set

$$\omega = -\Omega \qquad (82)$$

in such a way that for ω real and positive we are certainly off the spectrum and the solutions grow exponentially with n

$$Q_n \underset{n \to \infty}{\sim} \exp \gamma(\omega) n \qquad \omega > 0 \qquad (83)$$

A possible, but irrelevant, prefactor is omitted in (83).

Quantities such as $\gamma(\omega)$ are called Liapunov exponents, reference being made to the study of instabilities in classical dynamical systems (as previously, n may be thought of as an evolution variable). The knowledge of $\gamma(\omega)$ enables one to recover the spectral density. Assume Q_n to be given at two initial successive points, for instance corresponding to the values $n = -1$ and 0. Let us solve equation (81) iteratively. It follows that for $n > 0$, Q_n is an n-th degree polynomial in ω. In the large n limit, the roots of this polynomial tend to (minus) the eigenvalues Ω

$$Q_n \underset{n \to \infty}{\sim} \prod_{s=1}^{n} (\omega + \Omega_s) = \exp \sum_{s=1}^{n} \ln (\omega + \Omega_s) \qquad (84)$$

Thus

$$\gamma(\omega) = \int_0^\infty d\Omega \rho(\Omega) \ln(\omega + \Omega)$$
$$\rho(\Omega) = \frac{1}{\pi} \lim_{\varepsilon \to +0} \operatorname{Im} \frac{d\gamma}{d\omega}(-\Omega - i\varepsilon) \qquad (85)$$

The second formula (85) assumes an analytic continuation of γ (with a cut along the negative real axis). As above, we transform the linear second order difference equation (81) into a first order nonlinear one, a discrete analog of Ricatti's equation. With $-\omega$

replacing Ω, we set

$$R_n = \frac{Q_{n+1}}{Q_n}$$

$$R_n = 2 + \omega\ell_n - \frac{1}{R_{n-1}}$$

(86)

For $n \to \infty$, it follows from equation (83) that R_n is expected to behave as $\exp\gamma(\omega)$. This holds for a given realization of the set $\{\ell_n\}$. Thus on the average

$$\gamma(\omega) = \lim_{n\to\infty} \langle \ln R_n \rangle$$

$$\rho(\Omega) = \frac{1}{\pi}\mathrm{Im}\frac{\mathrm{d}}{\mathrm{d}\omega}\lim_{n\to\infty}\langle \ln R_n \rangle\bigg|_{\omega=-\Omega-\mathrm{i}\epsilon}$$

(87)

From equation (86), we can also derive an integral representation for the probability $P_n(R_n)\,\mathrm{d}R_n$ that R_n takes values in the interval $\mathrm{d}R_n$ around R_n, given R_0 (equal to one say). This results from the observation that R_{n-1} only depends on the variables ℓ_{n-1}, ℓ_{n-2}, ... which are independent from ℓ_n. Thus

$$P_n(R_n) = \int_0^\infty \mathrm{d}R_{n-1}P_{n-1}\left(R_{n-1}\right)$$

$$\times \int_0^\infty \mathrm{d}\ell e^{-\ell}\delta\left(R_n + \frac{1}{R_{n-1}} - 2 - \omega\ell\right) \quad (88)$$

a discrete analog of the Fokker–Planck equation. The condition $R_0 = 1$ and the restriction $\omega > 0$ entail that the support of P_n is in the interval $[1, \infty]$.

For $n \to \infty$ assume that P_n tends to a stationary distribution P. The latter is given by the fixed point equation

$$P(R) = \int_1^\infty \mathrm{d}R'P(R')\int_0^\infty \mathrm{d}\ell e^{-\ell}\delta\left(R + \frac{1}{R'} - 2 - \omega\ell\right)$$

$$= \int_1^\infty \mathrm{d}R'P(R')\frac{1}{\omega}\exp -\frac{1}{\omega}\left[R + \frac{1}{R'} - 2\right]\theta(R + R'^{-1} - 2)$$

(89)

The reasoning would be similar had the Poisson law for successive intervals be replaced by an arbitrary probability law. The difficulty in solving equation (89) is due to the θ-function under the integral sign. Once $P(R)$ is obtained from (89), the Liapunov exponent is given by

$$\gamma(\omega) = \int_1^\infty dR\, P(R) \ln R \tag{90}$$

Of course P depends parametrically on ω. Rather than attempting to find an explicit solution of equation (89), it is interesting to understand approximation methods which have a broader range of applicability.

Obtain $\gamma_0(\omega)$ and $\rho_0(\Omega)$ in the absence of disorder, i.e. when ℓ_n instead of being a fluctuating quantity is fixed to its mean value equal to unity. Thus $P(R) = \delta(R - A)$ where A is the largest root of the equation corresponding to (86), namely

$$A^2 - (2 + \omega)A + 1 = 0 \qquad A = 1 + \tfrac{1}{2}\omega + \left(\omega + \tfrac{1}{4}\omega^2\right)^{\frac{1}{2}} \tag{91}$$

Therefore

$$\gamma_0 = \ln A$$

$$\rho_0(\Omega) = \frac{1}{2\pi} \frac{1}{\left[\Omega(1 - \tfrac{1}{4}\Omega)\right]^{\frac{1}{2}}}$$

$$\underset{\Omega \to 0}{=} \frac{1}{2\pi\Omega^{\frac{1}{2}}} \left[1 + \tfrac{1}{8}\Omega + \cdots\right] \qquad 0 < \Omega < 4 \tag{92}$$

We shall study two opposite limits, the first one assumes weak disorder, the second a large frequency. If we pretend that the variables ℓ_n are weakly fluctuating, we expect to find an approximation valid for large wavelengths, hence small frequencies, which are the least sensitive to local disorder. We introduce a parameter λ by setting

$$\begin{aligned}
\ell_n &= \langle \ell_n \rangle + \lambda\left(\ell_n - \langle \ell_n \rangle\right) = \langle \ell_n \rangle + \lambda z_n \\
R_n &= A_n \exp\left(\lambda B_n + \lambda^2 C_n + \lambda^3 D_n + \lambda^4 E_n + \cdots\right)
\end{aligned} \tag{93}$$

with $\langle z_n \rangle = 0$ and, as $n \to \infty$, A_n tends to the amplitude A given by (91). With these definitions and $\langle B \rangle$, $\langle C \rangle$ ... standing for averages of B_n, C_n, ...,

$$\gamma(\omega) = \gamma_0(\omega) + \lambda \langle B \rangle + \lambda^2 \langle C \rangle + \cdots \tag{94}$$

Of course we are interested in the case $\lambda = 1$. Substitute the expression (93) (with A_n replaced by A) into equation (86). Using

(91) we find

$$A^2 \left[\exp \left(\lambda B_n + \lambda^2 C_n + \cdots \right) - 1 \right]$$
$$= \lambda (A - 1)^2 z_n - \left[\exp - \left(\lambda B_{n-1} + \lambda^2 C_{n-1} + \cdots \right) - 1 \right] \quad (95)$$

Identifying the coefficients of successive powers of λ, one derives the recursion relation

$$A^2 B_n = (A - 1)^2 z_n + B_{n-1}$$
$$A^2 \left(C_n + \tfrac{1}{2} B_n^2 \right) = C_{n-1} - \tfrac{1}{2} B_{n-1}^2 \quad (96)$$
$$\cdots$$

which show the usefulness of the expansion for small $(A - 1)$, i.e. small ω. Set

$$y = \frac{A - 1}{A + 1} z \quad (97)$$

then

$$\langle B \rangle = \langle y \rangle = 0 \qquad \langle C \rangle = -\tfrac{1}{2} \langle y^2 \rangle \qquad \langle D \rangle = \tfrac{1}{3} \langle y^3 \rangle$$

$$\langle E \rangle = -\tfrac{1}{4} \langle y^4 \rangle - \tfrac{1}{2} \frac{3 + 2A^2}{A^4 - 1} \langle y^2 \rangle^2 \qquad \cdots \quad (98)$$

i) Obtain equations (98)
ii) Show that if

$$X_n = B_n + C_n + D_n + E_n + \cdots \quad (99)$$

one has

$$\gamma(\omega) = \ln A + \langle X \rangle \quad (100)$$

$$\langle X \rangle = \langle \ln (1 + y) \rangle - \tfrac{1}{2} \frac{3 + 2A^2}{A^4 - 1} \langle y^2 \rangle^2 + \cdots \quad (101)$$

where the dots include terms involving products of two or more mean values of powers of the random variable y.

iii) Show that the most singular behaviour in $(A - 1)$ of terms of fifth and sixth order is

$$\langle F \rangle = \frac{3}{A^2 - 1} \langle y^2 \rangle \langle y^3 \rangle + \cdots$$

$$\langle G \rangle = -\tfrac{15}{2} \frac{1}{(A^2 - 1)^2} \langle y^3 \rangle^2 + \cdots \quad (102)$$

From these expressions we derive the behaviour of $\gamma(\omega)$ for ω tending to zero. Since

$$A - 1 = \omega^{\frac{1}{2}} + \tfrac{1}{2}\omega + \tfrac{1}{8}\omega^{\frac{3}{2}} + \cdots \tag{103}$$

we obtain up to and including terms of order ω^2

$$\gamma(\omega) = \omega^{\frac{1}{2}} - \tfrac{1}{8}\omega \left\langle z^2 \right\rangle - \tfrac{1}{384}\omega^{\frac{3}{2}} \left[16 - 16 \left\langle z^3 \right\rangle + 15 \left\langle z^2 \right\rangle^2 \right]$$
$$+ \tfrac{1}{512}\omega^2 \left[16 \left\langle z^2 \right\rangle - 8 \left\langle z^4 \right\rangle + 24 \left\langle z^2 \right\rangle^2 + 24 \left\langle z^2 \right\rangle \left\langle z^3 \right\rangle - 15 \left\langle z^2 \right\rangle^3 \right]$$
$$+ O(\omega^{\frac{5}{2}}) \tag{104}$$

This can now be specialized to the Poissonian case, setting $\lambda = 1$, with

$$\left\langle z^2 \right\rangle = 1 \qquad \left\langle z^3 \right\rangle = 2 \qquad \left\langle z^4 \right\rangle = 9 \qquad \cdots$$
$$\gamma(\omega) = \omega^{\frac{1}{2}} - \tfrac{1}{8}\omega + \tfrac{1}{384}\omega^{\frac{3}{2}} + \tfrac{1}{512}\omega^2 + \cdots \tag{105}$$

By analytic continuation

$$\mathrm{Re}\,\gamma(-\Omega - \mathrm{i}\varepsilon) = \tfrac{1}{8}\Omega + \tfrac{1}{512}\Omega^2 + O(\Omega^3)$$
$$\mathrm{Im}\,\gamma(-\Omega - \mathrm{i}\varepsilon) = \Omega^{\frac{1}{2}} - \tfrac{1}{384}\Omega^{\frac{3}{2}} + O(\Omega^{\frac{5}{2}}) \tag{106}$$

The leading correction to the average spectral density is therefore

$$\rho(\Omega) = \frac{1}{2\pi\Omega^{\frac{1}{2}}} \left[1 - \tfrac{1}{128}\Omega + O(\Omega^2) \right] \tag{107}$$

The method is a systematic one and would allow the computation of higher order terms in the expansion of $\rho(\Omega)$. The comparison between (107) and (92) is instructive. In both cases, the dominant term is the one corresponding to the spectrum of the Laplacian in the continuum. The first correction is of opposite sign, being sixteen times smaller for the random lattice. Thus, even in one dimension, the spectrum on a random lattice approximates the continuous spectrum at low frequency much better than on a regular one. One expects this effect to be much stronger the higher the dimension. This provides a definite motivation for the use of such lattices.

Equations (104) and (106) also reveal a new phenomenon. For a regular lattice the analytic continuation $\gamma(-\Omega - \mathrm{i}\varepsilon)$ is pure imaginary as ε tends to zero and Ω is positive. For a random lattice, there appears a positive real part. The natural interpretation of this quantity is that it is the inverse of finite localization length. With unit probability, for a frequency Ω belonging to the spectrum, not only do wavefunctions oscillate, but they also decrease exponentially. The behaviour of this localization length for small Ω is

$$L(\Omega) = \frac{8}{\Omega} - \frac{1}{8} + O(\Omega) \tag{108}$$

Naturally $L(\Omega)$ goes to infinity as $\Omega \to 0$, expressing that as one approaches the continuum limit the localization phenomenon disappears.

The magnitude of the localization length provides an empirical rule to obtain the range of validity of the weak disorder expansion. The latter makes sense as long as $L(\Omega)$ is large compared to the average spacing $\langle \ell_n \rangle$. We remark that, since ℓ_n can be arbitrarily small, the spectrum is unbounded.

To obtain some information for large Ω we need an alternative method. The above discussion suggests that, when Ω tends to infinity, states are strongly localized, so that γ is determined by small clusters of sites. We translate this for ω large by truncating the continued fraction expansion derived from equation (86)

$$R_n = 2 + \omega\ell_n - \cfrac{1}{2 + \omega\ell_{n-1} - \cfrac{1}{2 + \omega\ell_{n-1} - \cdots}} \tag{109}$$

We factor out $2 + \omega\ell_n$ and take the average of $\ln R_n$ over the Poisson distribution. If $\gamma = 0.57722\ldots$ stands for Euler's constant, we have

$$\langle \ln(2 + \omega\ell) \rangle = \ln 2 + \left\langle \frac{\omega}{2 + \omega\ell} \right\rangle = \ln 2$$

$$+ \, \mathrm{e}^{2/\omega} \left[\ln \tfrac{1}{2}\omega - \gamma + \frac{2}{\omega} - \frac{1}{2.2!}\left(\frac{2}{\omega}\right)^2 + \frac{1}{3.3!}\left(\frac{2}{\omega}\right)^3 - \cdots \right] \tag{110}$$

We set

$$C = \mathrm{e}^{\gamma} = 1.78107\cdots$$

and observe that for $p \geq 1$

$$\left\langle \frac{1}{(2 + \omega\ell)^{p+1}} \right\rangle = \frac{1}{p2^p\omega} + O\left(\frac{1}{\omega^2}\right) \qquad p \geq 1 \tag{111}$$

Thus

$$\gamma(\omega) = \ln 2 + \left\langle \frac{\omega}{2 + \omega\ell} \right\rangle - \sum_{p=1}^{\infty} \frac{1}{p} \left\langle \frac{1}{(2 + \omega\ell)^p} \right\rangle^2$$
$$- \left\langle \frac{1}{2 + \omega\ell} \right\rangle^2 \left\langle \frac{1}{(2 + \omega\ell)^2} \right\rangle + \cdots \tag{112}$$

The last term, the first one to involve the distribution of three successive intervals, is of order $(\ln\omega)^3/\omega^3$ and can be neglected in a computation up to and including terms of order ω^{-2}. We find therefore in the large ω limit

$$\gamma(\omega) = \ln \frac{\omega}{C} + \frac{2}{\omega}\left(\ln \frac{\omega}{2C} + 1 \right)$$
$$- \frac{1}{\omega^2}\left[\left(\ln \frac{\omega}{2C} - 1 \right)^2 + \eta - 4 \right] + O\left(\left[\frac{\ln\omega}{\omega}\right]^3 \right) \tag{113}$$

with a constant η given by

$$\eta = \sum_{1}^{\infty} \frac{1}{4^n n^2 (n+1)^2} = 0.13070\cdots \tag{114}$$

We conclude that as $\Omega \to \infty$

$$\rho(\Omega) = \frac{1}{\pi}\frac{d}{d\Omega}\mathrm{Im}\gamma(-\Omega - i\varepsilon) \underset{\Omega\to\infty}{\sim} \frac{2}{\Omega^2} + \frac{4}{\Omega^3}\left[\ln \frac{\Omega}{2C} - \frac{3}{2} \right] + \cdots$$

$$\frac{1}{L(\Omega)} = \mathrm{Re}\gamma(-\Omega - i\varepsilon) \underset{\Omega\to\infty}{\sim} \ln \frac{\Omega}{C} - \frac{2}{\Omega}\left(\ln \frac{\Omega}{C} + 1 \right)$$
$$- \frac{1}{\Omega^2}\left[\left(\ln \frac{\Omega}{2C} - 1 \right)^2 + \eta - 4 - \pi^2 \right] + \cdots \tag{115}$$

The logarithms occurring in these expressions are due to the finite limit of the probability distribution $p(\ell)$ as $\ell \to 0$. For Ω large, the localization length tends to zero as $1/\ln\Omega$, thus justifying the method. One can obtain more terms in the expansion at the price of rather heavy computation.

The results can be compared to a numerical simulation as shown on figure 2. The behaviour of $\rho(\Omega)$ in $2/\Omega^2$ as $\Omega \to \infty$ is in agreement with the fact that $\rho(\Omega)$ is integrable.

i) We have not made use of equation (89) which contains more information. It is possible to extract from this equation an

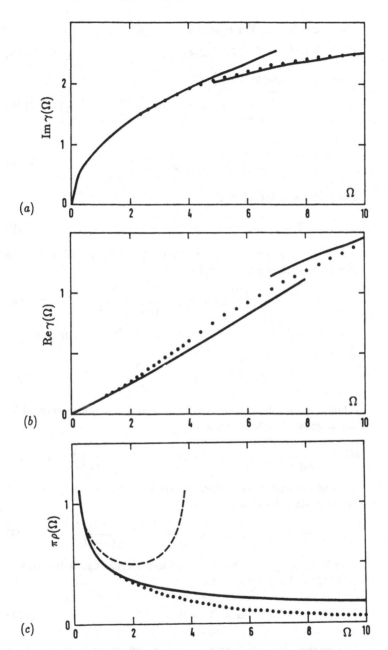

Fig. 2   (a) Comparison of small and large Ω expansions (full line) with the data (dots) for the integrated density. (b) Same comparison for the inverse correlation length. (c) The density of states in the continuum (full line), on a regular lattice (dashed line) and on a random lattice (dotted line).

analytic solution. For that purpose, set

$$R = 1 + x^{-1} \qquad P(R)\,\mathrm{d}R = p(x)\,\mathrm{d}x \qquad x \geq 0 \qquad (116)$$

With these notations we have

$$\omega x^2 p(x) = \int_{\mathrm{Sup}(0,x-1)}^{\infty} \mathrm{d}x' p(x') \exp -\frac{1}{\omega}\left[\frac{1}{x} - \frac{1}{1+x'}\right] \qquad (117)$$

and

$$\int_0^{\infty} \mathrm{d}x \; p(x) = 1 \qquad (118)$$

For x in the interval $(0,1)$

$$0 \leq x < 1 \qquad p(x) = \frac{1}{\alpha \omega x^2}\exp -\frac{1}{\omega x} \qquad (119)$$

where α is a positive constant to be determined from the normalization condition (118)

$$\frac{1}{\alpha} = \int_0^{\infty} \mathrm{d}x \; p(x) \exp\left[\frac{1}{\omega(1+x)}\right] \qquad (120)$$

To obtain $p(x)$ for $x \geq 1$, we trade $p(x)$ for the function $q(x)$ defined as

$$p(x) = \frac{1}{\alpha \omega x^2} \exp\left[-\frac{1}{\omega x}\right] q(x) \qquad (121)$$

Thus $q(x)$ is equal to 1 for $0 \leq x < 1$ and is given in general by the integrodifferential equation

$$\exp\left(-\frac{1}{\omega(x+1)}\right) \mathrm{d}q(x+1) + q(x)\,\mathrm{d}\left(\exp\left[-\frac{1}{\omega x}\right]\right) = 0 \qquad (122)$$

This shows that $q(x)$ is determined in the interval $n \leq x < n+1$ through its values for $0 \leq x < n$, from

$$q(x+1) = 1 - \int_0^x \mathrm{d}y q(y)\frac{1}{\omega y^2}\exp -\frac{1}{\omega y(y+1)} \qquad (123)$$

The function $q(x)$ is thus positive and decreasing. It behaves as $\omega^{1-x}/\Gamma(x)$ for $x \to \infty$. Also

$$\alpha = \sum_{n=1}^{\infty} q(n)\mathrm{e}^{-1/n\omega} \qquad (124)$$

Derive the expression for $\gamma(\omega)$ and re-obtain its high frequency expansion. Express the solution of equation (122) using a Laplace transform.

ii) Recover all the above results using the replica method.

10.2   Two-dimensional electron gas in a strong field

10.2.1   Landau levels – Quantum Hall effect

Let us illustrate the use of anticommuting variables on an example inspired by the study of the quantum Hall effect. The latter corresponds to the existence of plateaux in the Hall conductivity as a function of a (very strong) magnetic field at sufficiently low temperature. In such experiments, the electrons are trapped in a two-dimensional layer. The quantization of the Hall conductivity is accompanied by a sharp drop in the longitudinal resistance (or conductance!) showing that dissipative effects vanish at the center of the plateaux.

Let us recall the description of noninteracting electrons (charge e, mass m) constrained to move in the (x, y) plane, in the presence of a transversal field B. Energy levels are quantized and strongly degenerate (Landau). Choose a vector potential

$$A_x = -\tfrac{1}{2}By \qquad A_y = \tfrac{1}{2}Bx \qquad (125)$$

in the so-called symmetric gauge. The Hamiltonian reads

$$H_0 = \frac{1}{2m}(\mathbf{p} - e\mathbf{A})^2 = -\frac{\hbar^2}{2m}\left[\left(\partial_x + \frac{ie}{2\hbar}By\right)^2 + \left(\partial_y - \frac{ie}{2\hbar}Bx\right)^2\right]$$

$$(126)$$

Let us pretend that the electronic charge is positive, and introduce dimensionless units such that

$$X = \sqrt{\frac{eB}{2\hbar}}x \qquad Y = \sqrt{\frac{eB}{2\hbar}}y \qquad (127a)$$

and trade the magnetic field for the Larmor frequency

$$\omega = \frac{eB}{2m} \qquad (127b)$$

The Hamiltonian takes the form

$$H_0 = -\tfrac{1}{2}\hbar\omega\left[\left(\frac{\partial}{\partial X} + iY\right)^2 + \left(\frac{\partial}{\partial Y} - iX\right)^2\right]$$

$$= \hbar\omega\left[(-\partial + \tfrac{1}{2}\bar{z})(\bar{\partial} + \tfrac{1}{2}z) + \tfrac{1}{2}\right] \qquad (128)$$

where we use the complex notation

$$z = X + iY \qquad \partial = \frac{\partial}{\partial z} \qquad \bar{\partial} = \frac{\partial}{\partial \bar{z}} \qquad (129)$$

The operators

$$a = \bar{\partial} + \tfrac{1}{2}z \qquad\qquad a^+ = -\partial + \tfrac{1}{2}\bar{z} \qquad (130)$$

satisfy the bosonic commutation relations

$$[a, a] = [a^\dagger, a^\dagger] = 0 \qquad\qquad [a, a^\dagger] = 1 \qquad (131)$$

Hence the Landau spectrum reads

$$E_\ell = \hbar\omega \left(\ell + \tfrac{1}{2}\right) \qquad\qquad \ell = 0, 1, 2, \ldots \qquad (132)$$

corresponding to infinitely degenerate states

$$u_{\ell,m}(z, \bar{z}) = \frac{a^{\dagger\ell}}{\sqrt{\ell!}} u_{0,m}(z, \bar{z}) \qquad\qquad \ell, m = 0, 1, 2, \ldots \qquad (133)$$

The ground state wavefunctions satisfy

$$a u_{0,m} = \left(\bar{\partial} + \tfrac{1}{2}z\right) u_{0,m} = 0$$

with orthonormal solutions

$$u_{0,m} = \frac{z^m}{\sqrt{m!}} \frac{e^{-z\bar{z}/2}}{\sqrt{\pi}} \qquad\qquad m = 0, 1, 2, \ldots \qquad (134)$$

These functions generate the Fock–Bargmann space of entire functions, after factorization of $e^{-\frac{1}{2}z\bar{z}}/\sqrt{\pi} \equiv u_{0,0}$ (chapter 2 in volume 1). We absorb this factor into the measure and write

$$d\mu(z) = \frac{d^2 z}{\pi} e^{-z\bar{z}} \qquad\qquad u_{0,m} = \frac{e^{-\frac{1}{2}z\bar{z}}}{\sqrt{\pi}} v_m \qquad (135)$$

$$\int d\mu(z) \overline{v_m(z)} v_{m'}(z) = \delta_{m,m'}$$

where it is understood that $d^2 z = d\mathrm{Re}\, z\, d\mathrm{Im}\, z$.

  i) Give the explicit form of $u_{\ell,m}$ defined by equation (133).
 ii) Investigate the meaning of the quantum number m. What are the consequences of translational and rotational invariance? Pay attention to changes of gauge.
iii) Show that for large m, the state $u_{0,m}$ corresponds to a probability distribution concentrated in an annulus of width $\Delta(r^2) \simeq \sqrt{m}$ around a mean radius $\overline{r^2} = m$

$$\left| u_{0,m} \right|^2 = \frac{e^{-r^2} r^{2m}}{\pi m!} \underset{m \to \infty}{\sim} \frac{1}{\pi} \frac{e^{-(r^2 - m)^2/2m}}{\sqrt{2\pi m}} \qquad (136)$$

Deduce that the number of independent states in the fundamental Landau level concentrated inside a circular disk of area

$\mathcal{A} = \pi R^2$ is given asymptotically by the ratio of the flux $B\mathcal{A}$ to the quantum of flux $\varphi_0 = h/e$

$$N = \frac{eB\mathcal{A}}{2\pi\hbar} \tag{137}$$

if we return to original units. This result is independent of the choice of basis and of the shape of the surface provided N is large. Its interpretation is clear. The magnetic flux through the area $\mathcal{A}$, multiplied by the electric charge, is N times Planck's constant h. Alternatively, N is equal to the increase $(1/2\pi) \oint (e/\hbar)\mathbf{A} \cdot \mathbf{dx}$ in the phase of a wavefunction, measured in units of 2π, as one winds arounds the boundary of the sample. Show that these results extend to excited states.

What is significant, in the absence of any other interaction, is the macroscopic degeneracy of Landau levels whenever the magnetic length $\sqrt{\hbar/eB}$ (i.e. the square root of the area embracing a unit magnetic flux) is small compared to the overall scale (i.e. when N given by equation (137) is large). This quantum degeneracy leads to nontrivial quantum effects.

Consider an ideal situation where we neglect interactions among electrons as well as with the substrate. We also ignore magnetic contributions of electronic spins, the latter being assumed frozen along the strong magnetic field. Finally, pretend that the motion is constrained in the confining plane. Let us introduce an infinitesimal electric field $\mathbf{E}$ in this plane (hence perpendicular to the magnetic field $\mathbf{B}$). In classical terms, this field can be compensated for by a uniform displacement of the electron gas at a speed $\mathbf{v}$ adjusted in such a way that the Lorentz force $e(\mathbf{E}+\mathbf{v}\wedge\mathbf{B})$ vanishes. This is a purely kinematical effect, independent of the charge, such that

$$\mathbf{v} = \frac{\mathbf{E} \wedge \mathbf{B}}{\mathbf{B}^2} \tag{138}$$

and generates a Hall current given by

$$\mathbf{j}_H = en\mathbf{v} = \frac{en}{\mathbf{B}^2}\mathbf{E} \wedge \mathbf{B} \tag{139}$$

where n is the (surface) density of charges. This current does not lead to any dissipative effect ($\mathbf{j}_H.\mathbf{E} = 0$). In matrix notation,

$$\mathbf{j}_H = \overline{\overline{\sigma}}_H \mathbf{E} \tag{140}$$

with an antidiagonal conductivity matrix

$$\overline{\overline{\sigma}}_H = \begin{pmatrix} 0 & en/B \\ -en/B & 0 \end{pmatrix} \tag{141}$$

One can also add to this expression a phenomenological diagonal ohmic term σ_0 to restore the effects of impurities and interactions. Thus the total conductivity matrix looks like

$$\overline{\overline{\sigma}} = \begin{pmatrix} \sigma_0 & en/B \\ -en/B & \sigma_0 \end{pmatrix} \tag{142}$$

and its inverse, the resistivity, reads

$$\overline{\overline{\rho}} = \overline{\overline{\sigma}}^{-1} = \frac{1}{\sigma_0^2 + e^2 n^2 / B^2} \begin{pmatrix} \sigma_0 & -en/B \\ en/B & \sigma_0 \end{pmatrix} \tag{143}$$

In the limit where σ_0 is negligeable, we have the apparent paradox that both σ_{xx} and ρ_{xx} tend simultaneously to zero. Let us look at case where the chemical potential lies between two Landau levels, meaning that those levels with quantum number $\ell = 0, 1, \dots, L$ are totally occupied. According to equation (137), the electronic density is

$$n = \frac{eB}{2\pi\hbar}(L+1) \tag{144}$$

Hence the Hall constant, i.e. the antidiagonal term $\sigma_H = \sigma_{xy}$, is quantized according to

$$\sigma_H = \frac{en}{B} = \frac{e^2}{2\pi\hbar}(L+1) \tag{145}$$

Let ξ be the transverse section of a sample, and I the longitudinal current. The current density (ignoring edge effects) is $j = I/\xi$, while, if V_H is the transverse difference of potential, the electric field is $E = V_H/\xi$. As a result, and independently of the geometry or any other parameter,

$$\sigma_H = \frac{I}{V_H} = \frac{e^2}{h}(L+1) \tag{146}$$

These are the plateaux of the integral quantum Hall effect. Additional plateaux are observed at fractional occupation values of the Landau levels, and are believed to arise from interactions among electrons.

The quantized Hall effect, apart from being a spectacular manifestation of quantum mechanics at a macroscopic level, is important, if only because it involves fundamental constants and can lead to an improved accuracy on their measurement. What is truly remarkable is that the values (146) appear unmodified even in the presence of various impurities, (small) temperature effects, etc, and are observed with an amazing accuracy. The above description is of course somehow caricatural. One may naively think that a number of states become localized as a result of various kinds of defects – impurities, etc – and therefore cannot contribute to the conductivity. Arguments involving gauge invariance and topology show that the domain of validity of relation (146) extends beyond the ideal model and does apply to the real situation under much weaker assumptions (appendix A). However the width of the plateaux varies with experimental conditions and a complete theory should explain the observed behaviour.

We shall only study here how the density of states is affected by random impurities (Wegner 1983). It turns out that, with the help of simplifying assumptions, this problems admits an interesting analytic solution.

10.2.2   One particle spectrum in the presence of impurities

We want to understand how impurities affect the Landau levels. In practice, we shall limit ourselves to the lowest one with $\ell = 0$, and for a strong enough field, ignore excitations to higher levels, the gap between levels $\hbar\omega$ being much bigger than the perturbing potentials. Impurities are modelled by adding a random potential $V(\mathbf{x})$ to the Hamiltonian (126). Using reduced units defined in (127), we write

$$H = H_0 + V \qquad (147)$$

One assumes that correlations in the potential are of short-range, and in the limiting case that the potential is uncorrelated at distinct points. The origin of energies is chosen in general by requiring that

$$\langle V \rangle = 0 \qquad (148)$$

although in some instances it is more convenient to relax this condition. Our hypothesis on the locality of correlations translates into the characteristic functional

$$\left\langle \exp -i \int d^2x\alpha(\mathbf{x})V(\mathbf{x}) \right\rangle = \exp \int d^2x g(\alpha(\mathbf{x})) \qquad (149)$$

The function $g(\alpha)$ represents the Fourier transform of a distribution of potential at a site

$$\exp g(\alpha) = \int dV P(V)e^{-i\alpha V} \qquad (150)$$

Thus a Gaussian noise will correspond to

$$g(\alpha) = -\tfrac{1}{2}w\alpha^2 \qquad (151)$$

i) A Poisson model of random impurities corresponds to a uniform density ρ of zero-range scattering centers (Friedberg and Luttinger, 1975). The probability density to find N impurities at points $\mathbf{x}_1, \ldots, \mathbf{x}_N$, in an area $\mathcal{A}$ is given by

$$P(\mathbf{x}_1, \ldots, \mathbf{x}_N) = e^{-\rho\mathcal{A}}\rho^N/N! \qquad (152)$$

The corresponding potential of strength λ is

$$V(\mathbf{x}) = \lambda \sum_{i=1}^{N} \delta(\mathbf{x} - \mathbf{x}_i) \qquad (153)$$

Show that

$$g(\alpha) = \rho \left[e^{-i\lambda\alpha} - 1 \right] \qquad (154)$$

How should one modify this expression, when the coupling constant are also random variables ?

ii) A Lorentzian distribution of the potential

$$P(V) = \frac{\lambda}{\pi} \frac{1}{(V^2 + \lambda^2)} \qquad (155)$$

leads to a nonanalytic function

$$g(\alpha) = -\lambda |\alpha| \qquad (156)$$

As before, we obtain the average spectral density per unit area using the resolvent

$$\rho(E) = -\frac{1}{\pi}\mathrm{Im} \left\langle \left(\mathbf{x} \left| \frac{1}{E + io - H} \right| \mathbf{x} \right) \right\rangle \qquad (157)$$

The latter is in turn represented by a functional integral over a complex field φ through

$$\left(\mathbf{x}\left|\frac{1}{E+\mathrm{i}o-H}\right|\mathbf{x}'\right)$$

$$= \frac{1}{\mathrm{i}Z}\int \mathcal{D}(\varphi\bar\varphi)\varphi(\mathbf{x}')\bar\varphi(\mathbf{x}')\exp\mathrm{i}\int \mathrm{d}^2\mathbf{x}\,\bar\varphi(E+\mathrm{i}o-H)\varphi \quad (158)$$

with a normalization factor

$$Z = \int \mathcal{D}(\varphi\bar\varphi)\exp\mathrm{i}\int \mathrm{d}^2\mathbf{x}\,\bar\varphi(E+\mathrm{i}o-H)\varphi \quad (159)$$

This shorthand notation anticipates the fact that the effective action is the integral over a local density.   Henceforth the infinitesimal imaginary part in E will be understood, but not written explicitly.   It accounts for the choice of sign in the exponential.   Instead of using the replica trick, in order to average over disorder, let us present a variant, which looks more satisfactory from a mathematical point of view.   We represent Z^{-1}, which is nothing but $\det(E-H)$, by a path integral over an auxiliary complex fermionic field $(\psi,\bar\psi)$

$$Z^{-1} = \det(E-H) = \int \mathcal{D}(\psi\bar\psi)\exp\mathrm{i}\int \mathrm{d}^2\mathbf{x}\,\bar\psi(E-H)\psi \quad (160)$$

We may understand this by noting that a complex Fermi field counts for minus two real bosonic components.
    Therefore

$$\left(\mathbf{x}\left|\frac{1}{E-H}\right|\mathbf{x}'\right) = -\mathrm{i}\int \mathcal{D}(\varphi\bar\varphi\psi\bar\psi)\varphi(\mathbf{x})\bar\varphi(\mathbf{x}')$$

$$\times\exp\mathrm{i}\int \mathrm{d}^2\mathbf{x}\,[\bar\varphi(E-H_0)\varphi + \bar\psi(E-H_0)\psi - V(\bar\varphi\varphi + \bar\psi\psi)]$$

$$(161)$$

In this form we can compute the average over V using equation (149) with $\alpha(\mathbf{x})$ replaced by $\bar\varphi\varphi + \bar\psi\psi$.   Before doing so, we introduce the hypothesis that the potential should not excite transitions between distinct Landau levels, and that in (161) the only states to contribute correspond to the lowest level.   This amounts to integrating only over fields φ and ψ satisfying the conditions

$$\begin{aligned} (E-H_0)\varphi &= \mathcal{E}\varphi \\ (E-H_0)\psi &= \mathcal{E}\psi \end{aligned} \qquad \mathcal{E} = E - \tfrac{1}{2}\hbar\omega \qquad (162)$$

Thus

$$\varphi = e^{-\frac{1}{2}z\bar{z}}u(z) \qquad \psi = e^{-\frac{1}{2}z\bar{z}}v(z) \qquad (163)$$

where u (bosonic) and v (fermionic) are holomorphic in z ($\bar{u}$ and $\bar{v}$ are antiholomorphic). We do not include a factor $\pi^{-1/2}$ in (163) in order to maintain a symmetry between bosonic and fermionic integrations. The Jacobian corresponding to a transformation on the basis of all Landau levels is unity, with fermionic contributions compensating bosonic ones. Of course, we truncate this expansion by keeping only the lowest level. After averaging over V, we obtain

$$\left\langle \left(\mathbf{x} \left| \frac{1}{E-H} \right| \mathbf{x}' \right) \right\rangle$$

$$= -i \exp -\tfrac{1}{2} \left(z\bar{z} + z'\bar{z}' \right) \int \mathcal{D}(u\bar{u}v\bar{v})u(z)\bar{u}(z') \exp S \qquad (164)$$

where the action S reads

$$S = \int d^2 z\, e^{-z\bar{z}} \left\{ i\mathcal{E}(\bar{u}u + \bar{v}v) + g(e^{-z\bar{z}}(\bar{u}u + \bar{v}v)) \right\} \qquad (165)$$

Use translational invariance (accompanied by gauge transformations) to show that, within the approximation to the lowest Landau level, all the spatial dependence of the average of the resolvent can be factored out as

$$\left\langle \left(\mathbf{x} \left| \frac{1}{E-H} \right| \mathbf{x}' \right) \right\rangle = \exp\left\{ z\bar{z}' - \tfrac{1}{2}z\bar{z} - \tfrac{1}{2}z'\bar{z}' \right\} \left\langle \left(\mathbf{x} \left| \frac{1}{E-H} \right| \mathbf{x} \right) \right\rangle$$

$$(166)$$

while

$$\left\langle \left(\mathbf{x} \left| \frac{1}{E-H} \right| \mathbf{x} \right) \right\rangle = -i e^{-z\bar{z}} \int \mathcal{D}(u\bar{u}v\bar{v})u(z)\bar{u}(z) \exp S \qquad (167)$$

is $\mathbf{x}$-independent.

The action S possesses a supersymmetry which will become explicit if we introduce, besides the commuting coordinates $z, \bar{z}$, anticommuting ones θ and $\bar{\theta}$. We normalize the integral as

$$\int d\theta d\bar{\theta} e^{-\theta\bar{\theta}} = 1/\pi \qquad (168a)$$

to maintain the parallel with

$$\int d^2 z\, e^{-z\bar{z}} = \pi \qquad (168b)$$

We define a holomorphic superfield ϕ (and its antiholomorphic conjugate $\bar{\phi}$), a function of z and θ which unites u and v, as

$$\phi(z,\theta) = u(z) + \theta v(z) \qquad \overline{\phi(z,\theta)} = \overline{u(z)} + \overline{v(z)}\bar{\theta} \qquad (169)$$

Thus

$$S = i\pi\mathcal{E} \int d\theta d\bar{\theta} d^2 z e^{-(z\bar{z}+\theta\bar{\theta})}\bar{\phi}\phi + \int d^2 z g \left(e^{-z\bar{z}}\pi \int d\theta d\bar{\theta} e^{-\theta\bar{\theta}}\bar{\phi}\phi \right)$$

$$(170)$$

The first term is invariant under rotations in superspace, which contain transformations which do not mix z and θ as well as transformations which read in infinitesimal form

$$\begin{aligned} \delta z &= \bar{\omega}\theta & \delta\theta &= z\omega \\ \delta\bar{z} &= \bar{\theta}\omega & \delta\bar{\theta} &= z\bar{\omega} \end{aligned} \qquad (171)$$

and leave invariant the quadratic form $z\bar{z} + \theta\bar{\theta}$. Here $(\omega, \bar{\omega})$ is an "infinitesimal" anticommuting parameter. What is remarkable is that the second term of the action (170) is also invariant. To see this, let us assume that $g(\alpha)$ admits an expansion

$$g(\alpha) = \sum_{n=1}^{\infty} g_n \alpha^n \qquad (172)$$

Indeed $g(0)$ vanishes as a consequence of the normalization of the probability distribution of the potential. If an expansion of the form (172) is not available, we approximate $g(\alpha)$ by functions with this property. We have a series of interaction terms of the type

$$\sum_{1}^{\infty} g_n e^{-nz\bar{z}} \left[\pi \int d\theta d\bar{\theta} e^{-\theta\bar{\theta}}\bar{\phi}\phi \right]^n \qquad (173)$$

However the following identity holds

$$\left[\pi \int d\theta d\bar{\theta} e^{-\theta\bar{\theta}}\bar{\phi}\phi \right]^n = \frac{1}{n}\pi \int d\theta d\bar{\theta} e^{-n\theta\bar{\theta}}(\bar{\phi}\phi)^n \qquad (174)$$

which can be checked by an explicit computation

Prove relation (174).

This introduces a function $h(\alpha)$, associated to $g(\alpha)$, through

$$h(\alpha) = \sum_{n=1}^{\infty} \frac{g_n}{n}\alpha^n = \int_0^{\alpha} \frac{d\beta}{\beta} g(\beta) \qquad (175)$$

where the property $g(0) = 0$ is crucial to yield a finite $h(\alpha)$. With this notation, the action reads

$$S = \pi \int \mathrm{d}\theta \mathrm{d}\bar{\theta}\mathrm{d}^2 z \left[\mathrm{i}\mathcal{E}\bar{\phi}\phi \mathrm{e}^{-(z\bar{z}+\theta\bar{\theta})} + h(\bar{\phi}\phi \mathrm{e}^{(z\bar{z}+\theta\bar{\theta})}) \right] \qquad (176)$$

exhibiting explicitly rotational invariance in superspace. This is at the origin of spectacular simplifications in the evaluation of the path integral which follows.

Derive from translational invariance in superspace of the path integral, with

$$\phi(z,\theta) \to \phi(z - a, \theta - \omega) \exp\left\{ z\bar{a} + \theta\bar{\omega} - \tfrac{1}{2}a\bar{a} - \tfrac{1}{2}\omega\bar{\omega} \right\}$$

that the two-point function takes the form

$$\left\langle \phi(z,\theta)\overline{\phi(z',\theta')} \right\rangle = C \exp\left\{ z\bar{z}' + \theta\bar{\theta}' \right\} \qquad (177)$$

where C is the quantity, independent from configuration variables, that we wish to evaluate

$$C = \left\langle \left(\mathbf{x} \left| \frac{1}{E - H} \right| \mathbf{x} \right) \right\rangle \qquad (178)$$

Note that equation (177) means

$$\langle u(z)\bar{u}(z') \rangle = \langle v(z)\bar{v}(z') \rangle = C \exp z\bar{z}' \qquad \langle u(z)\bar{v}(z') \rangle = 0 \qquad (179)$$

To compute the two-point function $\langle \phi\bar{\phi}' \rangle$, we consider a perturbative expansion, making the best use of the supersymmetric holomorphic formalism. The propagator is the inverse of the operator occurring in the quadratic part of the action. For any superholomorphic function, we have the identity

$$f(z',\theta') = \int \mathrm{d}^2 z \, \mathrm{d}\theta \mathrm{d}\bar{\theta} f(z,\theta) \exp\left\{ z'\bar{z} + \theta'\bar{\theta} - z\bar{z} - \theta\bar{\theta} \right\} \qquad (180)$$

which generalizes the reproducing kernel in Bargmann's space. We deduce that the propagator is proportional to $\exp(z'\bar{z} + \theta'\bar{\theta})$. The interactions are given by the second term in the action (176). An interaction vertex of degree $2n$ contains a Gaussian factor $\exp -n\left\{ z\bar{z} + \theta\bar{\theta} \right\}$. We illustrate the Feynman rules by computing the contribution of the one-loop diagram to the mean value $\langle \phi\bar{\phi}' \rangle$ (figure 3).

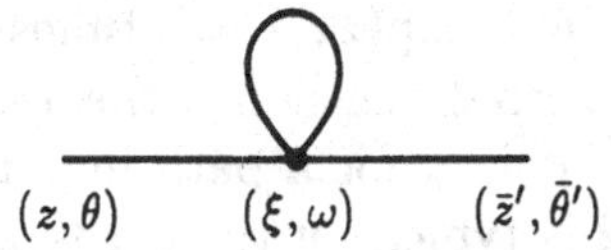

$$(z, \theta) \qquad (\xi, \omega) \qquad (\bar{z}', \bar{\theta}')$$

Fig. 3 One loop diagram contribution to the mean value $\langle \phi \bar{\phi}' \rangle$.

In the white noise case, i.e. when

$$g(\alpha) = -\tfrac{1}{2} w \alpha^2 \qquad\qquad h(\alpha) = -\tfrac{1}{4} w \alpha^2 \qquad (181)$$

the computation involves

i) the product of three propagators

$$(\mathrm{i}/\pi \mathcal{E})^3 \exp \left\{ z \bar{\xi} + \theta \bar{\omega} + \xi \bar{\xi} + \omega \bar{\omega} + \xi \bar{z}' + \omega \bar{\theta}' \right\}$$

ii) a vertex factor

$$-\pi w \exp -2 \left\{ \xi \bar{\xi} + \omega \bar{\omega} \right\}$$

iii) a symmetry factor equal to unity (the field is complex)
iv) finally an integral of the variables ξ, ω, leading to

$$(\mathrm{i}/\pi \mathcal{E})^3 \, (-\pi w) \exp \left\{ z \bar{z}' + \theta \bar{\theta}' \right\}$$

$$\times \int \mathrm{d}^2 \xi \, \mathrm{d}\omega \, \mathrm{d}\bar{\omega} \exp \left\{ -(\xi - z)(\bar{\xi} - \bar{z}') - (\omega - \theta)(\bar{\omega} - \bar{\theta}') \right\} \qquad (182)$$

We obtain the expected exponential prefactor containing all the dependence on external variables. What remains is to perform the Gaussian integral in superspace. Its value is unity. This is due to an exact cancellation between a bosonic and a fermionic determinant which are identical by virtue of the symmetry between the variables z and θ. Here we find the previous integrals

$$\int \mathrm{d}^2 \xi \exp -\xi \bar{\xi} = \pi \qquad \int \mathrm{d}\omega \mathrm{d}\bar{\omega} \exp -\omega \bar{\omega} = 1/\pi \qquad (183)$$

The phenomenon observed in this example extends to all orders. The combination of propagators and vertex factors leads to Gaussian integrals in superspace. In the exponential there appear quadratic and linear terms. When one factors out the expected prefactor, one recovers a supersymmetric Gaussian integral with value unity, as was the case above, and for the same reasons. In short, any diagram yields $\exp(z\bar{z}' + \theta\bar{\theta}')$ times a factor which would result in a similar zero-dimensional theory. We can therefore substitute a simple integral for a path integral. The dimensional reduction $d \to d - 2$ is typical of the existence of supersymmetry in such questions. It was first noticed by Parisi and Sourlas in a different context (section 5).

The conclusion of this discussion is the following expression

$$
\left\langle \left(\mathbf{x} \left| \frac{1}{E - H} \right| \mathbf{x} \right) \right\rangle = \left\langle \phi(z, \theta)\overline{\phi(z', \theta')} \right\rangle \exp - \left\{ z\bar{z}' + \theta\bar{\theta}' \right\}
$$
$$
= \frac{\int d\varphi d\bar{\varphi} \; \varphi\bar{\varphi} \exp \pi \left\{ i\mathcal{E}\bar{\varphi}\varphi + h(\bar{\varphi}\varphi) \right\}}{\int d\varphi d\bar{\varphi} \exp \pi \left\{ i\mathcal{E}\bar{\varphi}\varphi + h(\bar{\varphi}\varphi) \right\}} \qquad (184)
$$

The required quantity is obtained as the ratio of two integrals each one extending over the complex φ-plane. We recall that $\mathcal{E} = E - \frac{1}{2}\hbar\omega$. Since

$$
\left\langle \left(\mathbf{x} \left| \frac{1}{E - H} \right| \mathbf{x} \right) \right\rangle = \frac{1}{i\pi} \frac{\partial}{\partial \mathcal{E}} \ln \int_0^\infty dt \exp \pi \left\{ i\mathcal{E}t + h(t) \right\} \qquad (185)
$$

and the mean density of states is obtained in the explicit form

$$
\rho(E) = \frac{1}{\pi^2} \mathrm{Im} \frac{\partial}{\partial E} \ln \left[\int_0^\infty dt \exp \pi \left\{ i \left(E - \frac{1}{2}\hbar\omega \right) t + \int_0^t \frac{d\alpha}{\alpha} g(\alpha) \right\} \right] \qquad (186)
$$

where $g(\alpha)$ is given by equation (150) which defines the local disorder.

As an illustration, we evaluate the expression (186) for the examples introduced above. The reader will check the corresponding integrals. We also restore the dimensional factor

$$
K = \frac{eB}{\hbar} \qquad (187)
$$

which up to now was set equal to unity.

i) White noise (figure 4)

$$
g(\alpha) = -w\tfrac{1}{2}\alpha^2 \qquad \nu = \sqrt{\frac{2\pi}{wK^2}} \left(E - \tfrac{1}{2}\hbar\omega \right)
$$
$$
\rho(E) = \frac{K}{\pi^2} \sqrt{\frac{2}{w}} \frac{e^{\nu^2}}{1 + \left(2\pi^{-\frac{1}{2}} \int_0^\nu dx e^{x^2} \right)^2} \qquad (188)
$$

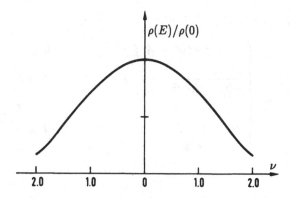

Fig. 4 Plot of the normalized density of states in the Gaussian case $\rho(E)/\rho(0)$ versus ν (equation (188)).

The total number of states is $\int dE \rho(E) = K^2/2\pi$. For large E, $\rho(E)$ decreases as

$$\rho(E) \underset{E \to \infty}{\sim} \frac{K}{\pi}\sqrt{\frac{2}{w}}\nu^2 e^{-\nu^2}\left(1 + O\left(\frac{1}{\nu^{-2}}\right)\right) \tag{189}$$

in agreement with a semiclassical approximation. At the center of the band ($\nu \to 0$), we have

$$\rho(E) \underset{E \to \hbar\omega/2}{\sim} \frac{K}{\pi^2}\sqrt{\frac{2}{w}}\left[1 + \left(1 - \frac{4}{\pi}\right)\nu^2 + O(\nu^4)\right] \tag{190}$$

giving a distribution which resembles a Gaussian but is flatter.

ii) Poisson distribution of short-range scatterers

$$g(\alpha) = \rho(e^{-i\lambda\alpha} - 1) \qquad \nu = \frac{2\pi}{\lambda K^2}(E - \tfrac{1}{2}\hbar\omega) \qquad f = \frac{2\pi}{K^2}\rho$$

$$\rho(E) = \frac{1}{\pi\lambda}\text{Im}F(\nu) \tag{191}$$

$$F(\nu) = \frac{\partial}{\partial\nu}\ln\left[\int_0^\infty dt \exp\left\{i\nu t - f\int_0^t \frac{d\alpha}{\alpha}(1 - e^{-i\alpha})\right\}\right]$$

We have assumed that the parameter λ, which characterizes the strength of the potential, is positive. The dimensionless ratio f is a measure of the impurity density in the natural unit of area embracing a unit magnetic flux.

One observes singularities when f goes through integral values, as shown on figure 5, where we have plotted the integrated density $N(E)$. The spectrum extends in the range

$\nu > 0$. As $\nu \to 0$, we find

$$\lambda \rho(E) \underset{\nu \to 0}{\sim} \begin{cases} (1 - f)\delta(\nu) + A(f)\nu^{-f} + \cdots & 0 < f < 1 \\[2mm] \dfrac{1}{\nu[(\ln \nu/\nu_0)^2 + \pi^2]} + \cdots & f = 1 \\[2mm] B(f)\nu^{2-f} + \cdots & 1 < f < 2 \\[1mm] B(2) + \cdots & f = 2 \\[1mm] B(f)\nu^{f-2} + \cdots & f > 2 \end{cases} \qquad (192)$$

Evaluate the constants appearing in the above singular behaviour at $\nu \to 0$,

$$A(f) = \frac{f \sin \pi f}{\Gamma(1-f)} \int_0^\infty \frac{dt}{t^f} \exp - \left\{ t + f \int_t^\infty \frac{d\alpha}{\alpha} e^{-\alpha} \right\} \qquad 0 < f < 1$$

$$\ln \nu_0 = \int_0^\infty dt \exp - \left\{ \int_t^\infty \frac{d\alpha}{\alpha} e^{-\alpha} + 1 e^{-t} \right\}$$

$$1/B(f) = \Gamma(f-1) \int_0^\infty \frac{dt}{t^f} \exp - f \int_t^\infty \frac{d\alpha}{\alpha} e^{-\alpha} \qquad f > 1 \qquad (193)$$

The most striking phenomenon occurs for $0 < f < 1$, where a fraction $(1 - f)$ of the states in the lowest Landau level is unaffected by the presence of the scatterers. There are weaker singularities in the density of states at integrals values of ν as shown on figure 5.

iii) Lorentzian distribution (equations (155)–(156)).

The density of states is again a Lorentzian as was $P(V)$

$$g(\alpha) = -\lambda \, | \, \alpha \, | \qquad \qquad \rho(E) = \frac{K^2}{2\pi^2} \frac{\lambda}{\lambda^2 + (E - \frac{1}{2}\hbar\omega)^2} \qquad (194)$$

The above discussion, even though it leads to interesting results for the density of states, does not illuminate the question of the robustness of the quantum Hall effect. It does not either clarify what is exactly meant by localization in the presence of a magnetic field. In appendix A we present a partial answer to the first question by exhibiting the Hall conductance as a topological invariant. This is an elaboration of an argument originally presented by Laughlin.

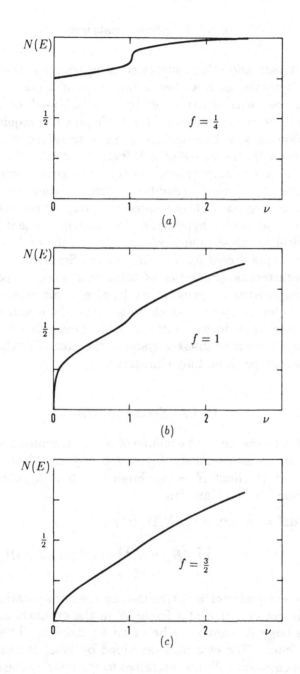

Fig. 5 The integrated density of states for the Poissonian model for various values of the density of impurities f. (a) $f = \frac{1}{4}$ (b) $f = 1$ (c) $f = \frac{3}{2}$.

10.3   Random matrices

Wigner, Dyson and others suggested comparing the spectrum of complex systems, such as heavy nuclei, with those of random Hamiltonians, with distribution laws as unbiased as possible, except for possible symmetries. The basic principle requires these distribution laws to be invariant under a transformation group derived from the existence of a Hilbert space. This implies in general the full unitary group, the real orthogonal group in the case where some reality condition (such as time reversal) is assumed, or else the unitary symplectic group. If one introduces an extra – ad hoc – hypothesis that matrix elements remain independent variables in any reference frame, the probability law becomes unique except for a scaling factor. Surprisingly enough, some characteristic properties of these ensembles, in particular the spacing between successive levels, give a fair representation of many observed spectra which range from atoms and nuclei to the zeroes of Riemann's ζ-function. Some universality is at work, of a nature specific to discrete systems, the precise mechanism of which has not yet been fully elucidated.

10.3.1   Semicircle law

We start with the simplest example of $N \times N$ Hermitian matrices. The matrix elements satisfy $H_{ij} = H_{ji}^*, 1 \leq i,j \leq N$. We are interested in the limit $N \to \infty$. From the above hypothesis, the distribution law is a Gaussian

$$P(H)\,\mathrm{d}H = \mathrm{cst}\ \exp-\left\{\tfrac{1}{2}\beta N \,\mathrm{Tr}\ H^2\right\}$$
$$\prod_{i=1}^{N} \mathrm{d}H_{ii} \times \prod_{1\leq i<j\leq N} \mathrm{d}\,\mathrm{Re}H_{ij}\ \mathrm{d}\,\mathrm{Im}H_{ij} \quad (195)$$

The constant prefactor is determined by the normalization condition. The introduction of a factor N in the exponential implies that the large N limit is to be taken for fixed β. This will be justified later. The ensemble described by (195) is the unitary Gaussian ensemble. We are interested by the distribution of eigenvalues $p(\Lambda)\mathrm{d}\Lambda$ where Λ is the real diagonal matrix of the eigenvalues λ_i, H being diagonalized with the help of a unitary matrix

U in the form

$$H = U^\dagger \Lambda U \tag{196}$$

To be specific, we can assume U to be of unit determinant although this still leaves $N - 1$ phases undetermined. To obtain $p(\Lambda)\mathrm{d}\Lambda$ from $p(H)\,\mathrm{d}H$, we express H in the vicinity of a diagonal Λ by writting $U = I + \mathrm{i}\delta K$, where δK is an infinitesimal Hermitian traceless matrix. To first order, $H = \Lambda + \mathrm{i}\,[\Lambda, \delta K]$. We introduce this parametrization in (195) and compute the corresponding Jacobian. Separating the integral over the (special) unitary group, we obtain

$$p(\Lambda)\mathrm{d}\Lambda = \mathrm{cst} \prod_{1 \leq i < j \leq N} (\lambda_i - \lambda_j)^2 \exp\left[-\tfrac{1}{2}\beta N \sum_{i=1}^{N} \lambda_i^2\right] \prod_{1 \leq i \leq N} \mathrm{d}\lambda_i \tag{197}$$

Show that in the real orthogonal case the eigenvalue distribution is

$$p(\Lambda)d\Lambda = \mathrm{cst} \prod_{1 \leq i < j \leq N} |\lambda_i - \lambda_j| \exp\left[-\tfrac{1}{2}\beta N \sum_{1 \leq \lambda \leq N} \lambda_i^2\right] \prod_{1 \leq i \leq N} \mathrm{d}\lambda_i \tag{198}$$

while for the unitary symplectic ensemble

$$p(\Lambda)d\Lambda = \mathrm{cst} \prod_{1 \leq i < j \leq N} (\lambda_i - \lambda_j)^4 \exp\left[-\tfrac{1}{2}\beta N \sum_{1 \leq \lambda \leq N} \lambda_i^2\right] \prod_{1 \leq i \leq N} \mathrm{d}\lambda_i \tag{199}$$

In all cases, we study the asymptotic properties of these distributions as N tends to infinity. These expressions suggest a statistical interpretation in terms of a one-dimensional Coulomb gas with logarithmic pair interactions. We define in general a partition function

$$Z(\alpha, \beta) = \frac{1}{N!} \int \prod_{1 \leq i \leq N} \mathrm{d}\lambda_i \exp -S = \int_{\lambda_1 \leq \lambda_2 \leq \cdots \leq \lambda_N} \prod_{1 \leq i \leq N} \mathrm{d}\lambda_i \exp -S$$

$$S = \tfrac{1}{2}\beta N \sum_{1 \leq i \leq N} \lambda_i^2 - \alpha \sum_{1 \leq i \leq j \leq N} \ln|\lambda_i - \lambda_j| \tag{200}$$

The values $\alpha = 1, 2, 4$ correspond respectively to the orthogonal, unitary and symplectic ensembles. The computation of Z gives the normalization of the distribution. Observe that $\alpha > 0$ corresponds to a short-distance repulsive potential.

Rather than trying to compute Z explicitly, we obtain its asymptotic behaviour using the familiar saddle point method. We order the λ_i in increasing order as indicated in (200) and introduce a variable $x = i/N$ ranging through equally spaced values in the interval $[0, 1]$. For N large, we approximate x by a continuous variable, and write

$$\frac{1}{N} \sum_{1 \le i \le N} F(\lambda_i) = \frac{1}{N} \sum_{1 \le i \le N} F\left(\lambda\left(\frac{i}{N}\right)\right)$$
$$\simeq \int_0^1 \mathrm{d}x F(\lambda(x)) = \int_{-\infty}^{+\infty} \mathrm{d}\lambda\rho(\lambda)F(\lambda) \tag{201}$$

defining a density $\rho(\lambda)$ of eigenvalues such that

$$\mathrm{d}\lambda\rho(\lambda) = \mathrm{d}x \qquad \int_{-\infty}^{+\infty} \mathrm{d}\lambda\rho(\lambda) = 1 \tag{202}$$

In this approximation the action takes the form

$$\frac{1}{N^2}S = \beta \int \mathrm{d}\lambda \tfrac{1}{2}\lambda^2\rho(\lambda) - \tfrac{1}{2}\alpha \int \int \mathrm{d}\lambda\,\mathrm{d}\lambda'\rho(\lambda)\rho(\lambda')\ln|\lambda - \lambda'| \tag{203}$$

The stationary point in ρ yields a thermodynamic approximation for Z as well as for the distribution of eigenvalues. Using a Lagrange multiplier for the normalization constraint (202), we obtain

$$\frac{\lambda^2}{\lambda_0^2} = \int \mathrm{d}\lambda'\rho(\lambda')\ln|\lambda - \lambda'| - \xi$$
$$\lambda_0^2 = 2\alpha/\beta > 0 \tag{204}$$

provided that λ belongs to the support of $\rho(\lambda)$. The condition of normalization suggests that this support be bounded. We now seek for such a solution. Taking the derivative of both sides of equation (204) with respect to λ gives

$$2\frac{\lambda}{\lambda_0^2} = \mathrm{PP} \int \mathrm{d}\lambda' \frac{\rho(\lambda')}{\lambda - \lambda'} \tag{205}$$

where the integration symbol means principal part and λ belongs to the support of ρ. It is convenient to introduce a function $F(z)$ analytic outside the support of ρ and behaving as z^{-1} at infinity, through

$$F(z) = \int \frac{d\lambda' \rho(\lambda')}{z - \lambda'} \qquad (206)$$

According to (205), when z tends to a point λ of the support of ρ with positive infinitesimal imaginary part, $F(\lambda + io)$ tends to $2\lambda/\lambda_0^2 - i\pi\rho(\lambda)$, while outside this support $F(\lambda)$ is real. Clearly $\text{Im} F(z) < 0$ for $\text{Im} z > 0$. The ensembles are invariant under reflection $H \leftrightarrow -H$, so that $\rho(\lambda) = \rho(-\lambda)$. A unique function satisfying all the above properties and analytic in a plane cut along $[-\lambda_0, \lambda_0]$ is

$$F(z) = \frac{2}{z + \sqrt{z^2 - \lambda_0^2}} = \frac{2}{\lambda_0^2} \left[z - \sqrt{z^2 - \lambda_0^2} \right] \qquad (207)$$

and the required asymptotic distribution with support in $[-\lambda_0, \lambda_0]$ is

$$\rho(\lambda) = \frac{2}{\pi\lambda_0} \sqrt{1 - \frac{\lambda^2}{\lambda_0^2}} \qquad \lambda^2 \leq \lambda_0^2 \qquad (208)$$

The solution (208) depends only on the ratio α/β, and in a reduced scale it is independent of the random ensemble, with $\alpha = 1, 2$ or 4. The initial choice of the factor N in the Boltzmann weight was designed to lead to this limiting distribution, called Wigner's semicircle law for obvious reasons, since

$$\left(\tfrac{1}{2}\pi\lambda_0\rho(\lambda) \right)^2 + \left(\frac{\lambda}{\lambda_0} \right)^2 = 1$$

i) Show that the distribution is correctly normalized.

ii) Verify that the saddle point equation (204) is satisfied and compute $Z(\alpha, \beta)$. Starting from equation (205) and integrating in the interval $[-\lambda_0, \lambda_0]$ one obtains equation (204) with a constant ξ given by

$$\xi = \int_{-\lambda_0}^{\lambda_0} d\lambda \rho(\lambda) \ln |\lambda|$$

$$= \ln \lambda_0 + \frac{2}{\pi} \int_{-1}^{+1} d\mu \sqrt{1 - \mu^2} = \ln \lambda_0 + \tfrac{1}{2} \left[\psi(\tfrac{1}{2}) - \psi(2) \right]$$

where $\psi(x)$ is the logarithmic derivative of Euler's $\Gamma(x)$ function

$$\psi(x) = -\gamma + \sum_{n=0}^{\infty} \left(\frac{1}{n+1} - \frac{1}{n+x} \right)$$

and γ is Euler's constant. It follows that $\psi(2) = 1 - \gamma$ and $\psi(\frac{1}{2}) = -\gamma + 2\ln\frac{1}{2}$. Thus

$$\xi = \ln\left(\frac{\lambda_0}{2e^{\frac{1}{2}}} \right) \tag{209}$$

Neglecting fluctuations around the saddle point value, the partition function is given by

$$Z(\alpha, \beta) \simeq \exp -S_0(\alpha, \beta) \tag{210a}$$

$$N^{-2}S_0(\alpha,\beta) = \tfrac{1}{2}\beta\lambda_0^2 \int \mathrm{d}\lambda\rho(\lambda) \left(\frac{\lambda^2}{\lambda_0^2} - \tfrac{1}{2} \int \mathrm{d}\lambda'\rho(\lambda')\ln|\lambda - \lambda'| \right)$$

$$= \tfrac{1}{4}\beta \int \mathrm{d}\lambda\rho(\lambda)(\lambda^2 - \xi\lambda_0^2) = \tfrac{1}{8}\alpha \left[3 - \ln\frac{\alpha}{2\beta} \right] \tag{210b}$$

10.3.2 *The fermionic method*

It is interesting to return to the previous discussion and show how one can make efficient use of Grassmann variables, one of the themes of this chapter. We consider as we did earlier the mean of the resolvent $(z - H)^{-1}$ in the unitary ensemble ($\alpha = 2$, $\lambda_0^2 = 4/\beta$). Set

$$\left\langle \left(\frac{1}{z - H} \right)_{jk} \right\rangle = \delta_{jk} F(z) \tag{211a}$$

The structure of the r.h.s. follows from the invariance under unitary transformations. Thus $NF(z)$ is the average of the trace. This entails

$$\rho(\lambda) = -\frac{1}{\pi} \mathrm{Im} F(\lambda + \mathrm{i}o) \tag{211b}$$

For $\mathrm{Im}\,z > 0$, we express $F(z)$ as a mean value of the propagator for an N-component complex scalar field $(\varphi, \bar{\varphi})$. The corresponding denominator is replaced by a Grassmannian integral over an N component fermionic field $(\psi, \bar{\psi})$. Consequently

$$F(z) = -\frac{\mathrm{i}}{N} \int \prod_{1 \leq j \leq N} \frac{\mathrm{d}\varphi_j \mathrm{d}\bar{\varphi}_j \, \mathrm{d}\psi_j \mathrm{d}\bar{\psi}_j}{\pi}$$

$$\bar{\varphi} \cdot \varphi \left\langle \exp \mathrm{i} \left\{ \bar{\varphi}(z - H)\varphi + \bar{\psi}(z - H)\psi \right\} \right\rangle \tag{212}$$

According to equation (197) in the Gaussian unitary ensemble

$$\langle \exp -i\,\mathrm{Tr}(HK)\rangle = \exp -\frac{1}{2\beta N}\,\mathrm{Tr}\,K^2 \qquad (213)$$

We apply this identity to the matrix

$$(K)_{jk} = \varphi_j\bar\varphi_k - \psi_j\bar\psi_k$$
$$\mathrm{Tr}\,K^2 = (\bar\varphi\cdot\varphi)^2 - (\bar\psi\cdot\psi)^2 + 2(\varphi\cdot\bar\psi)(\bar\varphi\cdot\psi)$$

Hence

$$F(z) = -\frac{i}{N}\int \prod_{1\le j\le N}\left(\frac{d\varphi_j d\bar\varphi_j\,d\psi_j d\bar\psi_j}{\pi}\right)\bar\varphi\cdot\varphi$$
$$\exp\Big\{iz(\bar\varphi\cdot\varphi + \bar\psi\cdot\psi)$$
$$-\frac{1}{2\beta N}\big[(\bar\varphi\cdot\varphi)^2 - (\bar\psi\cdot\psi)^2 + 2(\bar\psi\cdot\varphi)(\bar\varphi\cdot\psi)\big]\Big\} \qquad (214)$$

To perform the fermionic integral, it is easier to write an integral representation of the quartic term

$$\exp\frac{1}{2\beta N}(\bar\psi\cdot\psi)^2 = \left(\frac{\beta N}{2\pi}\right)^{\frac12}\int d\mu\,\exp -\{\mu\bar\psi\cdot\psi + \tfrac12\beta N\mu^2\}$$

which enables one to perform the Grassmannian integral in the form

$$\int \prod_{1\le i\le j\le N}d\psi_j d\bar\psi_j\,\exp -\bar\psi M\psi = \det M$$
$$M_{jk} = (\mu - iz)\delta_{jk} + \frac{1}{\beta N}\varphi_j\bar\varphi_k$$

and

$$\det M = (\mu - iz)^{N-1}\left(\mu - iz + \frac{1}{\beta N}\bar\varphi\cdot\varphi\right)$$

As a result

$$F(z) = -\frac{i}{N}\left(\frac{\beta N}{2\pi}\right)^{\frac12}\int d\mu \prod_{1\le j\le N}\frac{d\varphi_j d\bar\varphi_j}{\pi}(\bar\varphi\cdot\varphi)(\mu - iz)^{N-1}$$
$$\times\left(\mu - iz + \frac{1}{\beta N}\bar\varphi\cdot\varphi\right)$$
$$\times\exp\Big\{iz\bar\varphi\cdot\varphi - \tfrac12\frac{1}{\beta N}(\bar\varphi\cdot\varphi)^2 - \tfrac12\beta N\mu^2\Big\} \qquad (215)$$

We go over to polar coordinates in φ space using the angular integral

$$\int \prod_{1 \leq j \leq N} \left(\frac{d\varphi_j d\bar{\varphi}_j}{\pi} \right) \delta(\bar{\varphi} \cdot \varphi - 1) = \frac{2}{(N-1)!} \qquad (216)$$

and obtain for $F(z)$ the following expression as a double integral, valid for any N,

$$F(z) = -\frac{2i}{N!} \left(\frac{\beta N}{2\pi} \right)^{\frac{1}{2}} \int_{-\infty}^{+\infty} d\mu \int_0^\infty d\nu \, \nu^N (\mu - iz)^{N-1}$$
$$\left(\mu - iz + \frac{\nu}{\beta N} \right) \exp \left\{ iz\nu - \frac{1}{2\beta N} \nu^2 - \tfrac{1}{2}\beta N \mu^2 \right\} \qquad (217)$$

One of the two integrals can be performed in closed form. It is sufficient for our purpose to observe that for $N \to \infty$, we can use the saddle point method. We leave it to the reader to check that this leads readily to

$$F(z) = \tfrac{1}{2}\beta \left(z - \sqrt{z^2 - 4/\beta} \right) \qquad (218)$$

which coincides with the value found in (207) for $\alpha = 2$ and $\lambda_0^2 = 4/\beta$. The above method has the advantage that it enables one to study the corrections to the saddle point method and in particular to investigate the exponentially small tails in the distribution of levels.

i) Obtain the exponential tails of the level distribution.
ii) Using the fermionic method find the asymptotic value for

$$\left\langle \mathrm{Tr}(z_1 - H)^{-1} \, \mathrm{Tr}(z_2 - H)^{-1} \right\rangle$$

10.3.3 Level spacings

While the semicircle law does not give a realistic description of most spectra, the distribution of level spacings predicted by the Gaussian ensembles seems on the contrary to have a much wider validity. We shall discuss in some detail the unitary case, leaving as an exercise the other two cases (orthogonal and symplectic).

We return to the probability law for levels ($\alpha = 2$)

$$p_N(\lambda_1,\ldots,\lambda_N) = \frac{\prod_{j<k}(\lambda_j - \lambda_k)^2 \exp -\frac{1}{2}\beta N \Sigma_j \lambda_j^2}{\int \prod_j d\lambda_j \prod_{j<k}(\lambda_j - \lambda_k)^2 \exp -\frac{1}{2}\beta N \Sigma_j \lambda_j^2}$$

(219)

Since the scale is chosen in such a way that for fixed β as $N \to \infty$, the N levels remain with unit probability in a fixed interval, the level spacings are of order N^{-1}. If we are interested in the distribution of these spacings with a fixed average, we have to rescale the levels by a factor N.

Let $\chi(\lambda)$ be the characteristic function of an interval $[\lambda_0, \lambda_1]$

$$\chi(\lambda) = \begin{cases} 1 & \text{if} \quad \lambda \in [\lambda_0, \lambda_1] \\ 0 & \text{if} \quad \lambda \notin [\lambda_0, \lambda_1] \end{cases}$$

(220)

Therefore $1 - \chi$ is the characteristic function of the real axis with the interval $[\lambda_0, \lambda_1]$ deleted. The quantity

$$E(\lambda_0, \lambda_1) = \int \prod_{1 \le j \le N} \left[d\lambda_j (1 - \chi(\lambda_j)) \right] p(\lambda_1 \ldots, \lambda_{,N})$$

(221)

is the probability of finding no eigenvalue in the range $[\lambda_0, \lambda_1]$. The combination

$$E(\lambda_0, \lambda_1) - E(\lambda_0 - \delta\lambda_0 \lambda_1) - E(\lambda_0, \lambda_1 + \delta\lambda_1) + E(\lambda_0 - \delta\lambda_0, \lambda_1 + \delta\lambda_1)$$

$$\simeq -\delta\lambda_0 \delta\lambda_1 \frac{\partial^2}{\partial\lambda_0 \partial\lambda_1} E(\lambda_0, \lambda_1)$$

gives the probability of finding no level in the interval $[\lambda_0, \lambda_1]$ but to find at least one in $[\lambda_1, \lambda_1 + \delta\lambda_1]$ as well as in $[\lambda_0 - \delta\lambda_0, \lambda_0]$. In the limit where $\delta\lambda_0, \delta\lambda_1$ tend to zero, each of the boundary interval will only contain one level with unit probability. Let us assume that in the appropriate scale, as $N \to \infty$, $E(\lambda_0, \lambda_1)$ only depends on the difference $\lambda_1 - \lambda_0$ and not on the midpoint $\frac{1}{2}(\lambda_0 + \lambda_1)$ which we shall subsequently choose at the center of the spectrum to simplify matters. We see that

$$q(\lambda_1 - \lambda_0) = \frac{\partial^2}{\partial(\lambda_1 - \lambda_0)^2} E(\lambda_1 - \lambda_0)$$

(222)

is the probability of finding two successive levels at a distance $\lambda_1 - \lambda_0$. Here the key word is successive. It is possible also to study the two level probability distribution and it is likely that its short-distance behaviour coincides with the one for level spacings. The two quantities are unrelated at large separations.

To compute $E(\lambda_0, \lambda_1)$, we use an analogy between the probability law (219) and a one-dimensional gas of fermions in an harmonic potential. We stress that these fermions are unrelated to those appearing in the previous subsection. Indeed $p(\lambda_1, \ldots, \lambda_N)$ is the (modulus) square of an antisymmetric wavefunction

$$\psi(\lambda_1, \ldots, \lambda_N) = \mathcal{N}^{-1} \prod_{j<k} \left(\lambda_j - \lambda_k\right) \exp\left(-\tfrac{1}{4}\beta N \sum \lambda_i^2\right) \quad (223)$$

with $\mathcal{N}^{-1}$ a normalization factor. We can represent ψ as a Slater determinant constructed out of the first N eigenfunctions of an harmonic oscillator. They satisfy

$$(2a^\dagger a + 1)\psi_n = (2n+1)\psi_n \qquad [a, a^\dagger] = 1$$

$$a = \frac{1}{\sqrt{2}}\left(\sqrt{K}\lambda + \frac{1}{\sqrt{K}}\frac{\partial}{\partial\lambda}\right) \qquad a^\dagger = \frac{1}{\sqrt{2}}\left(\sqrt{K}\lambda - \frac{1}{\sqrt{K}}\frac{\partial}{\partial\lambda}\right)$$

$$K = \tfrac{1}{2}\beta N \qquad\qquad (224)$$

Hence

$$\psi_0(\lambda) = \left(\frac{K}{\pi}\right)^{\frac{1}{4}} e^{-K\frac{1}{2}\lambda^2}$$

$$\psi_n(\lambda) = \frac{a^{\dagger n}}{\sqrt{n!}}\psi_0(\lambda) = \left(\frac{K}{\pi}\right)^{\frac{1}{4}} \left(\frac{2^n K^n}{n!}\right)^{\frac{1}{2}} (\lambda^n + \cdots)e^{-K\frac{1}{2}\lambda^2}$$

$$\int_{-\infty}^{+\infty} d\lambda \psi_n(\lambda)\psi_m(\lambda) = \delta_{n,m} \qquad (225)$$

Thus we find

$$\psi(\lambda_1, \ldots, \lambda_N) = \frac{1}{(N!)^{\frac{1}{2}}} \begin{vmatrix} \psi_0(\lambda_N) & \cdots & \psi_{N-1}(\lambda_N) \\ \vdots & \ddots & \vdots \\ \psi_0(\lambda_1) & \cdots & \psi_{N-1}(\lambda_1) \end{vmatrix} \qquad (226)$$

$$\psi(\lambda_1, \ldots, \lambda_N) = \frac{1}{\left(\prod_1^N n!\right)^{\frac{1}{2}}} \left(\frac{K}{\pi}\right)^{\frac{1}{4}N} (2K)^{\frac{1}{2}N(N-1)}$$

$$\times \prod_{1\leq j<k\leq N} (\lambda_j - \lambda_k) \exp\left(-\tfrac{1}{2}K \sum_{1\leq j\leq N} \lambda_j^2\right) \quad (227)$$

Hence $p(\lambda_1, \ldots, \lambda_N)$ given by

$$p(\lambda_1, \ldots, \lambda_N) = |\psi(\lambda_1, \ldots, \lambda_N)|^2 \qquad (228)$$

$$= \frac{1}{\prod_1^N n!} \left(\frac{K}{\pi}\right)^{\frac{1}{2}N} (2K)^{N(N-1)} \prod_{j<k}(\lambda_j - \lambda_k)^2 \exp -K \sum_j \lambda_j^2$$

satisfies the correct normalization condition

$$\int \prod_j d\lambda_j p(\lambda_1, \ldots, \lambda_n) = 1 \tag{229}$$

Obtain the partition function $Z(\alpha = 2, \beta)$ using the above normalized probability.

Let us now turn to the computation of $E(\lambda_0, \lambda_1)$. From the fact that $(1 - \chi(\lambda))^2 = 1 - \chi(\lambda)$, it follows that if we set

$$f_n(\lambda) = (1 - \chi(\lambda))\psi_n(\lambda) \tag{230}$$

the expression for $E(\lambda_0, \lambda_1)$ reduces to

$$E(\lambda_0, \lambda_1) = \int \prod_{1 \le j \le N} d\lambda_j \frac{1}{N!} |\det f_k(\lambda_l)|^2$$

Expanding the determinant and integrating, one finds

$$E(\lambda_0, \lambda_1) = \det \left[\delta_{jk} - \int_{\lambda_0}^{\lambda_1} d\lambda \psi_j(\lambda)\psi_k(\lambda) \right] \Big|_{0 \le j,k \le N-1} \tag{231}$$

Gaudin has noticed that if one restricts the Darboux–Christoffel kernel

$$\mathcal{K}_N(\lambda, \lambda') = \sum_{0 \le j \le N-1} \psi_j(\lambda)\psi_j(\lambda')$$

$$= \sqrt{\frac{2N}{K}} \frac{\psi_N(\lambda)\psi_{N-1}(\lambda') - \psi_{N-1}(\lambda)\psi_N(\lambda')}{\lambda - \lambda'} \tag{232}$$

to the interval $[\lambda_0, \lambda_1]$, and if one looks for its N eigenvalues ξ and eigenvectors φ in the subspace generated by the functions $\psi_j(\lambda)$, the eigenvalue equation

$$\xi\varphi(\lambda) = \int_{\lambda_0}^{\lambda_1} d\lambda' \mathcal{K}_N(\lambda, \lambda')\varphi(\lambda') \tag{233a}$$

can be rewritten

$$\xi\varphi^j = \sum_{0 \le k \le N-1} \int_{\lambda_0}^{\lambda_1} d\lambda \psi_j(\lambda)\psi_k(\lambda)\varphi^k \tag{233b}$$

when one sets $\varphi = \sum_{0 \leq j \leq N-1} \varphi^j \psi_j(\lambda)$. The characteristic equation takes the form

$$\prod_{1 \leq s \leq N} (\xi - \xi_s) = \det \left[\xi \delta_{jk} - \int_{\lambda_0}^{\lambda_1} d\lambda \psi_j(\lambda) \psi_k(\lambda) \right] \tag{234}$$

Consequently, $E(\lambda_0, \lambda_1)$ can be expressed in terms of the eigenvalues ξ_s as

$$E(\lambda_0, \lambda_1) = \prod_{1 \leq s \leq N} (1 - \xi_s) \tag{235}$$

We look for an asymptotic form for the kernel $\mathcal{K}_N(\lambda, \lambda')$ as N gets large. This is obtained using a semiclassical approximation for the functions $\psi_j(\lambda)$. We restrict ourselves to a domain of order $1/N$ around the origin and recall that $K = \frac{1}{2}\beta N$ is of order N. In the equation satisfied by $\psi_N(\lambda)$

$$\left[\frac{\partial^2}{\partial \lambda^2} - K^2 \lambda^2 + K(2N+1) \right] \psi_N(\lambda) = 0$$

we neglect $K^2\lambda^2$ (of order one) as compared to $K(2N+1)$ (of order N^2). Thus for $\lambda(2NK)^{\frac{1}{2}}$ finite and $N \to \infty$, we find

$$\psi_N(\lambda) \sim \begin{cases} \psi_N(0) \cos \lambda (2NK)^{\frac{1}{2}} & N \text{ even} \\ \psi'_N(0)(2NK)^{-\frac{1}{2}} \sin \lambda (2NK)^{\frac{1}{2}} & N \text{ odd} \end{cases} \tag{236}$$

One can easily obtain $\psi_N(0)$ and $\psi'_N(0)$ in these expressions from the definitions (225). From $\sqrt{N+1}\psi_{N+1}(0) - \sqrt{N}\psi_{N-1}(0) = 0$ for N odd, we derive

$$\psi_{2s}(0) = (-1)^s \left(\frac{K}{\pi} \right)^{\frac{1}{4}} \frac{(2s!)^{\frac{1}{2}}}{2^s s!} \underset{s \to \infty}{\sim} (-1)^s \left(\frac{K}{s\pi^2} \right)^{\frac{1}{4}}$$

$$\psi'_{2s+1}(0) \underset{s \to \infty}{\sim} (-1)^s (4sK)^{\frac{1}{2}} \left(\frac{K}{s\pi^2} \right)^{\frac{1}{4}} \tag{237}$$

Inserting these evaluations in (232), one gets asymptotically, and close to the origin,

$$\mathcal{K}_N(\lambda, \lambda') \underset{N \to \infty}{\sim} \frac{\sin \left[(2NK)^{\frac{1}{2}} (\lambda - \lambda') \right]}{\pi(\lambda - \lambda')} \tag{238}$$

According to equation (208), the density of levels at the center of the distribution is $\rho(0) = 2/\pi\lambda_0$ with $\lambda_0^2 = 4/\beta = 4N/\beta N$, hence

$$N\rho(0) = \frac{(2KN)^{\frac{1}{2}}}{\pi} = \frac{1}{D} \tag{239}$$

where D is the average level spacing. We can rewrite the asymptotic kernel $\mathcal{K}$, which only depends on the difference $\lambda - \lambda'$, in the form

$$\mathcal{K}(\lambda, \lambda') = \frac{1}{D}\frac{\sin[\pi(\lambda - \lambda')/D]}{[\pi(\lambda - \lambda')/D]} \tag{240}$$

Choose D as a scale for eigenvalues by setting

$$\lambda = Dx \tag{241}$$

and denote by $\varphi(x)$ the eigenfunction of $\mathcal{K}$ in an interval

$$-\frac{1}{2}t \le x \le \frac{1}{2}t \quad (t > 0)$$

We have

$$\xi\varphi(x) = \int_{-\frac{1}{2}t}^{\frac{1}{2}t} dx' \frac{\sin\pi(x - x')}{\pi(x - x')}\varphi(x') \tag{242}$$

The kernel is even, meaning that we can split the eigenfunctions into even, $\varphi^{(+)}(x)$, and odd ones, $\varphi^{(-)}(x)$. In both cases with $\varepsilon = \pm 1$

$$\varepsilon\xi^{(\varepsilon)}\varphi^{(\varepsilon)}(x) = \int_{-\frac{1}{2}t}^{\frac{1}{2}t} dx' \frac{\sin\pi(x + x')}{\pi(x + x')}\varphi^{(\varepsilon)}(x') \tag{243}$$

The identity

$$\frac{\sin\pi(x + x')}{\pi(x + x')} = \int_{-\frac{1}{2}t}^{\frac{1}{2}t} dx'' \frac{\exp 2i\pi xx''/t}{\sqrt{t}} \cdot \frac{\exp 2i\pi x''x'/t}{\sqrt{t}} \tag{244}$$

shows that the kernel $\sin\pi(x + x')/\pi(x + x')$ is the square of the exponential kernel $t^{-\frac{1}{2}}\exp 2i\pi xx'/t$ in the interval $-\frac{1}{2}t \le x \le \frac{1}{2}t$. We can therefore choose $\varphi^{(\varepsilon)}(x)$ as an eigenvector of this second kernel with eigenvalues $\sqrt{t}\gamma^{(+)}$ or $i\sqrt{t}\gamma^{(-)}$

$$\int_{-\frac{1}{2}t}^{\frac{1}{2}t} \frac{dx'}{\sqrt{t}} \exp\frac{2i\pi xx'}{t}\varphi^{(\pm)}(x) = \int_{-\frac{1}{2}t}^{\frac{1}{2}t} \frac{dx'}{\sqrt{t}} \begin{pmatrix} \cos(2\pi xx'/t)\,\varphi^{(+)}(x') \\ i\sin(2\pi xx'/t)\,\varphi^{(-)}(x') \end{pmatrix}$$

$$= \begin{pmatrix} \sqrt{t}\gamma^{(+)}\varphi^{(+)}(x) \\ i\sqrt{t}\gamma^{(-)}\varphi^{(-)}(x) \end{pmatrix} \tag{245}$$

We have the relation

$$\xi^{(\pm)} = t\gamma^{(\pm)^2} \tag{246}$$

By a change of variable

$$x = \tfrac{1}{2}ty \qquad \varphi^{(\pm)}(x) = F^{(\pm)}(y) \tag{247}$$

this last system reads

$$\gamma^{(+)}F^{(+)}(y) = \int_0^1 dy' \cos \tfrac{1}{2}\pi t y y'\, F^{(+)}(y')$$
$$\gamma^{(-)}F^{(-)}(y) = \int_0^1 dy' \sin \tfrac{1}{2}\pi t y y'\, F^{(-)}(y') \tag{248}$$

and

$$E^{(\pm)}(t) = \prod_{s=1}^{\infty} \left(1 - t\gamma_s^{(\pm)^2}\right)$$
$$E(t) = E^{(+)}(t)E^{(-)}(t) \tag{249}$$

The γ's are also functions of t. The latter is the separation of eigenvalues (of the original random matrix) in relative units. In these units, the relation (222) between $E(t)$ and $q(t)$ remains valid. The eigenvalue equations (248) appear in various other contexts, in particular in communication analysis.

To summarize, in the scale $x = \lambda/D$, and with a kernel $\mathcal{K}_t(x, x') = \sin \pi(x - x')/\pi(x - x')$ restricted to the interval $[-\tfrac{1}{2}t, \tfrac{1}{2}t]$ we have

$$E(t) = \det(1 - \mathcal{K}_t) = \exp - \sum_1^{\infty} \frac{\operatorname{Tr} \mathcal{K}_t^n}{n}$$
$$q(t) = \frac{d^2 E(t)}{dt^2} \tag{250}$$

For small t we can expand $\operatorname{Tr} \mathcal{K}_t^n$ in powers of t, the leading term being t^n,

$$\operatorname{Tr} \mathcal{K}_t = t$$
$$\operatorname{Tr} \mathcal{K}_t^2 = t^2 - \tfrac{1}{18}\pi^2 t^4 + \tfrac{1}{675}\pi^4 t^6 - \tfrac{1}{8820}\pi^6 t^8 + \cdots$$
$$\operatorname{Tr} \mathcal{K}_t^3 = t^3 - \tfrac{1}{12}\pi^2 t^5 + \tfrac{1}{225}\pi^4 t^7 + \cdots$$
$$\operatorname{Tr} \mathcal{K}_t^4 = t^4 - \tfrac{1}{9}\pi^2 t^6 + \tfrac{121}{16200}\pi^4 t^8 + \cdots$$
$$\operatorname{Tr} \mathcal{K}_t^5 = t^5 - \tfrac{5}{36}\pi^2 t^7 + \cdots$$
$$\operatorname{Tr} \mathcal{K}_t^6 = t^6 - \pi^2 t^8 \tfrac{1}{6} + \cdots$$

$$\operatorname{Tr} \mathcal{K}_t^7 = t^7 + \cdots$$
$$\operatorname{Tr} \mathcal{K}_t^8 = t^8 + \cdots \tag{251}$$

Thus

$$E(t) = 1 - t + \tfrac{1}{36}\pi^2 t^4 - \tfrac{1}{675}\pi^4 t^6 + \tfrac{1}{17640}\pi^6 t^8 + O(t^{10})$$
$$q(t) = \tfrac{1}{3}(\pi t)^2 - \tfrac{2}{45}(\pi t)^4 + \tfrac{1}{315}(\pi t)^6 + O(t^8) \tag{252}$$

The behaviour of $q(t)$ in t^2 for small t reflects the quadratic repulsive factor of eigenvalues at short distance, as shown in the probability law (219). The convergence of the expansion (252) is limited by the first zero in the determinant $E(t)$.

i) The similar result for the Gaussian orthogonal ensemble is given by $E^{(+)}(t)$ (equation (249)) with an expansion

$$E^{(+)}(t) = 1 - t + \tfrac{1}{36}\pi^2 t^2 - \tfrac{1}{1200}\pi^4 t^5 + \tfrac{1}{8100}\pi^4 t^6 + \tfrac{1}{70560}\pi^6 t^7 + \cdots$$
$$q^{(+)}(t) = \tfrac{1}{6}\pi^2 t - \tfrac{1}{60}\pi^4 t^3 + \tfrac{1}{270}\pi^4 t^4 + \tfrac{1}{1680}\pi^6 t^5 + \cdots \tag{253}$$

Note that here $q^{(+)}(t)$ is linear in t for small t as expected.

ii) Show that asymptotically for $t \to \infty$

$$q(t) \sim e^{-\frac{1}{8}\pi^2 t^2} \qquad q^{(+)}(t) \sim e^{-\frac{1}{16}\pi^2 t^2} \tag{254}$$

and try to find the prefactors.

The understanding of the universality class of these spacing laws, and knowledge whether there might exist systems exhibiting interesting deviations is still unresolved. At any rate there exists a wide range of systems exhibiting quantum fluctuations where the techniques and results of the Gaussian ensembles find interesting applications.

10.4   The planar approximation

The study of Gaussian random matrices is a particular case of matrix valued field theories when the size N of the matrices grows indefinitely. One would *a priori* expect many simplifications to arise in this limit, on the basis of the analogy with the large n-limit of the vector model. This is precisely what 't Hooft (1974) suggested in the context of gauge fields. He discussed the nature of the corresponding simplifications, showing that it led to a new topological perturbative expansion, where the dominant term is the resummation of all diagrams which can be drawn on a plane (or by compactification on a sphere), hence the name planar approximation. Although this topic seems at first sight only loosely connected to the subject of this chapter, except by providing an extension of the Gaussian ensembles of random matrices, and extending some of the techniques used above, it turns out that it leads to an evaluation of the number of random triangulations of surfaces which ties in with another aspect of random systems (chapter 11). We shall limit ourselves to considerations in low dimension which enable one to exhibit the phenomena and to solve some combinatorial problems.

10.4.1   Combinatorics

We consider a matrix valued scalar field, denoted $M(\mathbf{x})$, an Hermitian $N \times N$ matrix. A general Lagrangian, invariant under the adjoint action of a global unitary group $SU(N)$ reads

$$\mathcal{L} = \tfrac{1}{2} \operatorname{Tr} \partial_\mu M \partial_\mu M + \tfrac{1}{2} \operatorname{Tr} M^2 + \sum_p g_p N^{1-\frac{1}{2}p} \operatorname{Tr} M^P \qquad (255)$$

We shall justify later the choice of a factor $N^{1-\frac{1}{2}p}$ associated to a term of order p. Let us show that, for fixed coupling constants and in the large N limit, only planar diagrams survive in the perturbative expansion. The crucial part in the Feynmann rules is the one giving the factors associated to the matrix nature of the field. The remaining rules are the familiar ones in a scalar theory, propagators, integrals, etc. It is convenient to represent the propagators of the field M by double lines, with an opposite orientation on each one, associated to the propagation of each of the indices. Let us study the N-dependence of a diagram with P propagators and V_p vertices of type p $\left(V = \sum_{3 \leq p} V_p \right)$. To simplify, we look at connected vacuum diagrams. If one counts each propagator twice, one counts p times the vertices of type p, hence

$$2P = \sum_{3 \leq p} p V_p \qquad (256)$$

Furthermore each of the B internal loops may be considered as the boundary of an oriented polyhedral surface. Joigning these faces along the propagators, one obtains a closed oriented connected surface, satisfying Euler's formula

$$\chi = 2 - 2H = V - P + B \qquad (257)$$

where χ is the Euler characteristic and we have denoted the genus by H (for handles), to avoid confusion with the coupling constants g_p. After summation over internal indices, each loop produces a factor N, so that, apart from the standard factors, the contribution of a diagram contains the product

$$\prod_{3 \le p} \left(g_p N^{1-\frac{1}{2}p} \right)^{V_p} N^B = \left(\prod_{3 \le p} g_p^{V_p} \right) N^{2-2H} \qquad (258)$$

In the limit $N \to \infty$, the leading contribution to the free energy (the generating function of connected vacuum diagrams) arises from diagrams drawn on a surface of genus zero ($\chi = 2$, $H = 0$), i.e. the planar ones. Corrections of order N^{-2} correspond to diagrams drawn on a torus, etc. The above reasoning can be generalized when source terms are added to the Lagrangian, provided that they carry the appropriate powers of N.

We set

$$E(g) = - \lim_{N \to \infty} \frac{1}{N^2} F(g) = - \lim_{N \to \infty} Z_N(g) \qquad (259)$$

As an illustration, we consider a φ^4 model ($g_4 \equiv g$), and enumerate the weights corresponding to the first few planar diagrams up to order g^3

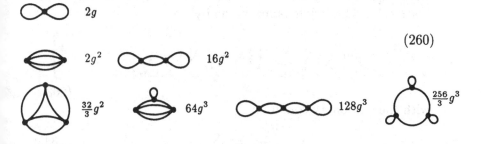

$$(260)$$

Extend the previous considerations to the (quartic) Lagrangian

$$\mathcal{L} = \operatorname{Tr} \partial_\mu M \partial_\mu M^\dagger + \operatorname{Tr} M M^\dagger + \frac{\alpha g}{2N} \operatorname{Tr} M M^\dagger M M^\dagger \qquad (261a)$$

where

$$\alpha = 1 \quad M \text{ real orthogonal}$$
$$\alpha = 2 \quad M \text{ Hermitian} \tag{261b}$$
$$\alpha = 4 \quad M \text{ complex}$$

with corresponding symmetry groups $SO(N)$, $SU(N)$ and $SU(N) \times SU(N)$. The leading terms are the same in all cases, but the dependence on N of subleading corrections varies according to the different choices of matrices.

In the zero-dimensional case, $E(g)$ is a generating function for combinatorial factors such as those shown in equation (260). The method used in the preceding section can be extended to evaluate this quantity. We present the computation for a quartic interaction, but there is no difficulty to include any polynomial interaction, even when it is unbounded from below. It is understood that the large N limit will be taken at the end of the calculation. Up to a normalizing g-independent constant, we have

$$\exp -N^2 E(g) = \text{cst} \int dM \exp - \left\{ \tfrac{1}{2} \text{Tr}\, M^2 + \frac{g}{N} \text{Tr}\, M^4 \right\}$$

$$= \int \prod_{1 \le i \le N} d\lambda_i \prod_{1 \le i < j \le N} (\lambda_i - \lambda_j)^2$$

$$\times \exp - \left\{ \sum_{1 \le i \le N} \tfrac{1}{2}\lambda_i^2 + \frac{g}{N}\lambda_i^4 \right\} \tag{262}$$

with $\{\lambda_i\}$ standing for the N real eigenvalues of the Hermitian matrix M. The saddle point method yields

$$E(g) = \lim_{N \to \infty} \frac{1}{N^2} \left\{ \sum_{1 \le i \le N} \left(\tfrac{1}{2}\lambda_i^2 + \frac{g}{N}\lambda_i^4 \right) - \sum_{1 \le i \ne j \le N} \left| \lambda_i - \lambda_j \right| \right\} \tag{263a}$$

with the stationary condition

$$\tfrac{1}{2}\lambda_i + \frac{2g}{N}\lambda_i^3 = \sum_{1 \le j \le N} \frac{1}{\lambda_i - \lambda_j} \tag{263b}$$

We order the eigenvalues and, in the large N limit, substitute to the discrete sequence of λ_i's the continuous function $\lambda_i =$

$\sqrt{N}\lambda(i/N)$. This translates into the approximation

$$E(g) = \int_0^1 dx\, \tfrac{1}{2}\lambda^2(x) + g\lambda^4(x) - \int_0^1 \int_0^1 dx\, dy \ln |\lambda(x) - \lambda(y)|$$

$$\tfrac{1}{2}\lambda(x) + 2g\lambda^3(x) = PP \int_0^1 dy \frac{1}{\lambda(x) - \lambda(y)}$$

$$(264)$$

We define the normalized density of eigenvalues $u(\lambda)$, according to

$$dx = u(\lambda)\, d\lambda$$

with $u(\lambda)$ even positive, with support in an interval $[-2a, 2a]$ satisfying

$$1 = \int_0^1 dx = \int_{-2a}^{+2a} d\lambda u(\lambda) \qquad (265)$$

Thus

$$\tfrac{1}{2}\lambda + 2g\lambda^3 = PP \int_{-2a}^{2a} dx \frac{u(\mu)}{\lambda - \mu}, \qquad -2a \le \lambda \le 2a \quad (266)$$

The analytic function

$$F(z) = \int_{-2a}^{2a} \frac{d\mu u(\mu)}{z - \mu} \qquad (267a)$$

is odd, defined in the complex plane cut along $[-2a, 2a]$, real on the real axis outside this interval. It behaves as z^{-1} as z goes to infinity and satisfies

$$F(\lambda \pm io) = \tfrac{1}{2}\lambda + 2g\lambda^3 \mp i\pi u(\lambda) \qquad -2a \le \lambda \le 2a \quad (267b)$$

The unique solution is

$$F(z) = \tfrac{1}{2}z + 2gz^3 - \left(\tfrac{1}{2} + 4ga^2 + 2gz^2\right)\sqrt{z^2 - 4a^2} \qquad (268)$$

The prefactor of the square root is such that $F(z)$ decreases as z^{-1} for z infinite. This requires that

$$12ga^4 + a^2 - 1 = 0 \qquad (269)$$

For small g, this means that a^2 has a convergent expansion

$$a^2 = \frac{1}{24g}\left[(1 + 48g)^{\frac{1}{2}} - 1\right] = 1 - 12g + 2(12g)^2 - 5(12g)^3 + \cdots$$

$$(270)$$

In equation (268), the square root is chosen to be positive for real z larger than $2a$. The discontinuity of F yields an expression for $u(\lambda)$ generalizing Wigner's semicircle law

$$u(\lambda) = \frac{1}{\pi} \left\{ \tfrac{1}{2} + 4ga^2 + 2g\lambda^2 \right\} \sqrt{4a^2 - \lambda^2} \qquad (271)$$

Returning to (264), one finds $E(g)$ as

$$E(g) - E(0) = \tfrac{1}{24}(a^2 - 1)(9 - a^2) - \tfrac{1}{2}\ln a^2 \qquad (272)$$

From the expression of a^2 as a function of g, this produces the series

$$
\begin{aligned}
E_0(g) \equiv E(g) - E(0) &= -\sum_{1}^{\infty} (-12g)^k \frac{(2k-1)!}{k!(k+2)!} \\
&= 2g - 18g^2 + 288g^3 - 6048g^4 + \cdots
\end{aligned}
\qquad (273)
$$

in agreement with the first few terms exhibited in equation (260).

In contradistinction to the complete perturbative series, the planar approximation enjoys the properties that the corresponding expansion is convergent, hence analytic around zero coupling $g = 0$. The nearest singularity arises from the expression of $a^2(g)$ and is located on the real negative axis at

$$g_c = -\tfrac{1}{48} \qquad (274)$$

As a result we obtain the asymptotic behaviour of the coefficients in the expansion (273)

$$E_0(g) = -\sum_{1}^{\infty} A_k(-g)^k \qquad A_k \sim \frac{1}{2\sqrt{\pi}}(48)^k k^{-\frac{7}{2}} \qquad (275)$$

i) The same method yields expressions for "Green functions" in the planar approximation. Thus

$$G_{2p}(g) = \left\langle \mathrm{Tr}\, M^{2p} \right\rangle = \int_{-2a}^{+2a} d\lambda\, u(\lambda)\lambda^{2p} \qquad (276)$$

Mean values of odd powers obviously vanish. From

$$zF(z) = \int_{-2a}^{2a} d\lambda\, u(\lambda)\frac{z}{z-\lambda} = \sum_{p=0}^{\infty} z^{-2p} G_{2p}(g)$$

we derive

$$G_{2p}(g) = \frac{(2p)!}{p!(p+2)!} a^{2p}(2p+2-pa^2)$$

$$= \frac{(2p)!}{p!(p-1)!} \sum_{k=0}^{\infty} \frac{(2k+p-1)!}{k!(k+p+1)!}(-12g)^k \qquad (277)$$

Similarly one can generate connected Green functions G^c_{2p}, defined recursively according to

$$G_{2p} = \sum_{\substack{r_q \geq 0 \\ \Sigma q r_q = p}} \frac{(2p)!}{(2p+1-\Sigma r_q)!} \frac{(G^c_2)^{r_1}}{r_1!} \frac{(G^c_4)^{r_2}}{r_2!} \cdots \frac{(G^c_{2q})^{r_q}}{r_q!} \cdots \qquad (278)$$

and one finds

$$G^c_{2p}(g) = \frac{(3p-1)!3^{1-p}}{(p-1)!(2p-1)!} \sum_{p=1}^{\infty} \frac{(-12g)^k(2k+p-1)!}{(k-p+1)!(k+2p)!} \qquad (279)$$

Finally, if the one-particle irreducible functions (the vertex functions) are expressed as

$$\Gamma_2 = [G^c_2]^{-1}$$

$$-\Gamma_4 = G^c_4 [G^c_2]^{-4}$$

$$-\Gamma_6 = G^c_6 [G^c_2]^{-6} - 3 [G^c_4]^2 [G^c_2]^{-7}$$

$$-\Gamma_8 = G^c_8 [G^c_2]^{-8} - 8 G^c_6 G^c_4 [G^c_2]^{-9} + 12 [G^c_4]^3 [G^c_2]^{-10}$$

$$\cdots \qquad (280a)$$

their generating function (analogous to the free energy expressed in terms of the magnetization after a Legendre transformation)

$$\Gamma(x) = \sum_{p=1}^{\infty} \Gamma_{2p} x^{2p} \qquad (280b)$$

satisfies the algebraic equation

$$3x^2(1-a^2)(1+\Gamma)^2 + 9a^4\Gamma(1+\Gamma-\Gamma^2/x^2) - a^2(2+a^2)^2\Gamma = 0 \quad (280c)$$

from which one can derive an expansion in powers of g.

ii) It is also possible to study systematically the corrections to the planar approximation. This amounts to enumerating the diagrams which are drawn on a torus, a double torus, etc. Let the corresponding contributions be denoted $E_1(g)$, $E_2(g)$, ... In the Hermitian case ($\alpha = 2$) we have

$$E(g) - E(0) = E_0(g) + \frac{1}{N^2} E_1(g) + \frac{1}{N^4} E_2(g) + \cdots \qquad (281)$$

A useful technique to obtaining the successive terms relies on the asymptotic expansion for large order of the orthogonal polynomials with respect to the measure $d\lambda \, \exp -(\frac{1}{2}\lambda^2 + (g/N)\lambda^4)$. These polynomials generalize the Hermite polynomials used in the previous section. One is led to the expressions

$$E_1(g) = \tfrac{1}{12} \ln(2 - a^2)$$

$$E_2(g) = \frac{1}{6!} \frac{(1 - a^2)^3}{(2 - a^2)^5}(82 + 21a^2 - 3a^4) \qquad (282)$$

$$\cdots$$

with a^2 given by equation (270). For $E_H(g)$ the leading term in g is, for any H,

$$E_H(g) = 2^{2H-2} \frac{(4H - 3)!}{H!(H - 1)!}g^{2H-1} + \cdots \qquad (283)$$

iii) We have discussed a quartic interaction, but the method obviously extends to any polynomial interaction in particular to a cubic one (to give a meaning to the original integral, we may assume the coupling constant to be pure imaginary). We sketch briefly the results in this case (useful by duality to study triangulations on surfaces). One wishes to evaluate for large N the integral

$$\exp -N^2 E(g) = \int \prod_{1 \le i \le N} d\lambda_i \prod_{1 \le i < j \le N} (\lambda_i - \lambda_j)^2$$

$$\times \exp - \sum_{1 \le i \le N} \left(\tfrac{1}{2}\lambda_i^2 + \frac{g}{\sqrt{N}}\lambda_i^3 \right) \qquad (284)$$

We apply the saddle point method with

$$\lambda_i = \sqrt{N}\lambda \left(\frac{i}{N} \right) \qquad\qquad d\lambda v(\lambda) = 2\, dx \qquad (285)$$

This leads to the equations

$$\lambda + 3g\lambda^2 = PP \int_{2a}^{2b} \frac{v(\mu)\, d\mu}{\lambda - \mu} \qquad 2a \le \lambda \le 2b \quad (286)$$

$$\int_{2a}^{2b} d\lambda v(\lambda) = 2$$

The support of $v(\lambda)$ is not symmetric except in the limit $g \to 0$. The function $F(z)$ having $v(\lambda)$ as its discontinuity along the cut $[2a, 2b]$, real outside this interval on the real axis, and vanishing as $2/z$ at infinity, is

$$F(z) = \int_{2a}^{2b} \frac{v(\lambda)\, d\lambda}{z - \lambda}$$
$$= z + 3gz^2 - [3gz + 1 + 3g(a + b)]\,[(z - 2a)(z - 2b)]^{\frac{1}{2}} \quad (287)$$

Set

$$\sigma = 3g(a + b) \quad\quad\quad (288)$$

The required behaviour of F at infinity gives the condition

$$18g^2 + \sigma(1 + \sigma)(1 + 2\sigma) = 0 \quad\quad\quad (289)$$

$$\sigma = -\tfrac{1}{4} \sum_{1}^{\infty} \frac{(72g^2)^k}{k!} \frac{\Gamma\left(\tfrac{1}{2}(3k - 1)\right)}{\Gamma\left(\tfrac{1}{2}(k + 1)\right)}$$

This yields for $E(g)$

$$E(g) - E(0) = -\frac{\sigma(3\sigma^2 + 6\sigma + 2)}{3(1 + \sigma)(1 + 2\sigma)^2} + \tfrac{1}{2}\ln(1 + 2\sigma)$$
$$= -\tfrac{1}{2} \sum_{k=1}^{\infty} \frac{(72g^2)^k}{(k + 2)!} \frac{\Gamma(3k/2)}{\Gamma(k/2 + 1)} \quad\quad\quad (290)$$

All these expressions are analytic around $g = 0$, the nearest singularity being

$$g_c^2 = \frac{1}{108\sqrt{3}}$$

The generating function of connected Green functions

$$\psi(j) = 1 + \sum_{1}^{\infty} j^p G_p^c \quad\quad\quad (291a)$$

obeys the equation

$$3g\psi^2 + (j - 3g)\psi - \tfrac{1}{2}j\frac{(1 + \sigma)(2 + 3\sigma)}{1 + 2\sigma} - j^3 = 0 \quad\quad\quad (291b)$$

One can also modify the Green functions by eliminating the tadpole contributions corresponding to $G_1 = G_1^c$. To this effect, one introduces an extra linear term $\rho \sum_{1 \le i \le N} \lambda_i$ in the action and one fixes the value of the parameter ρ by requiring the new one-point function G_1 to vanish. The modified generating function $\tilde{\psi}$ obeys

$$3g\tilde{\psi}^2 + (j - 3g)\tilde{\psi} - j(1 - \rho j + j^2) = 0 \quad\quad\quad (292a)$$

where ρ is related parametrically to g through

$$3g\rho = -\tau(1 - 3\tau)$$
$$3g^2 = \tau(1 - 2\tau)^2 \tag{292b}$$

Eliminating τ, we obtain for $\tilde{\psi}$ an explicit expansion in terms of graphs with E external lines, and V vertices (the remaining topological characteristics being fixed by the conditions that we deal with connected planar graphs with vertices of valence three)

$$\tilde{\psi}(j,g) = 1 + \sum_{E=2}^{\infty} j^E \sum_{\substack{V=E-2 \\ V-E \text{ even}}}^{\infty} (-3g)^V \tilde{\psi}_{E,V} \tag{293}$$

$$\tilde{\psi}_{E,V} = \frac{2E-2}{(E-1)!(E-2)!} 2^{\frac{1}{2}(V-E)+1} \frac{\left(\frac{1}{2}(3V+E)-1\right)!}{(E+V)!\left(\frac{1}{2}(V-E)+1\right)!}$$

To summarize, the method leads to algebraic equations in the pure combinatorial case (zero-dimensional field theory!), which provides the required quantities by an expansion in powers of g. Although it would be of great interest to carry out the same program in more realistic cases, this has not been fully successful, albeit in some phenomenological way, except for one-dimensional quantum field theory i.e. quantum mechanics.

10.4.2   *The planar approximation in quantum mechanics*

In general, the diagrams with a given topology are to be weighted by nontrivial Feynman integrals. It is therefore remarkable that, at least in the one-dimensional case, with a propagator $(p^2 + 1)^{-1}$ (p is a one-component momentum to be integrated from $-\infty$ to $+\infty$), one can obtain explicit expressions in the planar limit. As an illustration we study a set of N^2 coupled anharmonic oscillators, in such a way that the configuration variables can be arranged as the entries in an $N \times N$ hermitian matrix with the Hamiltonian invariant under the group $SU(N)$. We look at the ground state energy for

$$H = -\tfrac{1}{2}\Delta + V$$

$$\Delta = \sum_{1 \le i \le N} \frac{\partial^2}{\partial M_{ii}} + \tfrac{1}{2} \sum_{1 \le i < j \le N} \left[\frac{\partial^2}{(\partial \text{Re} M_{ij})^2} + \frac{\partial^2}{(\partial \text{Im} M_{ij})^2} \right]$$

$$V = \tfrac{1}{2} \text{Tr}\, M^2 + \frac{g}{N} \text{Tr}\, M^4 \tag{294}$$

This is only shown as an illustration and one could introduce more general potentials. For the ground state

$$H\psi = N^2 E(g)\psi \qquad (295)$$

the corresponding wavefunction is invariant under unitary transformations

$$\psi(M) = \psi(U^\dagger M U)$$

Thus ψ is a symmetric function of the eigenvalues of M, generalizing the notion of radial variables in "polar" coordinates. Hence $\psi \equiv \psi(\{\lambda_i\})$ and

$$E(g) = \lim_{N\to\infty} \frac{1}{N^2} \operatorname*{Inf}_{\psi} \mathcal{Z}_0^{-1} \mathcal{Z}$$

$$\mathcal{Z} = \int \prod_{1\le i\le N} d\lambda_i \prod_{1\le i<j\le N} (\lambda_i - \lambda_j)^2 \left\{ \frac{1}{2}\sum_{1\le i\le N} \left(\frac{\partial\psi}{\partial\lambda_i}\right)^2 + V\psi^2 \right\}$$

$$\mathcal{Z}_0 = \int \prod_{1\le i\le N} d\lambda_i \prod_{1\le i<j\le N} (\lambda_i - \lambda_j)^2 \psi^2 \qquad (296)$$

To the symmetric wavefunction $\psi(\{\lambda_i\})$, we associate an antisymmetric one denoted ϕ, defined as

$$\phi(\lambda_1,\ldots,\lambda_N) = \prod_{1\le i<j\le N} (\lambda_i - \lambda_j)\psi(\lambda_1,\ldots,\lambda_N) \qquad (297)$$

describing a gas of noninteracting fermions submitted to a quartic potential (in the present case). From equation (296) we derive the variational Schrödinger equation

$$\sum_{1\le i\le N} \left\{ -\frac{1}{2}\frac{\partial^2}{\partial\lambda_i^2} + \frac{1}{2}\lambda_i^2 + \frac{g}{N}\lambda_i^4 \right\} \phi = N^2 E(g)\phi \qquad (298)$$

The factor $1/N$ scaling the potential is included to ensure a correct limiting behaviour. Going over to polar coordinates did not generate additional terms in the potential. This would not be the case if we would have studied another ensemble of matrices. The antisymmetric behaviour of ϕ is the analog of the boundary conditions at the origin in cartesian polar coordinates. Call $e_1 < e_2 < e_3 < \cdots$ the successive energies corresponding to the one-body Hamiltonian

$$h = -\frac{1}{2}\frac{\partial^2}{\partial\lambda^2} + \frac{1}{2}\lambda^2 + \frac{g}{N}\lambda^4 \qquad (299)$$

and let e_F stand for the Fermi level, i.e. the highest occupied one. We have

$$N^2 E(g) = \sum_{1 \leq k} e_k \theta(e_F - e_k)$$

$$N = \sum_{1 \leq k} \theta(e_F - e_k) \tag{300}$$

These expressions are exact for any N (it would be more appropriate to write $E_N(g)$ instead of $N^2 E(g)$ for finite N, since the N^2 behaviour is only asymptotic). Of course this involves the values e_k which are not directly known. Luckily, for N large, the dominant contribution to the expressions (300) comes from very excited one-body states, and a semiclassical approximation becomes sufficient. In other words, we substitute to the above discrete sums, integrals over classical phase space (λ, p) according to

$$N^2 E(g) = N e_F$$
$$- \int \frac{\mathrm{d}\lambda \, \mathrm{d}p}{2\pi} \left(e_F - \tfrac{1}{2} p^2 - \tfrac{1}{2} \lambda^2 - \frac{g}{N} \lambda^4 \right)$$
$$\times \theta \left(e_F - \tfrac{1}{2} p^2 - \tfrac{1}{2} \lambda^2 - \frac{g}{N} \lambda^4 \right)$$

$$N = \int \frac{\mathrm{d}\lambda \, \mathrm{d}p}{2\pi} \theta \left(e_F - \tfrac{1}{2} p^2 - \tfrac{1}{2} \lambda^2 - \frac{g}{N} \lambda^4 \right) \tag{301}$$

Let us integrate over p and rescale λ and e_F as $\sqrt{N}\lambda$, $e_F \to N\varepsilon$. We obtain a pair of equations giving the solution in parametric form

$$E(g) = \varepsilon - \int_{-\infty}^{+\infty} \frac{\mathrm{d}\lambda}{3\pi} \left[2\varepsilon - \lambda^2 - 2g\lambda^4 \right]^{\frac{3}{2}} \theta \left(2\varepsilon - \lambda^2 - 2g\lambda^4 \right)$$

$$1 = \int_{-\infty}^{+\infty} \frac{\mathrm{d}\lambda}{2\pi} \left[2\varepsilon - \lambda^2 - 2g\lambda^4 \right]^{\frac{1}{2}} \theta \left(2\varepsilon - \lambda^2 - 2g\lambda^4 \right)$$

$$\tag{302}$$

which are elliptic integrals. The second equation expresses stationarity of E with respect to ε for fixed g. The equations (302) offer a striking example of the simplifications brought about by the planar approximation.

When we eliminate ε, we find E as a function of g, analytic in the vicinity of the origin. This is in contradistinction with the properties of the exact solution for which $g = 0$ is an essential singularity corresponding to a quantum instability by tunelling for $g < 0$. In the present approximation, the nearest singularity is for a negative value of g such that the Fermi level reaches the maximum of the potential, i.e.

$$g_c = -\frac{\sqrt{2}}{3\pi} \qquad \varepsilon = -\frac{1}{16g_c} \tag{303}$$

One would not expect that the planar approximation gives very sensible results for finite N and *a fortiori* for $N = 1$. It comes as a surprise to see that the comparison, as shown in the table, with the numerical data (called E_{exact}) obtained for the anharmonic oscillator (Hioe and Montroll (1975)), is quite favourable over a large range of couplings

g	E_{planar}	E_{exact}
0.01	0.505	0.507
0.1	0.547	0.559
0.5	0.651	0.696
1.0	0.740	0.804
50	2.217	2.500
1000	5.915	6.694

$$(304)$$

For large g, the agreement is less satisfactory. The behaviour of the planar approximation, as deduced from equation (302), is

$$E_{planar}(g) \underset{g \to \infty}{\sim} \frac{3}{7} \left(\frac{3}{2\pi^{\frac{1}{2}}} \Gamma^2(\tfrac{3}{4}) \right)^{\frac{4}{3}} g^{\frac{1}{3}} = 0.58993 g^{\frac{1}{3}} \tag{305}$$

while the exact solution behaves as

$$E_{exact}(g) \underset{g \to \infty}{\sim} 0.66799 g^{\frac{1}{3}} \tag{306}$$

10.5 Spin systems with random interactions

We present two further applications of anticommuting variables to disordered systems. The first is due to Parisi and Sourlas. Even though, in the final analysis, it does not allow a complete quantitative understanding of the behaviour of the system to be studied, close to its upper critical dimension, namely a spin model with a random, symmetry breaking, external magnetic field, for reasons to be clarified below, it gives a simple explanation for an observation made in the framework of perturbation theory. This relates the behaviour of the disordered system to a pure one in a lower dimension. In the second subsection, we present the work of Dotsenko and Dotsenko pertaining to another extreme situation. The goal is to look for the effect of weak bond impurities close to criticality, when disorder is expected to have a marginal effect according to a criterion due to Harris. Although the treatment may be open to some criticism, both the method of analysis as well as the results are interesting. These difficulties reflect once more the subtleties of disordered systems.

10.5.1   Random external field
and dimensional transmutation

Consider an Ising model in the presence of an external random field, in the continuous field φ^4 version, and assume a Gaussian distribution for the external field with zero-range correlations. Physical quantities are to be averaged over the quenched disorder. The thermal average is denoted $\langle A \rangle$ and its average over disorder $\overline{\langle A \rangle}$. With h the random external field, we write the free energy as

$$
F(h) = \ln \int \mathcal{D}\varphi \exp - \int d^d\mathbf{x} \left\{ \mathcal{L}(\mathbf{x}) + h(\mathbf{x})\varphi(\mathbf{x}) \right\}
$$
$$
\bar{F} = \int \mathcal{D}h\, F(h) \exp -\tfrac{1}{2} \int d^d\mathbf{x}\, h^2(\mathbf{x})
$$

(307)

with

$$
\mathcal{L}(\mathbf{x}) = \tfrac{1}{2}(\partial\varphi)^2 + V(\varphi) \qquad V = \tfrac{1}{2}m^2\varphi^2 + g\varphi^4 \qquad (308)
$$

To compute $\bar{F}$ one can use the replica trick, which replaces the computation of $\overline{F(h)}$ by the average $\overline{Z(h)^n}$, where $Z(h)$ is the partition function, as the number of replicas goes to zero. In a perturbative expansion the most divergent diagrams in the infrared limit are those involving the largest possible number of h^2 insertions. Resumming these most divergent diagrams leads to the conclusion that $F(h)$ is given by its saddle point approximation, with $\overline{F(h)}$ obtained by averaging this approximate value over disorder. This introduces the mean field φ_h, a solution of

$$-\Delta\varphi_h(\mathbf{x}) + V'(\varphi_h(\mathbf{x})) + h(\mathbf{x}) = 0 \qquad (309)$$

In this approximation the two-point function for instance is given by

$$\overline{\langle\varphi(\mathbf{x})\varphi(\mathbf{0})\rangle} = \int \mathcal{D}h\varphi_h(\mathbf{x})\varphi_h(\mathbf{0})\exp{-\tfrac{1}{2}}\int d^d x h^2(\mathbf{x})$$

$$= \int \mathcal{D}h\mathcal{D}\varphi \, |\det(-\Delta + V''(\varphi))| \, \varphi(\mathbf{x})\varphi(\mathbf{0}) \qquad (310)$$

$$\prod_{\mathbf{x}} \delta\left(-\Delta\varphi(\mathbf{x}) + V'(\varphi(\mathbf{x})) + h(\mathbf{x})\right)\exp{-\tfrac{1}{2}}\int d^d x h^2(\mathbf{x})$$

In the second expression we have used the definition of $\varphi_h(\mathbf{x})$ deduced from the mean field equation (309) in an indirect way short of exhibiting $\varphi_h(\mathbf{x})$ in closed form. The product of δ-functions is accompanied by a Jacobian, the absolute value of a determinant involving the second variation of the Lagrangian with respect to $\varphi(\mathbf{x})$. If the correspondence $h \leftrightarrow \varphi$ were one-to-one, the absolute value sign could be omitted, since $\det(-\Delta + V''(\varphi))$ would never vanish. This is implicitly assumed in a perturbative expansion. We will also use here this assumption, without a serious justification, as a means of resumming the leading logarithmic terms. Many disordered system are such that even the mean field approximation is of great complexity (this is for instance the case of spin glasses) and in such cases a similar assumption can be unwarranted.

After these apologies, we follow Parisi and Sourlas, ignore the absolute value sign in the determinant and proceed to investigate the consequences. The product of δ-functions is replaced by a parametric representation over a (pure imaginary) conjugate field $\alpha(\mathbf{x})$, and the determinant is expressed as a Grassmannian integral over a fermionic field $\psi, \bar\psi$. This gives the representation

$$\overline{< \varphi(\mathbf{x})\varphi(\mathbf{0}) >} = \int \mathcal{D}\varphi\mathcal{D}\alpha\mathcal{D}(\psi\bar\psi)\varphi(\mathbf{x})\varphi(\mathbf{0})\exp - \int d^d x \mathcal{L}_{\text{eff}}(\mathbf{x})$$

$$\mathcal{L}_{\text{eff}}(\mathbf{x}) = \bar\psi(-\Delta + V''(\varphi))\psi + \alpha(-\Delta\varphi + V'(\varphi)) - \tfrac{1}{2}\alpha^2 \qquad (311)$$

We expect that this Lagrangian satisfies a BRS supersymmetry. Let $\bar a$ denote an infinitesimal anticommutative parameter, and ε a d-dimensional vector, the transformation in question is of the type

$A \to A + \delta A$ for the set of fields $A \equiv (\varphi, \psi, \bar\psi, \alpha)$, with $\delta A = \bar a s A$

$$s\varphi = -\, \boldsymbol{\epsilon}.\mathbf{x}\psi \qquad\qquad s\alpha = 2\boldsymbol{\epsilon}.\boldsymbol{\partial}\psi$$
$$s\psi = 0 \qquad\qquad s\bar\psi = \boldsymbol{\epsilon}.\mathbf{x}\alpha + 2\boldsymbol{\epsilon}.\boldsymbol{\partial}\varphi \tag{312}$$

The variation of the action is the integral of a total derivative

$$\delta \int d^d \mathbf{x} \mathcal{L}_{\text{eff}}(\mathbf{x}) = 2\bar a \int d^d x \boldsymbol{\epsilon}.\boldsymbol{\partial} \left\{ \psi \left[-\Delta\varphi + V'(\varphi) \right] \right\} \tag{313}$$

which can be set equal to zero by requiring that the quantity in brackets vanishes at infinity.

Show that

$$s^2\varphi = s^2\psi = s^2\alpha = 0 \qquad\qquad s^2\bar\psi = -2\epsilon^2\psi \tag{314}$$

Could one modify the formalism to enforce $s^2 = 0$?

It is convenient to introduce a superfield Φ function of $\mathbf{x}$ and of two anticommuting variables θ and $\bar\theta$

$$\Phi(\mathbf{x}, \theta, \bar\theta) = \varphi(\mathbf{x}) + \bar\theta\psi(\mathbf{x}) + \bar\psi(\mathbf{x})\theta + \theta\bar\theta\alpha(\mathbf{x}) \tag{315}$$

We normalize the integration over θ and $\bar\theta$ by requiring that

$$\int d\bar\theta \, d\theta \, \theta\bar\theta = 1 \tag{316}$$

In superspace, we define the generalized Laplacian

$$\Delta_S = \Delta + \frac{\partial}{\partial\bar\theta} \frac{\partial}{\partial\theta} \tag{317}$$

where derivatives act to the left. One verifies that the effective action in the path integral (311) takes the form

$$S_{\text{eff}} = \int d^d x d\bar\theta \, d\theta \left\{ -\tfrac{1}{2}\Phi\Delta_S\Phi + V(\Phi) \right\}$$
$$= \int d^d x \left\{ \bar\psi(-\Delta + V''(\varphi))\psi + \alpha(-\Delta\varphi + V'(\varphi)) - \tfrac{1}{2}\alpha^2 \right\} \tag{318}$$

We leave it as an exercise to the reader to check this rather surprising fact, which implies invariance under rotations in superspace.

The interpretation of the two Grassmann coordinates is that they amount to subtracting (rather than to adding) two dimensions to d-dimensional space, in rotationally invariant integrals. To see this, let us look at an integral over a configuration variable

$(\mathbf{x}, \theta, \bar{\theta})$. Let f be an algebraic function with a sufficiently fast decrease at infinity. Consider the integral

$$-\frac{1}{\pi} \int d^d \mathbf{x} d\bar{\theta} \, d\theta \, f(\mathbf{x}^2 + \theta\bar{\theta}) = -\frac{1}{\pi} \int d^d x f'(\mathbf{x}^2)$$

Let us use polar coordinate with $\mathbf{x}^2 = r^2$, $S_d = 2\pi^{\frac{1}{2}d}/\Gamma(\frac{1}{2}d)$ the area of the unit sphere in d-dimensional space. We find by integration by parts for $d > 2$

$$-\frac{1}{\pi} \int d^d \mathbf{x} f'(r^2) = -\frac{S_d}{\pi} \int_0^\infty r^{d-1} \, dr f'(r^2)$$

$$= \frac{S_d}{2\pi} \left(\tfrac{1}{2}d - 1\right) \int_0^\infty dr^2 (r^2)^{\frac{1}{2}d-2} f(r^2) = \int d^{d-2} \mathbf{x} f(\mathbf{x}^2) \quad (319)$$

This verifies the claim for integrals over a single configuration variable. It extends to several variables, as the interested reader will find out, and leads to the astonishing dimensional reduction by two units, observed perturbatively and justified by this global analysis. The conclusion of Parisi and Sourlas is that the supersymmetric d-dimensional model is equivalent to a purely bosonic one in $d - 2$ dimensions. For the random field model, it applies at best close to its upper critical dimension (i.e. six for the disordered system, equivalent to a φ^3 theory). The behaviour in lower dimensions is not fully understood, and we shall not explore it further.

10.5.2 *The two-dimensional Ising model with random bonds*

We turn to another marginal effect of disorder on critical behaviour. Consider a system close to a continuous transition. Let θ denote the deviation from the critical temperature in reduced units. A relevant local operator K of dimension $\Delta \leq d$ contributes to the action an extra term of the form

$$\delta S = \int d^d x g(\mathbf{x}) K(\mathbf{x}) \quad (320)$$

with a coupling g of dimension $L^{\Delta-d}$. In the absence of perturbation the correlation length ξ behaves as $\theta^{-\nu}$. From scaling arguments, if we turn on a constant g, the singular part in the free energy per unit volume is of the form

$$F(\theta, g) = \theta^{2-\alpha} f\left(g/\theta^k\right) \quad (321)$$

with a nonnegative crossover exponent

$$k = \nu(d - \Delta) \qquad (322)$$

The scaling function is regular at the origin (this is the purpose of extracting the factor $\theta^{2-\alpha}$). When g is a quenched random external quantity, it is natural to expect that the would-be singular part of the total free energy takes a form generalizing equation (321)

$$\ln Z(\theta, g(\mathbf{x})) = \theta^{2-\alpha} \varphi \left(g(\mathbf{x})/\theta^k\right) \qquad (323)$$

where φ is a regular functional close to a vanishing argument. Assume that $g(\mathbf{x})$ has a zero mean. We have approximately

$$\overline{\ln Z(\theta, g(\mathbf{x}))} = \theta^{2-\alpha} \varphi_0$$
$$+ \tfrac{1}{2}\theta^{2-\alpha-2k} \int \int \mathrm{d}^d\mathbf{x}_1 \, \mathrm{d}\mathbf{x}_2 \overline{g(\mathbf{x}_1)g(\mathbf{x}_2)} \varphi_2(\mathbf{x}_1, \mathbf{x}_2) + \cdots \qquad (324)$$

If the integral on the r.h.s. is well-defined, the leading effect of disorder is expected to be of order $\theta^{2-\alpha-2k}$ with an exponent smaller or equal to the one pertaining to the pure system. The above argument breaks down when this exponent becomes negative

$$2 - \alpha - 2k < 0 \qquad (325a)$$

or, equivalently, by using hyperscaling, i.e. $2 - \alpha = \nu d$, when

$$\Delta < \tfrac{1}{2}d \qquad (325b)$$

as a naive perturbative expansion would suggest. This is the criterion due to Harris (1974) on the relevance of weak disorder.

It is therefore of interest to study a marginal situation corresponding to $\Delta = \tfrac{1}{2}d$. An example is offered by weak fluctuations of the bond couplings in a two-dimensional Ising model. The operator K is identified with the energy operator of dimension $\Delta = 1$. The marginality of disorder means that the associated field theoretic model will be just renormalizable. It turns out, as shown by Dotsenko and Dotsenko, that this model – first introduced in another context by Gross and Neveu – is infrared free in the regime of interest. As we increase the scale, the effect of disorder is weaker and weaker, to the extent that one can ascertain the leading large-distance behaviour in thermodynamic quantities. Even though some aspects of the derivation are still controversial, in particular the delicate use of the replica trick, the underlying field theoretic model is of interest in itself.

We start directly from the continuous version as a free real Majorana field theory, with a space dependent mass term $m(\mathbf{x})$ fluctuating around a zero mean value. We are interested in finding the average of the logarithm of the partition function Z_M (the index M referring to Majorana)

$$Z_M = \int \mathcal{D}(\psi, \bar{\psi}) \exp -S_M \qquad (326a)$$

$$S_M = \int \frac{\mathrm{d}^2 x}{2\pi} \left\{ \psi \bar{\partial} \psi - \bar{\psi} \partial \bar{\psi} + m(\mathbf{x}) \bar{\psi} \psi \right\} \qquad (326b)$$

There are several possible treatments. We choose to follow a presentation due to Shankar, which has the merit of succinctly exhibiting the main phenomenon. Since we are only interested in finding the effect on the free energy, we might as well double it, or equivalently square Z_M, to obtain a complex Dirac version, therefore introducing at first an $O(2)$ symmetry. Denote by $(\psi_1, \bar{\psi}_1)$, $(\psi_2, \bar{\psi}_2)$ the two copies of Majorana fields and define the two component Dirac fields through

$$u = \begin{pmatrix} \psi_1 + i\psi_2 \\ -\bar{\psi}_1 - i\bar{\psi}_2 \end{pmatrix} \qquad \bar{u} = \left(\bar{\psi}_1 - i\bar{\psi}_2, \psi_1 - i\psi_2 \right) \qquad (327)$$

An extra factor σ_3 has been inserted in taking the adjoint in such a way that the Dirac operator reads

$$D_m = \begin{pmatrix} \frac{1}{2}m & \partial \\ \bar{\partial} & \frac{1}{2}m \end{pmatrix} \qquad (328)$$

We shall freely interchange the notations $\mathbf{x}, (z, \bar{z})$ or even z alone, to denote the coordinates of a point. Thus

$$Z_D = Z_M^2 = \int \mathcal{D}(u, \bar{u}) \exp -S_D \qquad (329a)$$

$$S_D = \frac{1}{2\pi} \int \mathrm{d}^2 x \, \bar{u} D_m u \qquad (329b)$$

The appended index m on the Dirac operator is to remind us that the mass term varies in space. The evaluation of Z_D is in principle straightforward being the determinant of the operator D_m. The delicate question is as usual the role of modes close (or equal) to zero. We shall assume that they do not compromise the treatment discussed below. This is where the present approach is open to criticism, but we proceed nevertheless.

Before averaging over $m(\mathbf{x})$ let us recall the bosonization formulas of the Dirac field (chapter 2). They arise from the existence of an $O(2)$ symmetry between the two copies of Majorana fields, or in terms of the complex Dirac field of the invariance under

$$u \to e^{i\tau}u \qquad \bar{u} \to \bar{u}e^{-i\tau} \qquad (330)$$

Noether's theorem leads to a conserved current

$$j^z = \frac{1}{2i}\bar{u}\begin{pmatrix} 0 & 1 \\ 0 & 0 \end{pmatrix}u \qquad j^{\bar{z}} = \frac{1}{2i}\bar{u}\begin{pmatrix} 0 & 0 \\ 1 & 0 \end{pmatrix}u \qquad (331a)$$

$$\partial_z j^z + \partial_{\bar{z}} j^{\bar{z}} = 0 \qquad (331b)$$

Accordingly, in the massless theory, we represent the currents in terms of a free massless field φ, as

$$j^z = \bar{\psi}_2\bar{\psi}_1 \to \sqrt{2}\bar{\partial}\varphi \qquad j^{\bar{z}} = \psi_1\psi_2 \to -\sqrt{2}\partial\varphi \qquad (332)$$

These formulas have to be understood *cum grano salis*. They mean that the correlation functions computed in the corresponding theories do agree.

Recall that for $m = 0$, with a, b referring to points $\mathbf{x}_a$, $\mathbf{x}_b$

$$\langle\psi_a\psi_b\rangle = \frac{1}{z_{ab}} \qquad \langle\bar{\psi}_a\bar{\psi}_b\rangle = -\frac{1}{\bar{z}_{ab}}$$

$$\langle\varphi_a\varphi_b\rangle = -\tfrac{1}{2}\ln z_{ab}\bar{z}_{ab}$$

We verify that the two-point functions do agree

$$\langle j^z_a j^z_b\rangle = -\frac{1}{\bar{z}^2_{ab}} = 2\frac{\partial}{\partial\bar{z}_a}\frac{\partial}{\partial\bar{z}_b}\langle\varphi_a\varphi_b\rangle$$

$$\langle j^{\bar{z}}_a j^{\bar{z}}_b\rangle = -\frac{1}{z^2_{ab}} = 2\frac{\partial}{\partial z_a}\frac{\partial}{\partial z_b}\langle\varphi_a\varphi_b\rangle$$

Check the agreement of higher correlations.

The fluctuating field $m(\mathbf{x})$ is coupled to the energy operator which can also be represented in bosonic form as

$$\tfrac{1}{2}\bar{u}u = \bar{\psi}_1\psi_1 + \bar{\psi}_2\psi_2 \to 2\Lambda\cos\sqrt{2}\varphi = 2\varepsilon \qquad (333)$$

The ultraviolet cutoff $\Lambda = a^{-1}$, which takes care of the relative dimension of the operators, compensates for the absence of Wick ordering since

$$: e^{i\alpha\varphi} := \Lambda^{\frac{1}{2}\alpha^2} e^{i\alpha\varphi} \tag{334}$$

One of the virtues of the bosonic formulation is that, while in fermionic terms the spin field is a highly nonlocal operator, it takes a local form when expressed in terms of φ in the following sense. Taking the product of spins $\sigma_1\sigma_2$ in the two Majorana copies, we have the correspondence

$$\sigma_1\sigma_2 \to \sqrt{2}\Lambda^{\frac{1}{4}} \sin\varphi/\sqrt{2} \tag{335}$$

allowing us to compute squares of correlations in bosonic form. The prefactor $\Lambda^{\frac{1}{4}}$ is inserted to remedy for the absence of normal ordering and emphazises the dimension of the operator.

We are now prepared to take averages over $m(\mathbf{x})$. Suppose the latter fluctuates around a mean value m. Instead of averaging $\ln Z_M$ (or $\frac{1}{2} \ln Z_D^2$) let us rather compute the specific heat at $m = 0$, i.e. differentiate twice the average over m then set $m = 0$. This involves the integral over the two-point function of the energy operator ε i.e. the average of

$$\langle \varepsilon(\mathbf{x})\varepsilon(\mathbf{x'})\rangle = Z_B^{-1} \int \mathcal{D}(\varphi)\Lambda^2 \cos\sqrt{2}\varphi(\mathbf{x}) \cos\sqrt{2}\varphi(\mathbf{x'}) \exp -S_B$$

$$\tag{336a}$$

$$S_B = S_B^{(0)} + S_B^{(1)} \tag{336b}$$

$$S_B^{(0)} = \frac{1}{\pi} \int \mathrm{d}^2x \partial\varphi\bar{\partial}\varphi$$

$$S_B^{(1)} = \frac{1}{\pi} \int \mathrm{d}^2x\, m(\mathbf{x})\Lambda \cos\sqrt{2}\varphi \tag{336c}$$

For short we write : $\cos\alpha\varphi$: for $\Lambda^{\alpha^2/2} \cos\alpha\varphi$. We have still to characterize the probability law obeyed by $m(\mathbf{x})$. The only case amenable to a complete analysis is a Gaussian distribution with zero-range correlations

$$\overline{O(m)} = \frac{1}{2} \int \frac{\mathcal{D}m \ O(m(\mathbf{x})) \exp -(1/2g^2) \int \mathrm{d}^2x m^2(x)}{\int \mathcal{D}m \exp -\frac{1}{2} \int \mathrm{d}^2x m^2(x)} \tag{337a}$$

i.e. such that

$$\overline{m(x_1)m(x_2)} = g^2\delta^{(2)}(\mathbf{x}_1 - \mathbf{x}_2) \tag{337b}$$

Hence g^2 measures the intensity of disorder. To take the required average we have two options. Either to introduce n replicas of the bosonic field and let n go to zero, or to associate a Grassmannian partner. Results are formally the same and we choose the first option. Therefore

$$\langle \varepsilon(x)\varepsilon(x')\rangle = \lim_{n\to 0}\int \mathcal{D}(\varphi) : \cos\sqrt{2}\varphi^1 :: \cos\sqrt{2}\varphi^1 : \exp -S(\varphi)$$

$$(338a)$$

$$S(\varphi) = \int \mathrm{d}^2 x \frac{1}{\pi}\left\{\sum_c \partial\varphi^c \partial\varphi^c - \frac{g^2}{2\pi}\left(\sum_c : \cos\sqrt{2}\varphi^c :\right)^2\right\}$$

$$(338b)$$

The replica index c runs from 1 to n. If instead of using bosonic variables we had performed a similar treatment on the original Dirac action, we would have obtained the Gross–Neveu model with a symmetry group $O(2n)$. This symmetry is not easily expressed in the present context. The complete action with $m(\mathbf{x})$, u^c, $\bar{u}^c$, is the easiest one on which to carry completely the renormalization program. At any rate, equation (338) indicates that, apart from the limit $n \to 0$, we deal with a renormalizable theory. The obvious question is to investigate its infrared behaviour and to figure out how the effect of disorder, namely the effective g^2, behaves in this limit. An interesting aspect of the Gross–Neveu model is that it admits an exact treatment when $n = 1$, in which case its critical behaviour is one of those investigated in chapter 9 when dealing with theories with unit central charge and continuously varying exponents. To see this, observe that according to equations (332) and (333) we have, for a single component Dirac field, at a formal level, for the square of the current

$$\tfrac{1}{2}j^z j^{\bar{z}} = \tfrac{1}{2}\bar{\psi}_2\bar{\psi}_1\psi_1\psi_2 = \tfrac{1}{4}\left(\bar{\psi}_1\psi_1 + \bar{\psi}_2\psi_2\right)^2 \qquad (339)$$

Interpreted in bosonic language, in terms of the corresponding correlations, this means that we have the equivalence

$$-\partial\varphi\bar{\partial}\varphi \leftrightarrow \left(: \cos^2\sqrt{2}\varphi :\right)^2 \qquad (340)$$

This can also be confirmed by a more careful renormalization analysis which implies that the diagonal terms in the interaction

Lagrangian (338b) can be traded for a modified kinetic term as

$$S(\varphi) = \frac{1}{\pi} \int d^2x \left(1 + \frac{g^2}{2\pi}\right) \sum_c \partial \varphi^c \bar{\partial} \varphi^c$$

$$- \frac{g^2}{2\pi} \sum_{c \neq d} : \cos \sqrt{2} \varphi^c :: \cos \sqrt{2} \varphi^d : \quad (341)$$

When the number of components is equal to unity, the model is obviously equivalent to a massless free field theory where the field φ is rescaled by a factor $(1 + g^2/2\pi)^{\frac{1}{2}}$. This innocent looking modification alters of course the dimension of any operator of the form $K_\alpha =: \cos \alpha \varphi :$, from $\Delta_\alpha = \frac{1}{2}\alpha^2$ to $\Delta_\alpha(g) = \frac{1}{2}\alpha^2(1+g^2/2\pi)^{-1}$, thus leading to exponents depending continuously on g. For general n, the associated β-function has to vanish when $n = 1$, meaning that it has at least a factor $(n - 1)$. As we shall confirm below, this factor appears to the first power, hence $\beta(g)$ changes sign when n crosses unity. As a consequence, close to $g = 0$ the infrared and ultraviolet properties are interchanged. While for $n > 1$ the model is ultraviolet asymptotically free, with a spontaneously broken discrete symmetry through a dynamical generation of a fermion mass, in the opposite situation $n < 1$, we have infrared asymptotic freedom, the favourable circumstance for our present purposes.

Let us now proceed to the required computation. On the one hand, using the above observations and the form (341) of the action, we can rescale the fields φ^c according to

$$(1 + g^2/2\pi)^{\frac{1}{2}} \varphi^c = \tilde{\varphi}^c \qquad (342)$$

Set

$$\alpha = \sqrt{\frac{2}{1 + g^2/2\pi}} \qquad (343)$$

and recall that the dot prescription on exponentials implies an appropriate power of Λ. This leads to the action

$$S = \frac{1}{\pi} \int d^2x \left\{ \sum_c \partial \tilde{\varphi}^c \bar{\partial} \tilde{\varphi}^c \right.$$

$$\left. - \frac{1}{2\pi} \left[g\Lambda^{1-1/(1+g^2/2\pi)} \right]^2 \sum_{c \neq d} : \cos \alpha \tilde{\varphi}^c : : \cos \alpha \tilde{\varphi}^d : \right\} \qquad (344)$$

More accurately, we should have introduce an arbitrary finite scale μ to normalize the added power of Λ, which to leading order in g^2 should be interpreted as $(\Lambda/\mu)^{1-1/(1+g^2/2\pi)} \sim 1 + (g^2/2\pi)\ln(\Lambda/\mu)$. This first renormalization is however insufficient to insure a finite behaviour of all correlations, in particular of the energy–energy correlation (338). An additional renormalization is necessary for the operator $\sum_{c \neq d} \, : \cos\alpha\tilde{\varphi}^c \, :: \, \cos\alpha\tilde{\varphi}^d \,:$ as opposed to its individual factors. According to equation (344), the dimension of the interaction has been shifted from 2 to $2/(1 + g^2/2\pi)$. But again, this it to be interpreted in a perturbative sense though an expansion in g^2, and is justified a posteriori in that only the limit $g^2 \to 0$ will be of interest.

To see the additional divergences it is sufficient to investigate the perturbative expansion of the $\varepsilon\varepsilon$ correlation. We can carry this computation in two ways which can be usefully compared. Either we use the form (344), the $\tilde{\varphi}$ field and the corresponding Wick prescription. This requires an inequivalent treatment of the diagonal and off-diagonal terms in $\langle \varepsilon^a \varepsilon^b \rangle$. Or we retreat to the original more symmetrical expression (338b), with the field φ and only after understanding the first nontrivial counterterm do we apply the tricks eliminating the diagonal interaction. Indeed for $n \neq 1$, the value of g^2 appearing in the correspondence between φ and $\tilde{\varphi}$ is affected by renormalization. We choose to present the unsymmetrical version and urge the reader to follow also the alternative route, or else to return to the fermionic version. Although we only look at the off-diagonal terms in the $\langle \varepsilon\varepsilon \rangle$ expansion, it is also instructive to carry out the same discussion for the diagonal ones. Let us write

$$\varepsilon^a = \Lambda^{1-1/(1+g^2/2\pi)} E^a$$
$$E^a = \, : \cos\alpha\tilde{\varphi}^a :$$

(345)

Before taking the limit $n \to 0$, the averages can be written $\langle E^a(\mathbf{x}_1)E^b(\mathbf{x}_2) \rangle$. It is simpler to use the Feynman rules directly in configuration space, with

$$\langle E^a(\mathbf{x}_1)E^b(\mathbf{x}_2) \rangle = \delta^{ab} f(\mathbf{x}_1 - \mathbf{x}_2) \qquad f(\mathbf{x}) = \frac{1}{2\,|\mathbf{x}|^{\alpha^2}} \quad (346)$$

Choose $a \neq b$ and compute the successive perturbative contributions

$$\langle E^a(\mathbf{x}_1)E^b(\mathbf{x}_2) \rangle = 2 \left[\frac{g^2}{2\pi^2} \left(\frac{\Lambda}{\mu} \right)^{2-\alpha^2} \right] f(\mathbf{x}_1 - \mathbf{x}_2)$$

$$+4\left[\frac{g^2}{2\pi^2}\left(\frac{\Lambda}{\mu}\right)^{2-\alpha^2}\right]^2(n-2)$$

$$\int\int d^2x\, d^2y\, f(\mathbf{x}_1-\mathbf{x})f(\mathbf{x}_2-\mathbf{y})f(\mathbf{x}-\mathbf{y})+\cdots$$

$$(347)$$

The factor $(n-2)$ is specific to the off-diagonal correlation as can be seen from the interaction in equation (344). A subtraction takes care of the additional divergence if we set

$$\frac{g^2(\Lambda)}{2\pi^2}\left(\frac{\Lambda}{\mu}\right)^{2-\alpha^2}=\frac{g_R^2}{2\pi^2}-2\left(\frac{g_R^2}{2\pi^2}\right)^2(n-2)\pi\frac{\Lambda^{\alpha_R^2-2}-\mu^{\alpha_R^2-2}}{\alpha_R^2-2}+\cdots$$

$$(348)$$

which arises from the logarithmic divergent integral

$$\int_{\Lambda^{-1}<|\mathbf{x}|<\mu^{-1}}d^2x\, f(\mathbf{x})=\pi\int_{\Lambda^{-1}}^{\mu^{-1}}\frac{r\, dr}{r^{\alpha^2}}=\pi\frac{\Lambda^{\alpha^2-2}-\mu^{\alpha^2-2}}{\alpha^2-2}\simeq\ln\frac{\Lambda}{\mu}$$

To leading order in g_R^2,

$$g^2(\Lambda)=g_R^2\left[1-2(n-1)\frac{g_R^2}{2\pi}\ln\frac{\Lambda}{\mu}+\cdots\right]\qquad(349)$$

The two ingredients have conspired to change $n-2$ into the expected $n-1$. It is interesting to compare with what happens to the diagonal correlation, where a similar term (proportional to $(n-1)$ this time) has to be compensated by additional diagonal counterterms in the action, the latter being needed to introduce the correct coupling in the initial step rescaling φ to $\tilde{\varphi}$.

To leading order, the flow in coupling space close to the origin reads

$$\Lambda\frac{d}{d\Lambda}g(\Lambda)\equiv\beta(g(\Lambda))=-(n-1)\frac{g^3(\Lambda)}{2\pi}\qquad(350)$$

with the consequence that the theory is asymptotically free for $n>1$ in the ultraviolet regime. For $n<1$, the conclusion is reversed, at large distance the effective g decreases and perturbative computations become relevant.

In $\beta(g)$, not only is the coefficient of the g^3 term universal, but so is also the coefficient of g^5. According to Wetzel, one has

$$\beta(g)=-(n-1)\frac{g^3}{2\pi}+(n-1)\frac{g^5}{(2\pi)^2}+\cdots\qquad(351)$$

The r.h.s. vanishes again when $n=1$.

728 10 Disordered Systems and Fermionic Methods

The situation is similar to that with the φ^4 theory in dimension four presented in chapter 5, section 4.2 (volume 1). We can adapt the discussion given there, in particular equation (5.248), where one computes the specific heat, i.e. the integral of (338a). The most singular term arises again from the inhomogeneous piece called $b(g)$ in chapter 5, to lowest order a g-independent constant, proportional to the $\ln \Lambda$ term in the integral $\int d^2x \, \langle \varepsilon(\mathbf{x})\varepsilon(0) \rangle_0$ pertaining to the pure system. The specific heat singular term denoted $\Gamma_{0,2}$ in chapter 5, will be written $c(\theta, g)$ here

$$c(\theta, g) \sim \int_1^{\ell} \frac{d\ell'}{\ell'} b(g(\ell')) \exp \left\{ 2 \int_1^{\ell'} \frac{d\ell''}{\ell''} \gamma_2(g(\ell'')) \right\} \qquad (352)$$

where ℓ is an infrared length scale, measured in lattice units a for instance, and chosen in such a way that the homogeneous term in equation (5.248) be negligible. This means that, at the scale ℓ, the running temperature variable $\theta(\ell) \sim \theta_0$, where θ_0 is finite. The running temperature variable $\theta(\ell)$ is of the form $\theta(\ell) = \theta\ell$ to lowest order, since $\nu = 1$ for the pure system, hence $\ell \sim \theta_0/\theta$. As regards the other ingredients in equation (352), we need the solution of equation (350) corresponding to $n = 0$, which translated into the large length scale ℓ reads

$$g^2(\ell) = \frac{g^2}{1 + (g^2/\pi)\ln \ell} \qquad (353)$$

and the anomalous dimension $\gamma_2(g)$ obtained from the rescaling of the ε operator as

$$\gamma_2(g) = \frac{d}{d \ln \Lambda} \ln \left(\Lambda^{-1+1/(1+g^2/2\pi)} \right) \sim -\frac{g^2}{2\pi} \qquad (354)$$

Putting everything together, replacing $b(g)$ by $b(0)$

$$c(\theta, g) \sim \text{cst} \int_1^{\theta_0/\theta} \frac{d\ell'}{\ell'} \exp - \int_1^{\ell'} \frac{d\ell''}{\ell''} \frac{g^2}{\pi + g^2 \ln \ell''} \qquad (355)$$

The integrals are straightforward and yield

$$c(\theta, g) \approx \frac{\text{cst}}{g^2} \ln \left[1 + \frac{g^2}{\pi} \ln \frac{\theta_0}{\theta} \right] \qquad (356)$$

This is the surprising result found by Dotsenko and Dotsenko, showing a remaining weak singularity in $\ln \ln 1/\theta$ for exponentially small values of $\theta \leq \exp -\pi/g^2$. This subtle effect cannot be obtained by a perturbative weak disorder expansion. Assuming that the underlying hypothesis are correct, a direct observation confirming (356) would be an other evidence of the power of the renormalization group.

The bosonic version does not give access to the average of the spin–spin correlation. Show however, following Shankar, that using similar renormalization group arguments, one can obtain the average of its square at criticality in the form

$$\overline{\langle\sigma(\mathbf{x}_1)\sigma(\mathbf{x}_2)\rangle^2} \sim \frac{\left[1 + g^2/\pi \ln |\mathbf{x}_{12}|\right]^{\frac{1}{4}}}{|\mathbf{x}_{12}|^{\frac{1}{2}}} \qquad (357)$$

with logarithmic corrections typical of asymptotic freedom, correcting the $|\mathbf{x}|^{-\frac{1}{2}}$ behaviour in the pure case. Since

$$\left[\overline{\langle\sigma(\mathbf{x}_1)\sigma(\mathbf{x}_2)\rangle}\right]^2 \leq \overline{\langle\sigma(\mathbf{x}_1)\sigma(\mathbf{x}_2)\rangle^2}$$

this implies the inequality

$$\overline{\langle\sigma(\mathbf{x}_1)\sigma(\mathbf{x}_2)\rangle} \leq \mathrm{cst}\frac{\left[1 + g^2/\pi \ln |\mathbf{x}_{12}|\right]^{\frac{1}{8}}}{|\mathbf{x}_{12}|^{\frac{1}{4}}} \qquad (358)$$

Equation (357) is reminiscent of the spin–spin correlation of the XY-model at the end point of the critical line (see equation (4.184)) with vortices playing the role of disorder. However the logarithmic enhancement occuring on the right-hand side of equation (358) might disappear in a more sophisticated analysis of the average correlation. Investigate in a similar vein the averages of squares of higher spin correlations.

Appendix 10.A   The Hall conductance as a topological invariant

This appendix describes an elaboration of an argument due to Laughlin on the quantized Hall conductance for a minimally coupled interacting many body system, relating it to a topological invariant. We follow the presentation by Avron and Seiler. The idea is to combine gauge invariance and periodicity of wavefunctions when magnetic fluxes are increased by a quantum $\varphi_0 = h/e$, with a nontrivial topology including the wires and apparatus designed to measure (or to produce) the Hall voltage. The point of making such a connection with a topological invariant

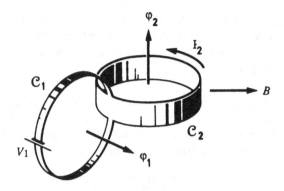

Fig. 6   The Hall setting including connecting wires and battery. The latter is replaced by a time dependent magnetic flux.

is that the latter will not be affected by smooth variations, impurities, etc.

One considers a many body Hamiltonian H_0 including the external strong (but fixed) magnetic field, as well as interparticle forces and interactions with the (possibly random) background. The discussion is carried at sufficiently low temperature that thermal effects are ignored and the crucial hypothesis is that, in the absence of current and voltage, the system is in a pure nondegenerate state. One can then switch on a very weak Hall voltage (through a battery say) and observe a current in the main circuit as shown on figure 6.

The Hamiltonian H_0 includes a vector potential $\mathbf{A}_0$ with a curl equal to the strong field $\mathbf{B}$ (and vanishing circulation around the loops C_1 and C_2 as indicated on the figure). To include the effect of the battery, as well as to have a conjugate variable for the current I_2, one adds to $\mathbf{A}_0$ an extra piece $\varphi_1 \mathbf{A}_1 + \varphi_2 \mathbf{A}_2$ such that

$$\oint_{C_a} \mathrm{dx}.\mathbf{A}_b = \delta_{ab} \qquad (A.1)$$

in such a way that the total magnetic flux is

$$\oint_{C_a} \mathrm{dx}.\mathbf{A} = \varphi_a \qquad (A.2a)$$

From Maxwell's equations, if φ_1 varies with time, the electromotive force induced in the loop is

$$\dot{\varphi}_1 = -\oint_{C_1} d\mathbf{x}.\mathbf{E} = -V_1 \qquad (A.2b)$$

The flux φ_1 is switched on adiabatically up to a linear regime corresponding to a very small V_1. Crudely $\varphi_1 = -tV_1$, or in a more sophisticated manner for negative times $\varphi_1 = -V_1 e^{\eta t}/\eta$ in the limit where η tends to zero.

If $\mathbf{A}$ denotes the sum $\mathbf{A}_0 + \varphi_1 \mathbf{A}_1 + \varphi_2 \mathbf{A}_2$, a typical Hamiltonian will have the structure

$$H = \sum_p \frac{1}{2m}(\frac{\hbar}{i}\,\boldsymbol{\nabla}_p - e\mathbf{A}(\mathbf{x}_p))^2 + V(\{\mathbf{x}\}) \qquad (A.3)$$

If we delete two cuts (to make the domain simply connected), $\mathbf{A}_1$ and $\mathbf{A}_2$ can be considered in a convenient gauge as gradients of two functions Λ_1 and Λ_2. As one winds around C_a, Λ_a increases by unity. One can therefore in this cut region perform a gauge transformation to reduce the vector potential to $\mathbf{A}_0$ and the Hamiltonian to H_0, at the price of writing wavefunctions as $\psi = \exp[ie/\hbar \sum(\varphi_1 \Lambda_1 + \varphi_2 \Lambda_2)]\psi_0$. We now have to account for a discontinuity in ψ_0, its value on one side of each cut being related to its value on the other side by a factor $\exp[ie\varphi_a/\hbar]$ for each coordinate variable $\mathbf{x}_p$. This shows the periodicity of the problem as φ_1 or φ_2 is increased by a quantum $\varphi_0 = h/e$, with H_0 (including the boundary conditions) defining a doubly periodic quantum mechanical system in the parameter plane φ_1, φ_2. Before applying this gauge transformation, the current density operator is a sum

$$\mathbf{J} = e\sum_p \mathbf{v}_p = \sum_p \mathbf{j}_p \qquad \mathbf{v}_p = \frac{1}{m}\left(\frac{\hbar}{i}\,\boldsymbol{\nabla}_p - e\mathbf{A}(\mathbf{x}_p)\right)$$

and

$$\frac{\partial H}{\partial \varphi_a} = -\tfrac{1}{2}\sum_p \left\{\mathbf{j}_p, \mathbf{A}_a(\mathbf{x}_p)\right\}_+$$

The hypothesis, made in the framework of the one body theory, that the Fermi level falls in a gap of the spectrum, is replaced here by the assumption that the system is in a normalized nondegenerate state $|\psi\rangle$. Therefore

$$\langle\psi|\frac{\partial H}{\partial\varphi_a}|\psi\rangle = -\int d^2\mathbf{x}_1\cdots d^2\mathbf{x}_N \sum_p \mathbf{A}_a(\mathbf{x}_p)\psi^*\mathbf{j}_p\psi$$

Expressing $\mathbf{A}_a$ as the gradient of Λ_a in the cut region, and using current conservation, we get only a boundary contribution arising from the unit discontinuity of Λ_a, of the form

$$I_a = -\langle\psi|\frac{\partial H}{\partial\varphi_a}|\psi\rangle \qquad (A.4)$$

This is a first justification for introducing the flux φ_2 in order to compute the current I_2 as a response function. To simplify the matter at the expense of rigour, we use $\varphi_1 = -tV_1$ to translate the time dependent Schrödinger equation satisfied by ψ into

$$-i\hbar V_1\frac{\partial}{\partial\varphi_1}|\psi\rangle = H(\varphi_1, \varphi_2)|\psi\rangle \qquad (A.5)$$

Thus differentiating this equation with respect to φ_2 and taking the scalar product with the adjoint bra $\langle\psi|$ and using finally the expression $(A.4)$ for the current I_2, we obtain

$$-i\hbar V_1\langle\psi|\frac{\partial^2}{\partial\varphi_1\partial\varphi_2}\psi\rangle = -I_2 + i\hbar V_1\left\langle\frac{\partial\psi}{\partial\varphi_1}|\frac{\partial\psi}{\partial\varphi_2}\right\rangle$$

The Hall conductance defined as

$$I_2 = \sigma_H V_1 \qquad (A.6)$$

is then given by a substitute for Kubo's formula, which reads

$$\sigma_H = i\hbar\left\{\frac{\partial}{\partial\varphi_2}\left\langle\psi|\frac{\partial\psi}{\partial\varphi_1}\right\rangle + \left\langle\frac{\partial\psi}{\partial\varphi_1}|\frac{\partial\psi}{\partial\varphi_2}\right\rangle - \left\langle\frac{\partial\psi}{\partial\varphi_2}|\frac{\partial\psi}{\partial\varphi_1}\right\rangle\right\} \qquad (A.7)$$

Instead of computing the derivatives and then letting V_1 (hence also φ_1) go to zero as well as φ_2, the trick is to replace the above expression by an average over each flux in an interval $[0, \varphi_0]$, the rationale being that φ_0 is very small as compared to macroscopic fluxes. This enables one to use the adiabatic approximation to leading order (for vanishingly small V_1). The latter asserts that, if for some initial value the state ψ is along an eigenstate of the φ-dependent Hamiltonian H, it remains so up to a phase. It is natural to take this state as being the normalized ground state

$$H(\varphi_1, \varphi_2)|\Omega(\varphi_1, \varphi_2)\rangle = E(\varphi_1, \varphi_2)|\Omega(\varphi_1, \varphi_2)\rangle \qquad (A.8)$$

where the energy eigenvalue is doubly periodic with period φ_0. Thus, in equation $(A.7)$, we approximate the first term on the right-hand side by $\partial/\partial\varphi_2 \langle\psi|H|\psi\rangle \approx \partial/\partial\varphi_2 E(\varphi_1,\varphi_2)$ with zero average, and identify the Hall conductance with

$$\langle\sigma_H\rangle = \frac{i\hbar}{\varphi_0^2} \int_0^{\varphi_0} \int_0^{\varphi_0} d\varphi_1 d\varphi_2 \left\{ \left\langle \frac{\partial\psi}{\partial\varphi_1} \Big| \frac{\partial\psi}{\partial\varphi_2} \right\rangle - \left\langle \frac{\partial\psi}{\partial\varphi_2} \Big| \frac{\partial\psi}{\partial\varphi_1} \right\rangle \right\}$$
$$(A.9)$$

Equivalently with a compact notation for the two-form integrated over the torus T in flux space

$$\langle\sigma_H\rangle = \frac{e^2}{h} \frac{i}{2\pi} \int\int_T \langle d\psi|d\psi\rangle = \frac{e^2}{h} \frac{1}{2i\pi} \int\int_T \mathrm{Tr}[\,dP\,P\,dP] \quad (A.10)$$

Here the projector $P = |\psi\rangle\langle\psi|$ is approximated to leading order by the ground state projector $|\Omega\rangle\langle\Omega|$, both being insensitive to a flux dependent (but coordinate independent) phase of the state.

The physics is doubly periodic in flux space, but this is neither the case for the Hamiltonian, nor for the state Ω. Under a shift through φ_0, H is changed into a unitary equivalent operator. To remedy this, we perform a unitary gauge transformation on the states corresponding to use $\mathbf{A}_a = \nabla \Lambda_a$ provided cuts are introduced in physical space. Thus with

$$F(\{\mathbf{x}\}) = \sum_p \frac{e}{\hbar} \left[\varphi_1\Lambda_1(\mathbf{x}_p) + \varphi_2\Lambda_2(\mathbf{x}_p) \right]$$
$$U = e^{iF}$$
$$|\Omega\rangle = U|\Omega_0\rangle \qquad\qquad (A.11)$$

the Hamiltonian H turns into H_0 and boundary conditions on $|\Omega_0\rangle$ are periodic in φ-space. The situation is analogous to the one in solid state physics for particles in a periodic potential and Bloch waves, with φ playing the role of quasimomentum. We now have $P = UP_0U^\dagger$. This transformation affects the integrand in equation $(A.10)$ by a total derivative with a vanishing contribution upon integration. As P_0 is now periodic in φ, this shows both that $\langle\sigma_H\rangle$ is sensible (being now invariant under shifts in φ-space), but also, under the assumption of no degeneracy, that it is e^2/h times an integer. Henceforth the subscript zero will be omitted, it being understood that $P(\varphi)$ is a doubly periodic smooth projector on a one-dimensional subspace. The theorem is then that the integral

$$K = \frac{1}{2i\pi} \int \int_T \text{Tr } dP\, P\, dP \qquad (A.12)$$

is both topologically invariant and an integer. It is a classical result due to von Neumann and Wigner that generically level crossing is of codimension three, meaning that it requires at least a three-dimensional parameter space. When the strong external field is kept fixed, the two-dimensional φ_1, φ_2 space is insufficient to allow degeneracies in general. To explain the fractional Hall effect, one would however require degenerate ground states for a generalization of the above argument. We limit ourselves here to the integral effect and to the hypothesis of no degeneracy. The invariant K is reminiscent of similar expressions in gauge theories and can be understood as an integral of curvature of an appropriate bundle.

We conclude this appendix with a proof that K is indeed an integer. For that purpose, consider an arbitrary parametrized loop in flux space and introduce along this loop the evolution equation

$$\frac{d}{ds}|X(s)\rangle = \left(\frac{dP}{ds}P - P\frac{dP}{ds} \right) |X(s)\rangle \qquad (A.13)$$

with $P(s) \equiv P(\varphi_1(s), \varphi_2(s))$, $\varphi_a(s+1) = \varphi_a(s)$. The r.h.s. operator is antihermitian so that the evolution along the loop is unitary. It is readily found that if $P(0)|X(0)\rangle = |X(0)\rangle$, then $|X(s)\rangle$ remains an eigenvector of P with eigenvalue one. The same holds true in the complementary orthogonal space. Under this assumption, after a complete turn, $|X(1)\rangle = e^{i\xi(\gamma)}|X(0)\rangle$. The phase $e^{i\xi(\gamma)}$ is of a type considered by Berry in his study of adiabatic transformations. To compute its value, recall that on each point of the loop $P = |\Omega\rangle\langle\Omega|$, with Ω the ground state of H_0 with discontinuities across the cuts. Thus

$$dP\, P - P\, dP = |d\Omega\rangle\langle\Omega| - |\Omega\rangle\langle d\Omega| + \{\langle d\Omega|\Omega\rangle - \langle\Omega|d\Omega\rangle\}|\Omega\rangle\langle\Omega|$$

Inserting this expression in equation $(A.13)$, we see that the projection $\langle\Omega|X\rangle$ satisfies

$$\frac{d}{ds}\langle\Omega|X\rangle = \langle\frac{d\Omega}{ds}|\Omega\rangle\langle\Omega|X\rangle \qquad (A.14)$$

with $|X(0)\rangle$ and $|X(1)\rangle$ both proportional to $|\Omega(\varphi_1(0), \varphi_2(0))\rangle$. Thus the phase is given by

$$e^{i\xi(\gamma)} = e^{\int_\gamma \langle d\Omega|\Omega\rangle} \qquad (A.15)$$

On the φ-torus $\mathcal{T}$, choose γ homologous to zero. It bounds two distinct domains S_1 and S_2, and the loop integral can be written in two ways as a surface integral which can be expressed in terms of the projector P due to the nondegeneracy hypothesis

$$\exp \int_\gamma \langle d\Omega|\Omega\rangle = \exp \int\int_{S_1} \mathrm{Tr} \; dP\, P \; dP = \exp \int\int_{S_2} \mathrm{Tr} \; dP\, P \; dP \qquad (A.16)$$

If γ shrinks to a point as well as S_1, the first two exponentials become unity, forcing the remaining integral over the whole torus to become a multiple of $2i\pi$, i.e. K defined by equation $(A.12)$ to be an integer, as was claimed.

One might argue that the definition taken for the Hall conductance as an average is not perhaps necessary. This is compensated for by the great generality of the argument based solely on the geometry of the setting and the consequences of gauge invariance in quantum mechanics. We finally note that Berry's phases such as the one introduced above occur in many significant circumstances when investigating the topology of the parameter space of quantum systems (or systems which admit a superposition principle).

Notes

The localization effect of disorder on wavefunctions is the subject of the classic paper by P.W. Anderson *Phys. Rev.* **109**,1492 (1958). One finds a number of references to the many works which have followed in the reviews by D. Thouless in *Ill Condensed Matter*, Les Houches, R. Balian, R. Maynard and G. Toulouse eds., North Holland (1979) and the articles by D. Thouless, E. Abrahams and F. Wegner in *Common Trends in Particle and Condensed Matter Physics*. E. Brézin, J.-L. Gervais, and G. Toulouse eds. *Phys. Reports* **67** (1980).

The treatment of a one-dimensional Hamiltonian with a Gaussian distributed potential is due to B. Halperin, *Phys. Rev.* **A139**, 104 (1965). This is closely connected with one-dimensional diffusion problems reviewed in S. Alexander, J. Bernasconi,

W.R. Schneider and R. Orbach, *Rev. Mod. Phys.* **53**, 175 (1981).
For another interesting solvable case see J.P. Bouchaud, A. Comtet, A. Georges, P. Le Doussal, *Europhys. Lett.* **3**, 653 (1987).
The replica trick was introduced in the study of random systems by S.F. Edwards and P.W. Anderson, *J. Phys.* **F5**, 965 (1975). The supersymmetry arising when representing Jacobians by Grassmann integrals was first discussed in the context of gauge theories by C. Becchi, A. Rouet, and R. Stora in *Renormalization Theory*, G. Velo and A.S. Wightman eds, Reidel, Dordrecht (1976).

The spectrum of random chains of oscillators was investigated by F.J. Dyson, *Phys. Rev.* **92**, 1331 (1953) and D.J. Thouless, *J. Phys.* **C5**, 77 (1972). The small disorder expansion is due to B. Derrida and R. Orbach, *Phys. Rev.* **B27**, 4694 (1983). The example discussed in the text is from a joint paper with E.J. Gardner and B. Derrida, *J. Phys.* **A17**, 1093 (1984).

For a review on the quantum Hall effect, see K. von Klitzing, *Rev. Mod. Phys.* **58**, 519 (1986) and *The Quantum Hall Effect*, R.E. Prange and S.M. Girvin eds, Springer Verlag, New York (1987). The computation of the one-particle spectrum in a strong magnetic field and a random potential is due to F. Wegner, *Z. Phys.* **B51**, 279 (1983). The generalization discussed in the text was worked out with E. Brézin and D. Gross, *Nucl. Phys.* **B235** [FS 11], 24 (1984). The insensitivity of the integral quantization of the Hall conductance to the presence of impurities, as a consequence of gauge invariance, is discussed in R.B. Laughlin, *Phys. Rev.* **B23**, 5632 (1981). We present in appendix A an elaboration of this argument due to J.E. Avron and R. Seiler, *Phys. Rev. Lett.* **54**, 259 (1985) following earlier work by several authors quoted in this paper. Phases occuring in adiabatic transitions are studied in M.V. Berry, *Proc. Roy. Soc. London* **A392**, 45 (1984).

Ensembles of random matrices introduced and studied by E.P. Wigner, F.J. Dyson, M.L. Mehta, M. Gaudin and others are presented by M.L. Mehta in *Random Matrices and the Statistical Theory of Energy Levels*, Academic Press, New York (1967). For the interpretation in terms of anticommuting variables, see the contribution by E. Brézin in the proceedings of the VIIIth Sitges conference, L. Garrido ed., Lecture Notes in Physics **216**, 115, Springer, Berlin (1985).

The planar approximation is due to G. 't Hooft, *Nucl. Phys.* **B72**, 461 (1974) *ibid*, **B75**, 461 (1974). Algebraic aspects were investigated by J. Koplik, A. Neveu and S. Nussinov, *Nucl. Phys.* **B123**, 109 (1977). Our presentation follows work done in collaboration with E. Brézin, G. Parisi and J.-B. Zuber, *Comm. Math. Phys.* **59**, 35 (1978) and D. Bessis and J.-B. Zuber, *Adv. in Appl. Math.* **1**, 109 (1980).

The behaviour of a spin system in a random external field was investigated by Y. Imry and S.K. Ma, *Phys. Rev. Lett.* **35**, 1399 (1975). Dimensional transmutation using supersymmetry was demonstrated by G. Parisi and N. Sourlas, *Phys. Rev. Lett.* **43**, 744 (1979). A general reference on the applications of anticommuting variables in the study of random systems is K.B. Efetov, *Adv. in Phys.* **32**, 53 (1983).

A criterion for assessing the effect of weak impurities on continuous phase transitions was put forward by A.B. Harris, *J. Phys.* **C7**, 1671 (1974). The application to the two-dimensional Ising model with random bonds was investigated by Vi.S. Dotsenko and Vl.S. Dotsenko in *Adv. in Phys.* **32**, 129 (1983), and R. Shankar, *Phys. Rev. Lett.* **58**, 2466 (1987). The underlying field theoretic model is due to D.J. Gross and A. Neveu, *Phys. Rev.* **D10**, 3235 (1974). For the two-loop renormalization of this model, see W. Wetzel, *Phys. Lett.* **153B**, 297 (1985).

11

RANDOM GEOMETRY

Statistical models with a geometrical basis arise in many circumstances, such as the theory of liquids, membranes, polymer net-
works, defects, microemulsions, interfaces, etc. Gauge theories
also lead to random surfaces, as does the theory of extended objects such as strings, and quantum gravity requires a generalization to four-manifolds. From a general point of view, local quantum field theory is rooted in the study of random paths. One
may wish to find such a universal model, generalizing Brownian
curves, to Brownian manifolds and in the first instance to surfaces (Polyakov, 1981). Despite many efforts, no such universal
archetype has been found, although the endeavour towards such a
model has uncovered a rich mathematical structure. By necessity,
our presentation will be limited to the most elementary aspects.

In the first section we discuss random lattices in Euclidean
space. The use of such lattices was advocated as a mean to restore
translational invariance while keeping the advantage of a short
distance cutoff (Christ, Friedberg and Lee, 1982). The formalism
could be generalized to other manifolds, but we refrain from doing
to, nor will we pursue the analysis of standard models on random
lattices, a difficult subject. Even free field theory on such a lattice
opens the Pandora's box of disordered systems. Concepts from
random lattices may be useful in the study of liquids or gases.

In the second section we look at random surfaces, both in
a discrete and continuous setting. Thus emerges the role of
reparametrization invariance, and its associated quantum anomaly
related to conformal field theory. We present in some detail the
Polyakov model for smooth surfaces, with its magic dimension 26,
as well as a discrete analog, pointing out the connection of the
latter with planar quantum field theory.

11.1   Random lattices

Random geometry belongs to an old mathematical tradition.
Recall for instance the problem of Buffon's needle. It is described
in a classical book by Santalo under the name of integral geometry.
In the early sixties, Bernal investigated random packing of spheres
and applied them to the study of liquid structure functions.
Here we describe a related topic, namely Poissonian random
lattices and their local properties. They provide a means of
introducing a short-distance cutoff in field theory, while hopefully
maintaining continuous geometrical symmetries, assuming self-
averaging of bulk quantities. This offers the opportunity to
describe in elementary terms some concepts both of topological
and geometrical interest. The study of models on random lattices
paves the way to an understanding of field theory in curved space,
since the use of a random lattice is a discrete analog of the choice
of an arbitrary coordinate system in continuous geometry.

11.1.1   *Poissonian lattices and cell statistics*

In a large regular region of volume Ω, in Euclidean d-dimensional
space, N points are chosen independently with uniform probability. We consider a limit with N and Ω tending to infinity keeping
the density ρ fixed

$$N \to \infty, \qquad \Omega \to \infty, \qquad N/\Omega \to \rho = a^{-d} \qquad (1)$$

The quantity $a = \rho^{-1/d}$ plays the role of an elementary length
scale. We choose units in such a way that $a = 1$, hence $\rho = 1$. As
we saw in chapter 10, in one dimension the points can be ordered
and the intervals ℓ between successive points become independent
variables in the thermodynamic limit, with a distribution

$$p(\ell)\mathrm{d}\ell = \mathrm{e}^{-\ell}\mathrm{d}\ell \qquad (2)$$

This justifies calling these lattices Poissonian.
   Consider the case of two dimensions. The generalization
to higher dimension will be straightforward. The following
construction is due to Dirichlet and Voronoi. To each point M_i
belonging to the lattice, we associate the closed cell C_i of those
points of the plane which are nearer to M_i than to any M_j, $j \neq i$.
The cell C_i is a closed convex polygon, as an intersection of half

planes. For each pair i, j the intersection $C_i \cap C_j$ is either empty or a segment lying on the bisecting line between M_i and M_j and belonging to the boundary of C_i and C_j. In the latter case we say that (M_i, M_j) form a pair of neighbours. We call q_i be the local coordination number, the number of neighbours of M_i, or equivalently the number of sides of the cell C_i. For a triplet $i \neq j \neq k$, if $C_i \cap C_j \cap C_j$ is not empty, it is a vertex of each of the corresponding cells, the center of the circle through M_i, M_j, M_j, and we call (M_i, M_j, M_k) an elementary 2-simplex, as (M_i, M_j) was an elementary 1-simplex. The planar region is now paved by 2-simplices as well as by cells C_i. The two sets are dual to each other.

Let $N_0 = N$ be the number of points M_i, equal to the number of cells, N_1 the number of links between neighbours and N_2 the number of 2-simplices, equal to the number of cell vertices. We have the relation

$$N_0 - N_1 + N_2 = \chi \tag{3}$$

where the Euler characteristic χ (2 for the sphere, 0 for a torus, etc.) is not extensive, i.e. in the infinite volume limit $\chi/N \to 0$. Cutting the plane along the edges of the cells we get N_0 pieces (the cells) with $2N_1 = \sum_i q_i$ edges and $3N_2$ vertices. Since each cell has as many vertices as edges (Euler's relation in dimension one) we find

$$2N_1 = 3N_2 = \sum_i q_i$$

$$N_0 = \sum_i 1 \tag{4}$$

Consequently, if the average coordination number is defined as

$$n_1 = \lim_{N \to \infty} \frac{1}{N} \sum_i q_i \tag{5}$$

the above relations imply

$$n_1 = 6 \qquad \lim \frac{N_1}{N_0} = 3 \qquad \lim \frac{N_2}{N_0} = 2 \qquad (6)$$

Equivalently, if n_a is the average number of a-simplices incident on a 0-simplex (a point) (with $n_1 \equiv q$)

$$n_1 = n_2 = 6 \tag{7}$$

The two-dimensional case is remarkable in that on the triangular regular lattice, the relations (6) and (7) are true locally. The random lattice may therefore be loosely considered as arising from defects in a regular triangular lattice, the defects being the points where the local coordination number are different from six. These points are the analogs of charges (the deviation from six), with equation (6) implying global neutrality.

Let us now introduce metric properties. We use the notation

$$\ell_{ij} = \text{distance between the neighbours } (M_i, M_j) \qquad \left\langle \ell_{ij} \right\rangle = \ell_1$$

$$\ell_{ijk} = \text{area of simplex } (M_i, M_j, M_k) \qquad \left\langle \ell_{ijk} \right\rangle = \ell_2$$

Similarly

$$\sigma_i = \text{area of cell } C_i \qquad \left\langle \sigma_i \right\rangle = \sigma_2$$

$$\sigma_{ij} = \text{length of edge perpendicular to the link } (ij) \qquad \left\langle \sigma_{ij} \right\rangle = \sigma_1$$

Since by convention the density is unity, we have

$$\sigma_2 = 1 \qquad \ell_2 = \tfrac{1}{2} \qquad (8)$$

The computation of ℓ_1 and σ_1 is more tricky and the reasoning is typical of random lattice problems. Let, for short, (i, j, k) be a 2-simplex with 0 as the dual vertex common to the cells C_i, C_j and C_k. The circle Γ of radius R centered in 0 is circumscribed to the triangle (i, j, k) (figure 1), and there can be no lattice point inside Γ, since for $\ell \neq i, j, k$ the distance $|OM_\ell|$ is greater or equal to $|OM_i|$, $\left|OM_j\right|$, and $|OM_k|$.

Given a point M_1 belonging to the lattice, the probability that two other points M_2 and M_3, lying within $d^2 x_2$ and $d^2 x_3$ respectively, form an elementary 2-simplex is therefore

$$dp = \frac{1}{n_2} \times \tfrac{1}{2}(N-1)(N-2) \times \left(1 - \frac{\pi R^2}{\Omega}\right)^{N-3} \frac{d^2 x_2}{\Omega} \frac{d^2 x_3}{\Omega}$$

where the first factor n_2 is a normalization of the average number of 2-simplices per 0-simplex, the second expresses the numbers of choices of M_2 and M_3 among the $N-1$ remaining points, and the third factor is the integrated probability of the $(N-3)$ remaining lattice points lying outside the circle Γ. In the limit $N \to \infty$, the required probability

$$dp = \tfrac{1}{12} e^{-\pi R^2} d^2 x_2 d^2 x_3 \qquad (9)$$

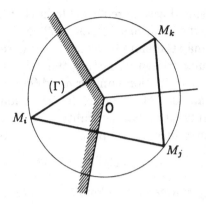

Fig. 1   The Dirichlet–Voronoi construction in two dimensions.

has a form reminiscent of the Poisson distribution (2).

Using as parameters the radius R of the circumscribed circle and the polar angles φ_1, φ_2, φ_3 of OM_1, OM_2, OM_3, with

$$d^2 x_2 d^2 x_3 = R^3 \left| \sin(\varphi_1 - \varphi_2) + \sin(\varphi_2 - \varphi_3) \right.$$
$$\left. + \sin(\varphi_3 - \varphi_1) \right| \, dR \, d\varphi_1 \, d\varphi_2 \, d\varphi_3$$

check that the probability (9) is correctly normalized

$$\int dp = \tfrac{1}{3}\pi \int_0^\infty dR R^3 e^{-\pi R^2} \int_0^{2\pi} d\varphi_3$$
$$\int_0^{\varphi_3} d\varphi_2 \left\{ \sin(\varphi_3 - \varphi_2) - \sin\varphi_3 + \sin\varphi_2 \right\}$$
$$= 2\pi^2 \int_0^\infty dR R^3 e^{-\pi R^2} = 1$$

To obtain ℓ_1, the average separation between neighbours, we average $|M_1 M_2| = 2R \left| \sin\tfrac{1}{2}(\varphi_2 - \varphi_1) \right|$ over dp, i.e.,

$$\ell_1 = \tfrac{2}{3}\pi \int_0^\infty dR \, R^4 e^{-\pi R^2}$$
$$\times \int_0^{2\pi} d\varphi_3 \int_0^{\varphi_3} d\varphi_2 \sin\tfrac{1}{2}\varphi_2 \left\{ \sin(\varphi_3 - \varphi_2) - \sin\varphi_3 + \sin\varphi_2 \right\}$$

Thus

$$\ell_1 = \frac{32}{9\pi} = 1.1317684 \tag{10}$$

i) In two dimensions the procedure followed above, to compute the average side of triangle and identifying it with the average link side is justified by the fact that each link belongs to a fixed number of triangles, namely two. Indeed one has

$$\ell_1 = \frac{1}{N_1} \sum_{\text{links}} \ell_{ij} = \frac{1}{2N_1} \sum_{\text{triangles } T} \sum_{\ell_{ij} \in T} \ell_{ij}$$

$$= \frac{3N_2}{2N_1} \times \text{average of a triangle side}$$

and $3N_2/2N \rightarrow 1$. In three dimensions, the number of tetrahedra sharing a link is not a constant throughout the lattice, so the average length of a tetrahedron side is not the average of a link of the lattice. But the procedure is correct for the average area of a triangular face. For lack of a better procedure the quantities referred to as $\ell_{(i)}$ in dimension d below are in fact computed as averages of the corresponding elements for a typical simplex.

ii) The quantity ℓ_1 for a random lattice may be compared to the equivalent on a regular triangular lattice with the same density of points i.e. $2^{\frac{1}{2}}3^{-\frac{1}{4}} = 1.0745699$ showing that the random lattice is somehow "looser".

iii) Using the probability density (9), check that ℓ_2 is $\frac{1}{2}$ as it should, and that its relative variance is quite large

$$\frac{\delta\ell_2}{\ell_2} = \left(\frac{\left\langle \ell_{ijk}^2 \right\rangle}{\left\langle \ell_{ijk} \right\rangle^2} - 1 \right)^{\frac{1}{2}} = \left(\frac{35}{2\pi^2} - 1 \right)^{\frac{1}{2}} = 0.8793$$

The average perimeter of cells is obtained using a clever argument due to Meijering. Let M be a lattice point and consider for any other lattice point M_k the bisecting line Δ_k of the segment MM_k. The number of such lines at distance between r and $r + \mathrm{d}r$ is the number of points M_k at distance between $2r$ and $2r + 2\,\mathrm{d}r$, namely $8\pi r\,\mathrm{d}r$. The fraction of length of such a line at a distance between R and $R + \mathrm{d}R$ is $(R^2 - r^2)^{-\frac{1}{2}}2R\,\mathrm{d}R$ (Figure 2). The total amount of would-be cell boundary length between R and $R + \mathrm{d}R$ is obtained by integrating the product of these two factors over r between the values 0 and R,

$$16\pi R\,\mathrm{d}R \int_0^R \frac{r\,\mathrm{d}r}{\sqrt{R^2 - r^2}} = 16\pi R^2\,\mathrm{d}R$$

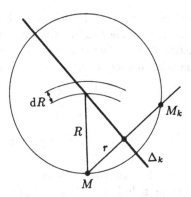

Fig. 2   Geometric construction involved in Meijering's argument.

An element of this length will actually belong to the boundary of the cell around M if the circle of radius R centered at the length element and (therefore going through M) does not contain any interior lattice point. This occurs with probability $e^{-\pi R^2}$. Thus the amount of perimeter of the cell boundary at a distance between R and $R + dR$ is

$$16\pi R^2 e^{-\pi R^2}\,dR$$

Integrating over R yields the mean perimeter

$$P = 16\pi \int_0^\infty R^2 e^{-\pi R^2}\,dR = 4 \qquad (11)$$

Since on average each cell has six sides, we find

$$\sigma_1 = \tfrac{2}{3} \qquad (12)$$

to be compared with the value $2^{\frac{1}{2}}3^{-\frac{3}{4}} = 0.62204032$ on a regular triangular lattice. Table I summarizes these two-dimensional local results.

It is possible to write an analytical (but untractable) expression for the probability p_n of finding a cell with n sides. Let M_0 at the origin of the coordinate system stand for a lattice point, and let M_1, ..., M_n with coordinates x_1, ..., x_n, be its neighbours. We denote by $\mathcal{A}$ the area of the union of circles centered at M_1, ..., M_n, and going through M_0, and by $\chi(M_1, \ldots, M_n)$ a characteristic function with value one if the bisecting lines of the segments (M_0, M_i)

Table I. Local averages in two dimensions corresponding to unit density.

Dimension of simplex	Number	Average number incident on site	Mean size	Mean size (dual)
0	$N_0 = N$	$n_0 = 1$		$\sigma_2 = 1$
1	$N_1 = 3N$	$n_1 = 6$	$\ell_1 = \frac{32}{9\pi}$	$\sigma_1 = \frac{2}{3}$
2	$N_2 = 2N$	$n_2 = 6$	$\ell_2 = \frac{1}{2}$	

Table II. Probability distribution for n-sided cells in a Poissonian two-dimensional lattice.

n	p_n
3	$(1.127 \pm 0.008)10^{-2}$
4	$(1.077 \pm 0.0011)10^{-1}$
5	0.258 ± 0.002
6	0.294 ± 0.003
7	0.198 ± 0.003
8	0.090 ± 0.020
9	$(2.88 \pm 0.07)10^{-2}$
10	$(6.95 \pm 0.20)10^{-3}$
20	$(1.5 \pm 0.8)10^{-13}$
35	$3.6.10^{-40}$
50	$1.5.10^{-73}$

generate an n-sided convex polygon, and zero otherwise. Then

$$p_n = \frac{1}{n!} \int \prod_{i=1}^{n} \mathrm{d}^2 x_i \chi(\mathbf{x}_1, \ldots, \mathbf{x}_n) e^{-\mathcal{A}} \qquad (13)$$

Integrating (numerically) this expression leads to the values quoted in table II. Although the average value $q = \sum_3^{\infty} n p_n = 6$ is also the most probable, occurring approximately in 30 percent of the cases, other small coordination numbers have an appreciable probability. For large n, p_n decreases very fast as $n^{-\alpha n}$, where α is between 1 and 2. The corresponding histogram is shown on figure 3.

Similarly one can study the mean distance λ_n to the perimeter of the cell or A_n the average area of an n-sided cell. One observes

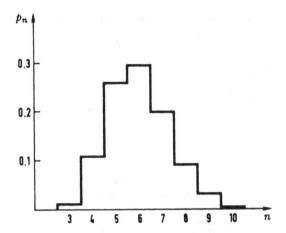

Fig. 3　Probability distribution of n-sided cells.

that A_n is almost linear in n with $A_n \sim \frac{1}{4}n + \text{cst}$ for large n, as if the cell was growing in an "unfriendly" environment. The ratio A_n/λ_n^2 seems to converge to a value close to π, suggesting that a very large cell behaves in certain respects as a circle.

The Dirichlet–Voronoi construction carries over in higher dimensions. Let us present some details in three dimensions. Let $N_0 = N$ be the number of points, N_1 of links, N_2 of triangles and N_3 of tetrahedra. The Euler characteristic vanishes in odd dimensions, thus

$$N_0 - N_1 + N_2 - N_3 = 0 \qquad (14)$$

Each triangle is shared by two tetrahedra, and there are four triangles on the boundary of a tetrahedron, thus

$$2N_2 = 4N_3 \qquad (15)$$

If as above n_a stands for the number of a-simplices incident on a point

$$n_a = (a + 1)N_a/N \qquad (16)$$

topology alone yields the relations

$$n_1 = 2 + \tfrac{1}{2}n_3$$
$$n_2 = \tfrac{3}{2}n_3 \qquad (17)$$

leaving one unknown, for instance n_3. The probability of finding a tetrahedron with one vertex M_1 at the origin and the three other ones M_2, M_3, M_4 at $\mathbf{x}_2$, $\mathbf{x}_3$, $\mathbf{x}_4$ within infinitesimal volumes $\mathrm{d}^3 x_i$ is

$$\mathrm{d}p = \frac{1}{3!n_3} \mathrm{e}^{-\frac{4}{3}\pi R^3} \prod_{i=2}^{4} \mathrm{d}^3 x_i \qquad (18)$$

In this relation R is the radius of the sphere circumscribed to the tetrahedron. Thus

$$n_3 = \frac{1}{3!} \int \mathrm{e}^{-\frac{4}{3}\pi R^3} \prod_{i=2}^{4} \mathrm{d}^3 x_i \qquad (19)$$

To evaluate this integral one can use the following coordinates. Let 0 be the center of the circumscribed sphere and $\mathbf{u}_1, \mathbf{u}_2, \mathbf{u}_3, \mathbf{u}_4$ unit vectors along OM_1, OM_2, OM_3 and OM_4. Call $R^3 w$ the volume of the tetrahedron with

$$w = \pm\frac{1}{3!} \det \begin{vmatrix} 1 & 1 & 1 & 1 \\ \mathbf{u}_1 & \mathbf{u}_2 & \mathbf{u}_3 & \mathbf{u}_4 \end{vmatrix} \qquad (20)$$

with a sign chosen in such a way that w be positive. Then

$$\frac{1}{3!} \prod_{i=2}^{4} \mathrm{d}^3 x_i = R^8 w \, \mathrm{d}R \mathrm{d}^2 \hat{u}_1 \mathrm{d}^2 \hat{u}_2 \mathrm{d}^2 \hat{u}_3 \mathrm{d}^2 \hat{u}_4 \qquad (21)$$

with $\mathrm{d}^2 \hat{u}$ the measure on the sphere. Then

$$n_3 = 72\pi \langle w \rangle \qquad (22)$$

where w stands for the rotational average

$$\langle w \rangle = \int w \prod_{i=1}^{4} \frac{\mathrm{d}^2 \hat{u}_i}{4\pi} \qquad (23)$$

We shall obtain below (equation (36)) the d-dimensional generalization of this integral. Restricted to $d = 3$ it yields

$$n_3 = \tfrac{96}{35}\pi^2 = 27.0709\cdots$$

$$n_1 = 2 + \tfrac{48}{35}\pi^2 = 15.53546\cdots \qquad n_2 = \tfrac{144}{35}\pi^2 = 40.6064 \, (24)$$

where use has been made of equation (17). Alternatively this means that

$$N_3/N = \tfrac{24}{35}\pi^2 \qquad N_2/N = \tfrac{48}{35}\pi^2 \qquad N_1/N = 1 + \tfrac{24}{35}\pi^2 \qquad (25)$$

Of course these numbers do not correspond to any regular lattice.
The average volume of a tetrahedron in natural units is the
reciprocal of N_3/N, i.e.

$$\ell_3 = 35/24\pi^2 = 0.1478 \cdots \tag{26}$$

From the probability density (18), one can also compute the
average area of a triangle

$$\ell_2 = \left(\frac{3}{4\pi}\right)^{\frac{2}{3}} \tfrac{875}{243}\pi^{-1}\Gamma(\tfrac{2}{3}) = 0.5973 \cdots \tag{27}$$

and the mean length of a 1-simplex, i.e. the average distance
between neighbours, with the caveat explained above

$$\ell_1 = \left(\frac{3}{4\pi}\right)^{\frac{1}{3}} \tfrac{1715}{2304}\Gamma(\tfrac{1}{3}) = 1.2371 \tag{28}$$

For the dual lattice $\sigma_3 = 1$. The average area of cell boundary
is obtained using Meijering's reasoning as

$$\text{average area of cell boundary} = \left(\frac{4\pi}{3}\right)^{\frac{1}{3}} \tfrac{8}{3}\Gamma(\tfrac{2}{3}) = 5.821 \tag{29}$$

Thus σ_2 follows by dividing out by the connectivity n_1

$$\sigma_2 = \frac{\left(\frac{4\pi}{3}\right)^{\frac{1}{3}} \frac{4}{3}\Gamma\left(\frac{2}{3}\right)}{1 + \frac{24}{35}\pi^2} = 0.3747 \cdots \tag{30}$$

These data are collected in table III while table IV present similar
data pertaining to four dimensions.

The average coordination number n_1 grows very fast with
dimension being already $\frac{340}{9} = 37.778$ in four dimensions as
opposed to 8 on a regular hypercubical lattice.

The above computations are rather arduous, and fewer results
are known in general dimension d. Let us quote some of them.

i) Since each $(d-1)$-simplex belongs to two d-simplices and since
$d+1$ $(d-1)$-simplices are incident on a d-simplex

$$2N_{d-1} = (d+1)N_d \qquad n_{d-1} = \tfrac{1}{2}dn_d \tag{31}$$

Euler's relation applied to the boundary of a cell, homeomorphic
to a sphere yields

$$n_1 - n_2 + \cdots (-1)^{d-1}n_d = 1 - (-1)^d \tag{32}$$

Table III. Cell data in three dimensions

Dimension of simplex	Number	Number incident on a point	Average size	Average size (dual)
0	$N_0 = N$	$n_0 = 1$		
1	$N_1 = \frac{1+24\pi^2}{35}N$	$n_1 = 2 + \frac{48}{35}\pi^2$	$\ell_1 = \left(\frac{3}{4\pi}\right)^{\frac{1}{3}} \frac{1715}{2304}\Gamma\left(\frac{1}{3}\right)$	$\sigma_2 = \dfrac{\left(\frac{4\pi}{3}\right)^{\frac{1}{3}}\frac{4}{3}\Gamma\left(\frac{2}{3}\right)}{1+\frac{24\pi^2}{35}}$
2	$N_2 = \frac{48}{35}\pi^2 N$	$n_2 = \frac{144}{35}\pi^2$	$\ell_2 = \left(\frac{3}{4\pi}\right)^{\frac{2}{3}} \frac{875}{243}\frac{1}{\pi}\Gamma\left(\frac{2}{3}\right)$	
3	$N_3 = \frac{24}{35}\pi^2 N$	$n_3 = \frac{96}{35}\pi^2$		

Table IV. Cell data in four dimensions

Dimension of simplex	Number	Number incident on a point	Average size	Average size (dual)
0	$N_0 = N$	$n_0 = 1$		$\sigma_4 = 1$
1	$N_1 = \frac{170}{9}N$	$n_1 = \frac{340}{9}$	$\ell_1 = \left(\frac{2}{\pi^2}\right)^{\frac{1}{4}} \frac{3^3}{7\times11}\frac{16}{15}\frac{1}{\pi}\Gamma\left(\frac{1}{4}\right)$	$\sigma_3 = \frac{9}{225}(2\pi^2)^{\frac{1}{4}}\frac{1}{\pi}\Gamma\left(\frac{3}{4}\right)$
2	$N_2 = \frac{590}{9}N$	$n_2 = \frac{1770}{9}$	$\ell_2 = 2^{2\frac{1}{2}}\frac{2^7}{11^2\times13}\frac{17}{16}\Gamma\left(\frac{1}{2}\right)$	
3	$N_3 = \frac{715}{9}N$	$n_3 = \frac{2860}{9}$	$\ell_3 = \left(\frac{2}{\pi^2}\right)^{\frac{3}{4}} \frac{2^7 7^2}{3\times13^2\times57}\frac{18}{17}\Gamma\left(\frac{3}{4}\right)$	
4	$N_4 = \frac{286}{9}N$	$n_4 = \frac{1470}{9}$	$\ell_4 = \frac{9}{286}$	

More generally if $n_{k,m}, k > m$, $(n_{k,0} \equiv n_k)$ denotes the number of k-simplices incident on an m-simplex

$$\sum_{k=m+1}^{d} (-1)^{k-m-1} n_{k,m} = 1 - (-1)^{d-m} \qquad (33)$$

Equivalently since

$$n_{k,m} = \frac{(k+1)!}{(k-m)!(m+1)!} \frac{N_k}{N_m} \qquad (34)$$

$$\sum_{k=m}^{d} \frac{(k+1)!}{(k-m)!(m+1)!} (-1)^{d-k} N_k = N_m \qquad (35)$$

The case $m = d$ is trivial, $m = d - 1$ or $d - 2$ give equivalent information, and so on. Hence equations (35) leave only $\frac{1}{2}d - 1$ (d even) or $\frac{1}{2}(d-1)$ (d odd) unknown quantities among the d ratios $N_1/N, \ldots, N_d/N$. Thus topology alone fixes them in two dimensions, leaves one unknown in three and four dimensions etc...

ii) It is possible to obtain the average number n_d of d-simplices incident on a point in closed form. It is given by the integral, generalizing the expression found in (19) for the three-dimensional case, as

$$n_d = \frac{1}{d!} \int d^d x_2 \cdots d^d x_{d+1} \exp -V$$

where V is the volume of the sphere circumscribed to the d-simplex $(0, x_2, \ldots, x_{d+1})$ the radius of which we denote by R, thus $V = d^{-1} R^d S_d$ with S_d the area of the unit sphere

$$S_d = \frac{2\pi^{\frac{1}{2}d}}{\Gamma(\frac{1}{2}d)}$$

Taking the origin 0 at the center of the sphere with $\hat{u}_1, \ldots, \hat{u}_{d+1}$ the unit vectors pointing along $OM_1, \ldots, OM_{d+1}$ and w_d standing for the obvious generalization of (20), one finds

$$n_d = \int_0^\infty dR\, R^{d^2-1} S_d^{d+1} e^{-R^d S_d/d} \langle w_d \rangle = \frac{S_d d!}{d^2} \langle w_d \rangle$$

with $\langle w_d \rangle$ the average of a $(d+1) \times (d+1)$ determinant

$$\langle w_d \rangle = \int \prod_0^d \frac{\mathrm{d}^{d-1}\hat{u}_k}{S_d} \frac{1}{d!} \left| \det \begin{vmatrix} 1 & \cdots & 1 \\ \hat{\mathbf{u}}_0 & \cdots & \hat{\mathbf{u}}_d \end{vmatrix} \right|$$

$$= \int \prod_0^d \frac{\mathrm{d}^{d-1}\hat{u}_k}{S_d} \frac{1}{d!} (\det A_{\alpha\beta})^{\frac{1}{2}}$$

The indices α, β range from 0 to d, $A_{\alpha\beta} = 1 + \hat{\mathbf{u}}_\alpha \cdot \hat{\mathbf{u}}_\beta$, and the integration is over $d+1$ unit vectors. For a regular simplex $\hat{\mathbf{u}}_\alpha \cdot \hat{\mathbf{u}}_\beta = \delta_{\alpha\beta} - (1 - \delta_{\alpha\beta})/d$ and $w_{\mathrm{reg}} = \left[d! d^{\frac{1}{2}d} \right]^{-1} (1+d)^{\frac{1}{2}(1+d)}$. The end points of the d vectors $\hat{\mathbf{u}}_1, \ldots, \hat{\mathbf{u}}_d$ build a $(d-1)$-simplex of volume Δ in a hyperplane at a distance h from $\hat{u}_0$ and $w = h\Delta/d$. Let $\hat{u}$ stand for a unit vector along the normal of this hyperplane with two possible orientations. In general

$$\int f(\hat{u}_1, \ldots, \hat{u}_d) \prod_1^d \mathrm{d}^{d-1}\hat{u}_k = \int \tfrac{1}{2} \mathrm{d}^{d-1}\hat{u}(d-1)! \Delta \prod_1^d \mathrm{d}^{d-1}\hat{u}_k$$

$$f(\hat{u}_1, \ldots, \hat{u}_d) \, \delta(\hat{u} \cdot [\hat{u}_2 - \hat{u}_1]) \cdots \delta(\hat{u} \cdot [\hat{u}_d - \hat{u}_1])$$

Define $\hat{\mathbf{u}}_k = \cos\theta_k \hat{u} + \sin\theta_k \hat{\mathbf{v}}_k$ where $\hat{\mathbf{v}}_k$ is a unit $(d-1)$-dimensional vector orthogonal to $\hat{u}$, and

$$\mathrm{d}^{d-1}\hat{u}_k = \sin\theta_k^{d-3}\mathrm{d}\cos\theta_k \mathrm{d}^{d-2}\hat{v}_k$$

The arguments of the δ-functions are $\cos\theta_k - \cos\theta_1, k \geq 2$, and force $\hat{u}$ to be orthogonal to all $\hat{\mathbf{v}}_k$'s. We substitute $f \equiv h\Delta/d$, and observe that $\Delta = \sin\theta^{d-1} w_{d-1}$. Thus performing the integration over $\hat{u}_0$

$$\langle w_d \rangle = \frac{1}{d} \left(\frac{S_{d-1}}{S_d} \right)^d \tfrac{1}{2}(d-1)! S_d \langle w_{d-1}^2 \rangle$$

$$\times \int_0^\pi \mathrm{d}\cos\theta (\sin\theta)^{d(d-3)+2(d-1)}$$

$$\times \frac{\int_0^\pi \mathrm{d}\varphi \sin\varphi^{d-2} |\cos\varphi - \cos\theta|}{\int_0^\pi \mathrm{d}\varphi \sin\varphi^{d-2}}$$

The last ratio comes from the average over the height h. Consequently

$$\langle w_d \rangle = \frac{(d-2)!(d+1)}{d^2} \frac{S_{d-1}^{d+1}}{S_d^d} \frac{\Gamma\left(\frac{1}{2}\right)\Gamma\left(\frac{1}{2}(d^2-1)\right)}{\Gamma\left(\frac{1}{2}d^2\right)} \langle w_{d-1}^2 \rangle$$

$$\langle w_{d-1}^2 \rangle = \frac{1}{(d-1)!^2} \langle \det B_{ab} \rangle \qquad B_{ab} = 1 + \hat{\mathbf{v}}_a \cdot \hat{\mathbf{v}}_b \qquad 1 \leq a, b \leq d$$

Expanding the $d \times d$ determinant as a sum over products of cycles, observing that each cycle contributes a factorized term, using

$$\left\langle \hat{v}_a^k \hat{v}_b^{k'} \right\rangle = \delta^{kk'} \delta_{ab} \frac{1}{d-1}$$

$$\langle B_{12} B_{23} \cdots B_{\ell 1} \rangle = 1 + \langle \hat{\mathbf{v}}_1 . \hat{\mathbf{v}}_2 \ \hat{\mathbf{v}}_2 . \hat{\mathbf{v}}_3 \ \cdots \ \hat{\mathbf{v}}_\ell . \hat{\mathbf{v}}_1 \rangle = 1 + \frac{1}{(d-1)^{\ell-1}}$$

we find

$$\eta_d = \langle \det B_{ab} \rangle$$

$$= \sum_{\substack{\text{permutations} \mathcal{P}}} (-1)^{\mathcal{P}} \prod_{\substack{\text{cyclic decomposition of } \mathcal{P}}} \left(1 + \frac{1}{(d-1)^{\ell-1}} \right)$$

We group together all terms with the same cyclic decomposition. Let $\alpha_\ell \geq 0$ be the number of cycles of length ℓ with $\sum_\ell \ell \alpha_\ell = d$. The number of permutations having this cyclic decomposition is

$$\frac{d!}{\prod_{\ell=1}^d \ell^{\alpha_\ell} \alpha_\ell!}$$

and the signature is $(-1)^{\mathcal{P}} = (-1)^{\Sigma_\ell (\ell-1)\alpha_\ell}$. Therefore

$$\frac{1}{d!} \eta_d = \sum_{\{\alpha_\ell\}} \frac{1}{\prod_1^d \ell^{\alpha_\ell} \alpha_\ell!} \prod_\ell (-1)^{\alpha_\ell(\ell-1)} \left[1 + \frac{1}{(d-1)^{\ell-1}} \right]^{\alpha_\ell}$$

$$= \text{coefficient of } t^d \text{ in } \exp \sum_{\ell=1}^\infty (-1)^{\ell-1} \left[1 + \frac{1}{(d-1)^{\ell-1}} \right] \frac{t^\ell}{\ell}$$

$$= \text{coefficient of } t^d \text{ in } (1+t) \left(1 + \frac{t}{d-1} \right)^{d-1} = \frac{1}{(d-1)^{d-1}}$$

Combining all this information, we get

$$\langle w_d \rangle = \frac{d+1}{d} \frac{S_{d-1}^{d+1}}{[(d-1)S_d]^d} \frac{\Gamma\left(\frac{1}{2}\right) \Gamma\left(\frac{1}{2}(d^2-1)\right)}{\Gamma\left(\frac{1}{2}d^2\right)} \tag{36a}$$

and as a result,

$$n_d = \frac{2}{d} \frac{\Gamma(d)}{\Gamma\left(\frac{1}{2}(d+1)\right)^2} \left[\frac{\Gamma\left(\frac{1}{2}\right) \Gamma\left(\frac{1}{2}d+1\right)}{\Gamma\left(\frac{1}{2}(d+1)\right)} \right]^{d-1} \frac{\Gamma\left(\frac{1}{2}\right) \Gamma\left(\frac{1}{2}(d^2+1)\right)}{\Gamma\left(\frac{1}{2}d^2\right)} \tag{36b}$$

Asymptotically for large d

$$n_d \underset{d\to\infty}{\sim} \frac{2}{d^{\frac{1}{2}}} e^{\frac{1}{4}} (2\pi d)^{d-\frac{1}{2}} \tag{37}$$

The average volume of a d-simplex is

$$\ell_d = \frac{d+1}{n_d} \tag{38}$$

Similarly generalizing the reasoning in low dimensions one can obtain the average area of a cell boundary, call it P, as

$$P = 2^{d+1}\frac{1}{d}\Gamma\left(2 - \frac{1}{d}\right)\frac{1}{\Gamma\left(d - \frac{1}{2}\right)}\left[\Gamma\left(\tfrac{1}{2}d + 1\right)\right]^{2-1/d} \tag{39}$$

Dividing out by n_{d-1} given by equations (31) and (36) we obtain what was called above ℓ_{d-1}

$$\ell_{d-1} = \frac{d^2}{\pi^{\frac{1}{2}}}\frac{\Gamma\left(d + (d-1)/d\right)}{\left[\Gamma(d)\right]^2}\left[\frac{\Gamma\left(\tfrac{1}{2}d + 1\right)}{\pi^{d/2}}\right]^{(d-1)/d}$$

$$\times \frac{\left[\Gamma\left(\tfrac{1}{2}(d+1)\right)\right]^d}{\left[\Gamma\left(\tfrac{1}{2}d + 1\right)\right]^{d-1}}\left[\frac{\Gamma\left(\tfrac{1}{2}d^2\right)}{\Gamma\left(\tfrac{1}{2}(d^2 + 1)\right)}\right]^2 \frac{\Gamma\left(\tfrac{1}{2}(d^2 + 2)\right)}{\Gamma\left(\tfrac{1}{2}(d^2 + d - 1)\right)} \tag{40}$$

Up to now we have only studied local properties. Even though the points are chosen at random and independently, the Dirichlet construction correlates all of them. It would be of interest to understand the nature of correlations among distant cells. This is an intricate problem, to which no conclusive answer seems to be known. Other interesting constructions involve random packings, in particular of spheres.

As an application of the power of elementary topological considerations such as those developed in this section, let us derive the five regular polyhedra (or Platonic solids) in three dimensions. A finite subgroup of rotations acts transitively as substitutions on the N_0 vertices, N_1 links and N_2 faces, which means that all vertices, all links and all faces respectively, are equivalent. Let q denote the coordination number, i.e. the number of neighbours of a vertex, and $\tilde{q}$ the number of edges of each face, i.e. the number of neighbouring faces. Each link is common to two faces and joins two vertices, and the Euler characteristic χ is equal to 2, thus

$$2N_1 = \tilde{q}N_2 = qN_0 \qquad N_0 - N_1 + N_2 = 2$$

Equivalently

$$2(N_0 + N_2) = 4 + \tilde{q}N_2 = 4 + qN_0$$

Consequently $(4 - q)N_0 + (4 - \tilde{q})N_2 = 8$, showing that q or $\tilde{q}$ has to be smaller than 4. Since both are larger or equal to three (otherwise we have no solid), one concludes that the faces are either triangles ($\tilde{q} = 3$) or the vertices have a coordination number equal to 3 ($q = 3$). From the above relations one also deduces that $(6 - q)N_0 = 12 + 2(\tilde{q} - 3)N_2$ showing that q can only take the values 3,4 and 5 since the r.h.s. is larger or equal to 12. Similarly $\tilde{q}$ can only take the same values. Thus we have (with a complete duality) either $q = 3$ and $\tilde{q} = 3, 4$ or 5 or vice versa. From the original equalities we compute N_1 as

$$N_1 = \frac{2}{(2/q) + (2/\tilde{q}) - 1}$$

which has to be an integer. For $q = 3$ and $\tilde{q}$ respectively equal to 3,4 or 5 we find $N_1 = 6$, 12 and 30. Similarly for $\tilde{q} = 3$ and $q = 3, 4, 5$. This leads to the table of regular solids, where in the first column we recall the nature of the faces (given by the number $\tilde{q}$ of their edges) and in the last column the order of the corresponding rotation group, equal to $2N_1$, since it acts transitively on the oriented links. The solids are divided into three groups, with the octahedron dual to the cube, and the icosahedron dual to the dodecahedron.

Table V. Regular solids in three dimensions.

Faces	$\tilde{q}$	q	N_0	N_1	N_2	Order of rotation group
4 triangles (Tetrahedron)	3	3	4	6	4	12
8 triangles (Octahedron)	3	4	6	12	8	24
6 squares (Cube, hexahedron)	4	3	8	12	6	24
20 triangles (Icosahedron)	3	5	12	30	20	60
12 pentagons (Dodecahedron)	5	3	20	30	12	60

Can the reader extend these arguments in higher dimension ?

If, to the above list of finite rotation groups, we add the two infinite series of cyclic and dihedral groups, we recover the $A-D-E$ classification encountered in section 3.4 of chapter 9.

11.1.2   Field equations

Given an arbitrary lattice we want to set up a statistical, or field theoretical model, using some of the metrical properties attached to points, links, etc. As a matter of fact we have constructed two lattices. The first one is a simplicial lattice L made of 0, 1, 2,... simplices. A 0-simplex is a point i (we set $\ell_i \equiv 1$), a 1-simplex is a link joining two neighbours (ij) at a distance $\ell_{ij} = \ell_{ji}$, crossing the hyperplane of face common to the two cells around i and j, and so on. The dual lattice $\tilde{L}$ is made of d-, $(d-1)$-, $\cdots$ cells. We call i the d-cell pertaining to point i and σ_i its (hyper-) volume, (i,j) is then the $(d-1)$-cell dual to the link (ij) and common to cells i and j, with σ_{ij} its "area" etc.

Define $\mathcal{F}_0$ as the set of 0-forms, i.e. functions defined on sites, $i \rightarrow \varphi_i$. Similarly $\mathcal{F}_1$ is the set of antisymmetric 1-forms, defined on oriented 1-simplices, i.e. links, $\varphi_{ij} = -\varphi_{ji}$; $\mathcal{F}_2$ is made of antisymmetric 2-forms assigned to 2-simplices $\varphi_{ijk} = -\varphi_{jik} = \cdots$, and so on. In parallel, let us call $\tilde{\mathcal{F}}_d$ the set of d-densities, namely functions ψ_i defined on positively oriented d-cells. The $(d-1)$-densities ψ_{ij}, defined on oriented $(d-1)$-cells build $\tilde{\mathcal{F}}_{d-1}$, and so on. The orientation on the cell $(ij\ldots)$ and on the dual simplex $(ij\ldots)$ are chosen to be compatible, i.e. their product is unity.

The forms φ_i, φ_{ij}, φ_{ijk}, $\ldots$ may be considered as restrictions on the lattice of scalar, vector, antisymmetric tensors fields and so on, defined in the continuum. The correspondence is as follows

$$
\varphi_i \leftrightarrow \varphi(\mathbf{x}_i)
$$

$$
\varphi_{ij} \leftrightarrow \frac{1}{\ell_{ij}} \int_i^j \mathrm{d}x^\mu \varphi_\mu(\mathbf{x})
$$

$$
\varphi_{ijk} \leftrightarrow \frac{1}{\ell_{ijk}} \int \int_{(ijk)} \mathrm{d}x^\mu \wedge \mathrm{d}x^\nu \varphi_{\mu\nu}(\mathbf{x}) \tag{41}
$$

$$
\cdots
$$

Between the dual vector spaces $\mathcal{F}_p$ and $\tilde{\mathcal{F}}_{d-p}$ there exists a natural scalar product. With φ in $\mathcal{F}_p$, ψ in $\tilde{\mathcal{F}}_{d-p}$, we have

$$
p = 0 \qquad \langle \varphi | \psi \rangle = \frac{1}{1!} \sum_i \varphi_i \psi_i
$$

$$
p = 1 \qquad \langle \varphi | \psi \rangle = \frac{1}{2!} \sum_{ij} \varphi_{ij} \psi_{ij}
$$

$$p = 2 \qquad \langle \varphi \,|\, \psi \rangle = \frac{1}{3!} \sum_{ijk} \varphi_{ijk} \psi_{ijk}$$

$$\ldots \tag{42}$$

The factorial compensates for overcounting and the sum extends over sites, links, triangles etc. Should one wish it, the scalar product among real quantities could be extended to a Hermitian one among complex ones. The obvious duality between $\mathcal{F}_p$ and $\tilde{\mathcal{F}}_{d-p}$ is set up as

$$\varphi \leftrightarrow \psi$$

$$\mathcal{F}_p \leftrightarrow \tilde{\mathcal{F}}_{d-p}$$

$$\ell_{ijk\ldots} \varphi_{ijk\ldots} = \frac{\psi_{ijk\ldots}}{\sigma_{ijk\ldots}} \tag{43}$$

where no summation is understood. For any p, this entails the following relation among dimensions

$$[\psi] = [\varphi]\,[\text{length}]^d \tag{44}$$

The dimensional factor is the volume of a cell when $p = 0$. When $p = 1$, it is d-times the sum of the volumes of two pyramids of apex i and j respectively and base σ_{ij}, and so on. We denote by $\psi = \tilde{\varphi}$ or $\varphi = \tilde{\psi}$ the above correspondences.

On $\mathcal{F}_p$ we can therefore set up a quadratic potential (or mass) term as

$$V(\varphi) = \tfrac{1}{2} \langle \varphi \,|\, \tilde{\varphi} \rangle \tag{45}$$

$$p = 0 \qquad V(\varphi) = \tfrac{1}{2} \sum_i \sigma_i \varphi_i^2$$

$$p = 1 \qquad V(\varphi) = \tfrac{1}{2} \sum_{(ij)} \ell_{ij} \sigma_{ij} \varphi_{ij}^2$$

$$p = 2 \qquad V(\varphi) = \tfrac{1}{2} \sum_{(ijk)} \ell_{ijk} \sigma_{ijk} \varphi_{ij}^2 \tag{46}$$

$$\ldots$$

To construct kinetic terms, we need the analogs of gradient and divergence. Define the operator d (the analog of the exterior derivative) as a mapping from $\mathcal{F}_p$ to $\mathcal{F}_{p+1}$ (giving zero on $\mathcal{F}_d$) as

$$\mathrm{d} : \; \mathcal{F}_p \to \mathcal{F}_{p+1}$$

$$(\mathrm{d}\varphi)_{ij} = \frac{\varphi_i - \varphi_j}{\ell_{ij}}$$

$$(\mathrm{d}\varphi)_{ijk} = \frac{\ell_{ij}\varphi_{ij} + \ell_{jk}\varphi_{jk} + \ell_{ki}\varphi_{ki}}{\ell_{ijk}}$$

$$(\mathrm{d}\varphi)_{i_0 i_1 \ldots i_q} = \sum_{s=0}^{q} \frac{(-1)^s \ell_{i_0 \ldots \hat{i}_s \ldots i_q} \varphi_{i_0 \ldots \hat{i}_s \ldots i_q}}{\ell_{i_0 i_1 \ldots i_q}} \tag{47}$$

Clearly the operator d has a vanishing square

$$\mathrm{d}^2 = 0 \tag{48}$$

while from equation (42), its transpose $\tilde{\mathrm{d}}$ maps $\tilde{\mathcal{F}}_p$ into $\tilde{\mathcal{F}}_{p+1}$ according to

$$\left\langle \varphi \left| \tilde{\mathrm{d}}\psi \right.\right\rangle = \left\langle \mathrm{d}\varphi \left| \psi \right.\right\rangle \tag{49}$$

Explicitly

$$\tilde{\mathcal{F}}_{d-1} \to \tilde{\mathcal{F}}_d \qquad (\tilde{\mathrm{d}}\psi)_i = \sum_{j(i)} \frac{\psi_{ij}}{\ell_{ij}} \tag{50}$$

$$\tilde{\mathcal{F}}_{d-2} \to \tilde{\mathcal{F}}_{d-1} \qquad (\tilde{\mathrm{d}}\psi)_{ij} = \sum_{k(ij)} \frac{\ell_{ij}\psi_{ijk}}{\ell_{ijk}} \tag{51}$$

$$\ldots$$

Obviously

$$\tilde{\mathrm{d}}^2 = 0 \tag{52}$$

The divergence operator d^* is then obtained by pulling back $\tilde{\mathrm{d}}$ using the duality map (43) according to the scheme

$$
\mathrm{d}^* \begin{array}{ccc}
\mathcal{F}_p & \overset{\text{duality}}{\to} & \tilde{\mathcal{F}}_{d-p} \\
\downarrow & & \downarrow \\
\mathcal{F}_{p-1} & \overset{\text{duality}}{\leftarrow} & \tilde{\mathcal{F}}_{d-p+1}
\end{array} \tilde{\mathrm{d}} \tag{53}
$$

For instance d^* maps $\mathcal{F}_1$ on $\mathcal{F}_0$. We start from φ_{ij}, apply duality, and get $\psi_{ij} = \tilde{\varphi}_{ij} = \ell_{ij}\sigma_{ij}\varphi_{ij}$. Acting with $\tilde{\mathrm{d}}$ we obtain

$$(\tilde{\mathrm{d}}\psi)_i = \sum_{j(i)} \frac{\psi_{ij}}{\ell_{ij}} = \sum_{j(i)} \sigma_{ij}\varphi_{ij}$$

Returning by duality to $\mathcal{F}_0$, the final expression is

$$(\mathrm{d}^*\varphi)_i = \frac{1}{\sigma_i} \sum_{j(i)} \sigma_{ij}\varphi_{ij} \tag{54}$$

which is seen to be the discrete analog of the divergence. As an example, equation (54) clearly exhibits Gauss's relation between the flux of the electric field and the integral of the charge density. Again

$$d^{*2} = 0 \tag{55}$$

and d^* annihilates $\mathcal{F}_0$. One can also construct a similar nilpotent operator $\tilde{d}^*$.

A kinetic part in an action can then be defined as

$$K(\varphi) = V(\mathrm{d}\varphi) = \tfrac{1}{2} \left\langle \mathrm{d}\varphi \middle| \widetilde{\mathrm{d}\varphi} \right\rangle$$

$$= \tfrac{1}{2} \sum_{i_0,\dots,i_p} \frac{\sigma_{i_0\dots i_p}}{\ell_{i_0\dots i_p}} \left(\sum_{s=0}^{p} (-1)^s \ell_{i_0\dots \hat{i}_s\dots i_p} \varphi_{i_0\dots \hat{i}_s\dots i_p} \right)^2 \tag{56}$$

Specifically, for scalars and vectors respectively

$$\varphi \in \mathcal{F}_0 \qquad K(\varphi) = \tfrac{1}{2} \sum_{(ij)} \frac{\sigma_{ij}}{\ell_{ij}} (\varphi_i - \varphi_j)^2 \tag{57a}$$

$$\varphi \in \mathcal{F}_1 \qquad K(\varphi) = \tfrac{1}{2} \sum_{(ijk)} \frac{\sigma_{ijk}}{\ell_{ijk}} (\ell_{ij}\varphi_{ij} + \ell_{jk}\varphi_{jk} + \ell_{ki}\varphi_{ki})^2 \tag{57b}$$

As in the continuum theory, the dimension of K is

$$[K] = [\varphi]^2 \, [\text{length}]^{d-2} \tag{58}$$

To obtain free massless fields, we use the kinetic term as an action. The classical field equations are obtained by looking for extrema of K in the form

$$\delta K = \left\langle \cdot \middle| \widetilde{\delta \varphi} \right\rangle = 0$$

therefore

$$d^* \mathrm{d}\varphi = 0 \tag{59}$$

The operator $d^* d$ has a dimension of $[\text{length}]^{-2}$ and maps $\mathcal{F}_p$ on $\mathcal{F}_p$.

Let us look first at the scalar case. Since $d^* \mathcal{F}_0 = 0$, we can replace $d^* d$ by $(d^*d + dd^*) = (d + d^*)^2 = -\Delta$, where Δ is the Laplacian on scalars, such that

$$(-\Delta\varphi)_i = (d^* \mathrm{d}\varphi)_i = \frac{1}{\sigma_i} \sum_{j(i)} \frac{\sigma_{ij}}{\ell_{ij}} (\varphi_i - \varphi_j) \tag{60}$$

Define

$$p_{j,i} = \frac{1}{2d} \frac{\sigma_{ij}\ell_{ij}}{\sigma_i} \qquad (61)$$

a dimensionless positive hopping probability from a site i to a neighbouring site j satisfying

$$\sum_j p_{j,i} = 1 \qquad (62)$$

Hence

$$\frac{1}{2d}(-\Delta\varphi)_i = \sum_{j(i)} p_{j,i} \frac{\varphi_i - \varphi_j}{\ell_{ij}^2} \qquad (63)$$

Show that the functions

$$\varphi_i = a + \mathbf{k}.\mathbf{x}_i \qquad (64)$$

where $\mathbf{k}$ is a constant vector, are harmonic, i.e. satisfy

$$(-\Delta\varphi)_i = 0 \qquad (65)$$

For $p \geq 1$, the field equations (59) are not equivalent to the Laplace equation, very much as Maxwell's equations for the (four-) vector potential, which correspond to the case $p = 1$, do not entail without a special choice of gauge that each component be harmonic. There also exists here a gauge invariance of the form $\varphi \to \varphi' = \varphi + \mathrm{d}\varphi''$ since $\mathrm{d}^*\mathrm{d}\varphi = \mathrm{d}^*\mathrm{d}\varphi'$. For instance, when $p = 1$, we can rewrite equation (59) explicitly as

$$(\mathrm{d}^*\mathrm{d}\varphi)_{ij} \equiv \frac{1}{\sigma_{ij}} \sum_{k(ij)} \frac{\sigma_{ijk}}{\ell_{ijk}} (\ell_{ij}\varphi_{ij} + \ell_{jk}\varphi_{jk} + \ell_{ki}\varphi_{ki}) = 0 \qquad (66)$$

Show that if one assumes that φ_{ij} results from an underlying vector field $A_\mu(\mathbf{x})$ defined in the continuum

$$\varphi_{ij} \leftrightarrow \frac{1}{\ell_{ij}} \int_i^j \mathrm{d}x^\mu A_\mu(\mathbf{x})$$

then equation (66) is a discretized version of Maxwell's equation

$$\partial^\mu F_{\mu\nu} \equiv \Delta A_\nu - \partial_\nu(\partial.A) = 0$$

where $F_{\mu\nu} = \partial_\mu A_\nu - \partial_\nu A_\mu$. In the continuum we have

$$\Delta A_\nu = \partial^\mu(\partial_\mu A_\nu - \partial_\nu A_\mu) + \partial_\nu(\partial.A)$$

and more generally a similar definition of the Laplacian extends to any antisymmetric tensor field. In discretized form, the Laplacian is defined as

$$-\Delta = d^*d + dd^* = (d + d^*)^2 \qquad (67)$$

Thus $-\Delta$ appears as the square of the so-called Dirac–Kähler operator defined on a set $\Phi \equiv \{\varphi_i, \varphi_{ij}, \varphi_{ijk}, \ldots\}$ of forms of any degree as

$$(d + d^*)\Phi = 0 \qquad (68)$$

This has been suggested as a possible extension of Dirac's equation on a lattice, although the tensorial transformation properties do not correspond to the standard ones in the continuum.

One can study on random lattices other types of field equations than the free massless linear ones. Attempts have been made at numerical investigations of four-dimensional gauge theories on such lattices. The task is formidable. Even two-dimensional models present serious difficulties. One has to face the question of how the geometric and topological disorder interact with the intrinsic dynamics, meaning a competition between several length scales. In all cases, we are looking for the way in which the large distance properties of the random lattice allow the restoration of continuous symmetries.

11.1.3 The spectrum of the Laplacian

The simplest example of a comparison between the behaviour of a dynamical system on a random lattice and a regular one, is the study of the spectrum of the scalar Laplacian given by equations (60) and (63). We have already presented in chapter 10, section 1.4, the easiest one-dimensional case. Let us look now at some general properties in arbitrary dimension. The difficulty is that we can not use Fourier analysis anymore. Let us consider a large, but finite box, with N sites, and let $\varphi^{(\alpha)}$ and ω_α stand for the eigenfunctions and eigenvalues of the Laplacian. The eigenfunctions are normalized according to

$$\left\langle \varphi^{(\alpha)} | \tilde{\varphi}^{(\beta)} \right\rangle \equiv \sum_i \sigma_i \varphi_i^{(\alpha)} \varphi_i^{(\beta)} = \delta^{\alpha\beta} \qquad (69)$$

An arbitrary function φ such that

$$\gamma_\alpha = \left\langle \varphi \left| \tilde{\varphi}^{(\alpha)} \right. \right\rangle \equiv \sum_i \sigma_i \varphi_i \varphi_i^{(\alpha)} \qquad (70)$$

admits an expansion of the form

$$\varphi_i = \sum_\alpha \gamma_\alpha \varphi_i^{(\alpha)} \qquad (71)$$

with

$$(-\Delta\varphi)_i = \sum_\alpha \gamma_\alpha \omega_\alpha \varphi_i^{(\alpha)} \qquad (72)$$

Here the functions $\varphi^{(\alpha)}$ which replace the plane waves, and the corresponding eigenvalues ω_α are still stochastic variables. We look for the most likely, or rather average distribution of eigenvalues.

As the number of sites N tends to infinity, with the index α running through N values, let $N\rho(\omega)$ denote the number of eigenvalues in an interval $d\omega$ around ω. We have

$$\int_0^\infty d\omega \rho(\omega) = 1 \qquad (73)$$

We expect that the thermodynamic limit ensures that $\rho(\omega)$ is already the most likely distribution. Our aim will be to ascertain the behaviour of $\rho(\omega)$ for small as well as large values of ω. For $\omega \to 0$, if eigenfunctions can be approximated by plane waves over large distances, it is likely that $d\omega\rho(\omega) \sim d^d\mathbf{k}/(2\pi)^d$ with $\omega = \mathbf{k}^2$, i.e.

$$\rho(\omega) \underset{\omega \to 0}{\sim} \frac{\omega^{\frac{1}{2}d-1}}{(4\pi)^{\frac{1}{2}d}\Gamma(\frac{1}{2}d)} \qquad (74)$$

To lend credence to this estimate, let us study how the Laplacian acts on the functions

$$\varphi_i(\mathbf{k}) = \exp i\mathbf{k} \cdot \mathbf{x}_i \qquad (75)$$

Consider the local quantities

$$\omega_i(\mathbf{k}) = \varphi_i^*(\mathbf{k})(-\Delta\varphi(\mathbf{k}))_i = \frac{1}{\sigma_i} \sum_{j(i)} \frac{\sigma_{ij}}{\ell_{ij}} \left[1 - e^{i\mathbf{k}\cdot(\mathbf{x}_j - \mathbf{x}_i)} \right] \qquad (76)$$

For small $\mathbf{k}$, $\omega_i(\mathbf{k})$ is of order $\mathbf{k}^2$, according to equations (64) and (65). The average

$$\overline{\omega(\mathbf{k})} = \frac{\sum_i \sigma_i \omega_i(\mathbf{k})}{\sum \sigma_i} \tag{77}$$

can be interpreted as the expectation value of $(-\Delta)$ in the state $\varphi(\mathbf{k})$. For $N \to \infty$, there exists no preferred direction in the lattice, so the expression (77) can be averaged over the directions of the wavevector $\mathbf{k}$

$$\overline{\omega(\mathbf{k})} = \sum_{s=1}^{\infty} (-1)^{s-1} \frac{\mathbf{k}^{2s}}{2^{2s}} \frac{1}{s!} \frac{\Gamma(\frac{1}{2}d)}{\Gamma(\frac{1}{2}d+s)} \overline{\ell^{2s-1}} \tag{78}$$

The last average stands for

$$\overline{\ell^{2s-1}} = \frac{1}{N} \sum_i \sum_{j(i)} \sigma_{ij} \ell_{ij}^{2s-1} \tag{79}$$

It can be computed following Meijering's argument of section 1.1, with the result that

$$\overline{\omega(\mathbf{k})} = \frac{2^d}{d\pi^{\frac{1}{2}}} \Gamma(1 + \tfrac{1}{2}d)^2 \sum_{s=1}^{\infty} \left[\frac{-\Gamma(1 + \tfrac{1}{2}d)^{2/d}}{\pi} \right]^{s-1}$$
$$\times \frac{\Gamma(s + \frac{d-1}{2})\Gamma(2 + \frac{2(s-1)}{d})}{\Gamma(s + \frac{d}{2})\Gamma(s+d-1)} \frac{\mathbf{k}^{2s}}{s!} \tag{80}$$

This rather cumbersome formula reduces to simple expressions in low dimensions, $d = 1$ or 2. We have

$$d = 1 \qquad \overline{\omega(\mathbf{k})} = \ln(1 + \mathbf{k}^2) \tag{81a}$$
$$d = 2 \qquad \overline{\omega(\mathbf{k})} = 2\pi \left[1 - e^{-\mathbf{k}^2/2\pi} I_0\left(\mathbf{k}^2/2\pi\right) \right] \tag{81b}$$

where $I_0(z)$ is the modified Bessel function

$$I_0(z) = \sum_{n=0}^{\infty} \left(\frac{z^2}{4}\right)^n \frac{1}{(n!)^2} \underset{z\to\infty}{\sim} \frac{e^z}{\sqrt{2\pi z}} \left(1 + \frac{1}{8z} + \cdots\right) \tag{82}$$

In all cases $\overline{\omega(\mathbf{k})}$ behaves as $\mathbf{k}^2$ for small $\mathbf{k}$. This comforts the estimate (74). The computation of $\overline{\omega(\mathbf{k})}$ for any $\mathbf{k}^2$ allows us also to draw some interesting conclusions. In dimension one, we observe that it is unbounded, giving a proof that the support of $\rho(\omega)$ is also unbounded. However, in dimension larger that one, $\overline{\omega(\mathbf{k})}$ is bounded as $\mathbf{k}^2$ tends to infinity. For instance for $d = 2$,

$\overline{\omega(\mathbf{k})} \to 2\pi(1 - 1/|\mathbf{k}| + \cdots)$. In general, this limit can be obtained by dropping the oscillatory term in averaging (76)

$$\lim_{\mathbf{k}\to\infty} \overline{\omega(\mathbf{k})} = \pi d \frac{\Gamma(2 - 2/d)}{[\Gamma(1 + \frac{1}{2}d)]^{2/d}} \tag{83}$$

showing that the support of $-\Delta$ extends at least up to this value.

It is most likely that for small ω, in large enough dimension, the eigenstates are extended ones, with unit probability. At the other extreme, we expect localized states for large ω. A very crude argument, based on this expectation, yields an estimate of $\rho(\omega)$ for ω tending to infinity. Looking at the form of the Laplace operator (60) or (63), we see that "weak links", where σ_{ij}/ℓ_{ij} is exceptionally large, play a significant role in localizing the wavefunctions. Assume that σ_{ij}/ℓ_{ij} is large due to the unfrequent occurrence of two points i, j very close to each other. A bold and extreme approximation is that an eigenstate has a wavefunction φ essentially concentrated on the two neighbouring sites i, j, with

$$\begin{aligned} \omega\varphi_i &= \frac{1}{\sigma_i} \frac{\ell_{ij}}{\sigma_{ij}}(\varphi_i - \varphi_j) \\ \omega\varphi_j &= \frac{1}{\sigma_j} \frac{\ell_{ij}}{\sigma_{ij}}(\varphi_j - \varphi_i) \end{aligned} \tag{84}$$

Therefore

$$\omega = \frac{\sigma_{ij}}{\ell_{ij}} \left(\frac{1}{\sigma_i} + \frac{1}{\sigma_j} \right) \tag{85}$$

It is of course assumed that the right-hand side of this expression is very large (with a small probability). Since i and j are very close, we can think, in the limit, of the two neighbouring cells as a unique cell, call it $\mathcal{C}$, partitioned by a plane bisecting the infinitesimal segment (ij), or equivalently an arbitrary plane through the center of $\mathcal{C}$, cutting it into two parts of volumes σ_+ and σ_-, and having an area σ_0 inside $\mathcal{C}$. Then

$$\rho(\omega)\mathrm{d}\omega \underset{\ell\to 0}{\sim} p(\ell)\mathrm{d}\ell \tag{86a}$$

$$\omega \sim \frac{1}{\ell}\sigma_0 \left(\frac{1}{\sigma_+} + \frac{1}{\sigma_-} \right) \tag{86b}$$

where $p(\ell)d\ell$ behaves as $S_d\ell^{d-1}d\ell$ and the average in (86b) is both on the cell $\mathcal{C}$ and the direction of the plane. This results in

$$\rho(\omega) \geq \frac{2\pi^{\frac{1}{2}d}}{\Gamma(\frac{1}{2}d)}\frac{1}{\omega^{d+1}} \qquad (87)$$

This power law decrease is in agreement with the unit integral of $\rho(\omega)$, equation (73), and is verified for $d = 1$. A tail in the spectrum of the type (87) is of course absent in the case of a regular lattice. The above argument shows that it is related to localization effects. To determine the localization threshold ω_c is a delicate matter. While for $d = 1, \omega_c = 0$, it is not known what is even the lower critical dimension (possibly $d = 2$) where ω_c departs from zero.

i) Show that from equation (83) one can derive the exact sum rule

$$\int_0^\infty d\omega\rho(\omega)\omega = \pi d\frac{\Gamma(2 - 2/d)}{\left[\Gamma(1 + \frac{1}{2}d)\right]^{2/d}} \qquad (88)$$

This is infinite for $d = 1$ due to a divergence at large ω, in agreement with the estimate (87), but eventually grows like d, due to the suppression of small and large ω's in large dimension.

ii) Derive the inequality

$$\frac{\langle\omega^2\rangle}{\langle\omega\rangle^2} - 1 \geq \frac{1}{q} \qquad (89)$$

where q is the average coordination number of the lattice, and prove that it becomes an equality on any regular lattice. As q grows very rapidly with d on a random lattice, and $\rho(\omega)$ tends to be strongly peaked, it is likely that (89) becomes asymptotically an equality.

11.2   Random surfaces

We will not repeat the motivations for attempting to find models of random surfaces. A compact and basic formulation is still missing. As a consequence the following presentation will necessarily be fragmentary. We have to give a suitable definition of a surface since it involves various degrees of internal or external structure. We first have topological properties such as connectedness, orientability, Euler characteristics, number of boundaries.

The generalization in higher dimension implies homology, and if a differentiable structure is assumed, dual cohomology groups. We have then the metric properties, parallel transport and associated curvature, with generalizations to other fibre bundles and their connections on surfaces. One can also study complex structures. These are intrinsic properties, but one can also wish to investigate the embeddings in other manifolds, in practical terms mostly Euclidean (or Minkovskian) space. An important circumstance is the one of interfaces, mostly in $\mathcal{R}^3$ of course.

An aspect, which distinguishes surfaces from curves, is the possible existence of nontrivial boundaries. Typical of these questions is the Plateau problem of finding minimal surfaces bounding a given curve, or the study of Wilson loops in gauge theories.

Yet another question, which was not on the forefront for random curves, but could also have been discussed in this case, is the one of reparametrization invariance, or general covariance as it is called in the context of gravitation theory. A correlative point of view is the one of dynamical generation of surfaces. A curve could be viewed as the path traced by a structureless point as it evolves in time, but a surface is generated by a string with infinitely many degrees of freedom (hence the name *string field theory* now in common use).

To sum up, the much richer set of possibilities afforded by the concept of surface opens a vast new domain. We start with a short description of piecewise linear manifolds as exemplified by triangulated surfaces.

11.2.1 Piecewise linear surfaces

We consider an assembly of triangles endowed with a flat Euclidean metric and compatible incidence relations among the edges (and therefore the vertices). This implies of course the equality of lengths for identified sides. We assume for simplicity the resultant manifold to be connected, without boundary, and orientable. Its topology is then dictated by its Euler characteristic $\chi = 2 - 2g$ where g (denoted H in chapter 10 section 4) is the genus, or number of handles. The incidence relations define the notion of nearest neighbour pairs (or links) and 2-simplices (the original triangles) and the construction allows us to compute distances. The manifold has an obvious differentiable structure except possibly at singular points (the vertices). As usual N_0, N_1, N_2 are the numbers of vertices, links and triangles related by

$$2N_1 = 3N_2 \qquad \chi = N_0 - N_1 + N_2 \qquad (90a)$$

Thus in terms of the number N of vertices

$$N_0 = N$$
$$N_1 = 3(N - \chi) \qquad (90b)$$
$$N_2 = 2(N - \chi)$$

We shall discuss later the question of inequivalent triangulations and their statistical weights.

A practical way of implementing the above construction is to pick a model, or parameter space, as a given piecewise flat triangulated connected and orientable manifold, to map its N points in Euclidean space $\mathcal{R}^d$ and to interpolate linearly between the images of neighbours. This produces in the target space another triangulated surface (with possible self-intersections), or equivalently we may consider that it endows the original model with a different metric structure. As an example, we may start from a finite region in a Euclidean plane carrying a regular triangular lattice with the shape of a torus ($g = 1$) and allow for various maps in $\mathcal{R}^d$. There are of course numerous other ways of defining surfaces such as those encountered in chapter 6 when discussing lattice gauge theories, as collections of plaquettes from regular lattices with specific rules for assembling them.

The triangles are Euclidean, hence their internal angles θ_1, θ_2, θ_3 add to π. For each triangle we can therefore split unity into $1 = (\theta_1 + \theta_2 + \theta_3)/\pi$, with each θ between 0 and π. Summing over triangles yields N_2, but we can also collect first all θ 's pertaining to the same vertex, then sum over vertices. Call θ_i the sum at each vertex. Then $N_2 = \sum_i \theta_i/\pi$. From equation (90) this is also $2N - 2\chi = -2\chi + 2\sum_i 1$. Identification leads to

$$\chi = \sum_i \left(1 - \frac{\theta_i}{2\pi}\right) \qquad (91)$$

which is the discrete form of the Gauss–Bonnet formula, in terms of deficit angles $2\pi - \theta$. When the deficit angle vanishes at a vertex the corresponding triangles fit in flat space and do not contribute to χ. We have the following interpretation : deficit angle ⇔ frustration from a planar situation ⇔ curvature. Positive curvature is related to a positive deficit angle ($\theta_i < 2\pi$), negative curvature means overabundance of "matter" and negative deficit angle ($2\pi < \theta_i$).

This can be understood in the example of a sphere of radius r, with curvature $R = 1/r^2$, the product of the two inverse principal radii of curvature, here equal. For a spherical triangle with inner angles α_1, α_2, α_3, the area A is given by $RA = (\alpha_1 + \alpha_2 + \alpha_3 - \pi)$ The total amount that a tangent vector has rotated in one circumnavigation (rounding vertices) is

$$\theta = (\pi - \alpha_1) + (\pi - \alpha_2) + (\pi - \alpha_3)$$
$$= 2\pi - RA \quad \text{i.e.} \quad RA = 2\pi - \theta$$

the total angular deficit. On the triangulated surface, the curvature is concentrated at the vertices, with RA taking a finite limit as A shrinks to zero and R tends to have a δ-function singularity. With the present normalization (which may differ from the one found in other sources), it follows that the Gauss–Bonnet formula has the following form in the continuum

$$\chi = \frac{1}{2\pi} \int \mathrm{d}A\, R \tag{92}$$

i.e. $\chi = 2$ for a sphere where $A = 4\pi r^2, R = r^{-2}$.

For surfaces, curvature is a scalar quantity. In a discrete setting, equation (91) implies a dissymmetry between positive and negative curvature, as we cannot have more than a 2π deficit ($\theta = 0$), while one can have as much as one wants of "extra matter" for large θ.

One can envision higher dimensional analogs of piecewise flat manifolds, with d-dimensional simplices replacing triangles. The Euler characteristics $\chi = N_0 - N_1 + N_2 - N_3 + \cdots$ remains a topological invariant. Indeed, adding a vertex will produce the substitutions $N_0 \to N_0 + 1$, $N_1 \to N_1 + C_{d+1}^1$, $N_2 \to N_2 + C_{d+1}^2$, ..., with the last change being $N_d \to N_d + C_{d+1}^d - 1$. Therefore $\chi \to \chi + (1 - 1)^{d+1} = \chi$. One would like to get a relation as simple as (91) relating χ to metric properties, with a smooth limit in the continuum case and such that contributions from regions where the simplices fit in flat space vanish (as was true for the expression

(91)). This goal can apparently not be fulfilled. The following is the best substitute.

Consider a p-simplex in the triangulation incident on a given vertex and round off the corresponding corner, to a spherical shape, which by rescaling corresponds to a fraction $\varphi^{(p)}$ of the area of the unit $(p-1)$-dimensional sphere S_{p-1}. The case of a one-simplex – a link – corresponds to the sphere S_0, a set of two points, in which case we assign $\varphi^{(1)} = \frac{1}{2}$ to each end point. For a vertex itself, we define $\varphi^{(0)} = 1$. If we sum the $\varphi^{(p)}$'s over the $p+1$ vertices of a given p-simplex, we obtain the full area of S_{p-1}, thus the $\varphi^{(p)}$ add to unity. The sum of unity over the p-simplices being N_p, we have, by collecting first the contributions pertaining to a vertex i denoted $\varphi_i^{(p)}$

$$\varphi_i^{(p)} = \sum_{\substack{p \text{ simplices} \\ \text{incident on } i}} \varphi^{(p)}$$

the obvious relation

$$N_p = \sum_i \varphi_i^{(p)}$$

Thus for the Euler characteristic

$$\chi = \sum_i \sum_{p=0}^{d} (-1)^p \varphi_i^{(p)} \tag{93}$$

When $d = 2$ we recover formula (91), but in higher dimension this expression does not exhibit explicitly that contributions from flat regions vanish. It is possible to re-express (93) in terms of interior angles, but in any case the relation with the curvature tensor is hard to see (Cheeger, Müller and Schrader).

We also note here that the four-dimensional Hilbert–Einstein action for gravity has been discretized by Regge for a simplicial complex in the form

$$\sum a_2 \left(1 - \frac{\theta}{2\pi}\right)$$

generalizing equation (92) where the sum runs over 2-simplices of area a_2 with θ the corresponding sum of dihedral angles. The conditions under which this approximates the continuous version

$$\int \mathrm{d}^4 x \sqrt{g} R$$

require some careful analysis (Feinberg, Friedberg, Lee and Ren).

Returning to triangulated surfaces with their metric properties, we can easily generalize the field equations for random lattices. In the following, we shall be interested mostly by free massless scalar fields. If the surface were smooth with local parameters α^1, α^2, and a metric positive definite tensor $g_{ab}(\alpha)$ such that $g = \det g_{ab}$, with g^{ab} the inverse of g_{ab}, we would write a reparametrization invariant action as

$$S = \int \mathrm{d}^2\alpha \sqrt{g} \tfrac{1}{2} g^{ab}(\alpha) \partial_a \varphi \partial_b \varphi \qquad (94)$$

We assume for simplicity that the surface has no boundary and observe that in general one would require several overlapping coordinate patches to be able to define the above integral. If instead of taking real values, φ is the image of a map from the surface to a Riemannian d-dimensional manifold M, with local coordinates φ^μ, and a metric $G_{\mu\nu}(\varphi)$, the action would be replaced by

$$S^{(M)} = \int \mathrm{d}^2\alpha \sqrt{g} \tfrac{1}{2} g^{ab}(\alpha) \partial_a \varphi^\mu \partial_b \varphi^\nu G_{\mu\nu}(\varphi) \qquad (95)$$

This defines a general σ-model. As a particular case, if the manifold M coincides with the Euclidean space $\mathcal{R}^d$, with a flat metric $G_{\mu\nu}(\varphi) = \delta_{\mu\nu}$, the action (95) amounts to the sum of d-copies of the scalar action (94).

Let us look first at the structureless one-component action and extend its definition to a piecewise linear triangulated surface embedded in $\mathcal{R}^d$. Take on each triangle with vertices $\mathbf{x}_1, \mathbf{x}_2, \mathbf{x}_3$ a set of barycentric coordinates

$$\mathbf{x}(\alpha) = \alpha^1 \mathbf{x}_1 + \alpha^2 \mathbf{x}_2 + \alpha^3 \mathbf{x}_3, \qquad \alpha_i \geq 0, \qquad \alpha^1 + \alpha^2 + \alpha^3 = 0 \quad (96)$$

If φ takes values φ_i at the vertices we extend it linearly inside the triangle through

$$\varphi(\alpha) = \alpha^1 \varphi_1 + \alpha^2 \varphi_2 + \alpha^3 \varphi_3 \qquad (97)$$

this being a natural harmonic function. Computing the action (94) as a sum over triangles, we find

$$S_{\text{discrete}} = \sum_{(ijk)} \tfrac{1}{8} \left[\varphi_i(\mathbf{x}_j - \mathbf{x}_k) + \varphi_j(\mathbf{x}_k - \mathbf{x}_i) + \varphi_k(\mathbf{x}_i - \mathbf{x}_j) \right]^2 / \ell_{ijk}$$

$$(98)$$

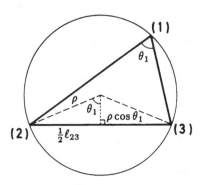

Fig. 4   Construction of the dual quantities.

The sum is over triangles of area ℓ_{ijk}, related to the lengths of the sides through

$$16\ell_{ijk}^2 = (\ell_{ij}+\ell_{jk}+\ell_{ki})(\ell_{ij}+\ell_{jk}-\ell_{ki})(\ell_{ij}-\ell_{jk}+\ell_{ki})(-\ell_{ij}+\ell_{jk}+\ell_{ki})$$
$$(99)$$

Translations, rotations or constant shifts of the field leave the action invariant. Henceforth we will omit the suffix "discrete", since the context will clearly indicate when we deal with triangulated surfaces.

In spite of the fact that an *a priori* notion of duality seems to be missing, it is possible to give to the r.h.s. in (98) a form similar to the one corresponding to flat random lattices (equation (57a)).

The quantities $\ell_i = 1$, ℓ_{ij}, and ℓ_{ijk} have obvious definitions the second being the length of the side (ij), the third the area of the triangle ℓ_{ijk} as above. Define now $\sigma_{ijk} = 1, \sigma_{ij}, \sigma_i$, corresponding to a (virtual) dual lattice as follows. Pick one triangle and call it (1,2,3) for short with interior angles $\theta_1, \theta_2, \theta_3$ between 0 and π. The corresponding contribution to the action (98) is

$$S_{123} = \tfrac{1}{4}\left[\cot\theta_1(\varphi_2 - \varphi_3)^2 + \cot\theta_2(\varphi_3 - \varphi_1)^2 + \cot\theta_3(\varphi_1 - \varphi_2)^2\right]$$
$$(100)$$

The coefficients can take either sign, but the whole expression is positive. Let ρ denote the radius of the circle of center O circumscribed to the triangle. We have

$$\tfrac{1}{2}\cot\theta_i = (\rho\cos\theta_i)/\ell_{jk}$$

with $\rho \cos \theta_i$ the (algebraic) distance of O to the edge jk (positive if i and O are on the same side of the chord jk). Two and only two triangles $(23i_1)$ and $(23i_2)$ share a given edge (23). We define the indefinite quantity σ_{23}, the (algebraic) length of the virtual link dual to (23), as the sum of the two quantities $\rho_1 \cos \theta_{i_1}$ and $\rho_2 \cos \theta_{i_2}$ pertaining to these two triangles. Thus

$$\sigma_{ij} = \rho_1 \cos \theta_{i_1} + \rho_2 \cos \theta_{i_2} \tag{101}$$

in such a way that $\frac{1}{2}\sigma_{ij}/\ell_{ij}$ will be the coefficient of $(\varphi_i - \varphi_j)^2$ in the action. This is consistent with the flat lattice definition and yields

$$S = \frac{1}{2} \sum_{(ij)} \frac{\sigma_{ij}}{\ell_{ij}} (\varphi_i - \varphi_j)^2 \tag{102}$$

The positivity of the action entails inequalities such as

$$\sum_{j(i)} \frac{\sigma_{ij}}{\ell_{ij}} > 0$$

The remaining quantity to be defined is σ_i, the area of the virtual cell i, which should satisfy $\sum_i \sigma_i = \sum_{(ijk)} \ell_{ijk} = A$, where A stands for the total area. A natural choice fulfilling this condition is

$$\sigma_i = \tfrac{1}{4} \sum_{j(i)} \sigma_{ij} \ell_{ij} \tag{103}$$

which reduces, in the flat case, to area of the Voronoi cell. In the present context, some σ_i's can however be negative. It should be observed that the construction is well-defined once the lengths ℓ_{ij} of the edges are given. Thus an abstract collection of metric triangles with compatible incidence relations is all that is required, independently of any embedding.

With the ℓ 's and σ 's as above, the definition of the operators d and d* goes through as in equations (47), (53) and (54), with the obvious restrictions to a two-dimensional manifold. Thus

$$\begin{aligned}
\varphi_i &\to (\mathrm{d}\varphi)_{ij} = (\varphi_i - \varphi_j)/\ell_{ij} \\
\varphi_{ij} &\to (\mathrm{d}\varphi)_{ijk} = (\ell_{ij}\varphi_{ij} + \ell_{jk}\varphi_{jk} + \ell_{ki}\varphi_{ki})/l_{ijk}
\end{aligned} \tag{104a}$$

and

$$\begin{aligned}
\varphi_{ij} &\to (\mathrm{d}^*\varphi)_i = \frac{1}{\sigma_i} \sum_{j(i)} \sigma_{ij}\varphi_{ij} \\
\varphi_{ijk} &\to (\mathrm{d}^*\varphi)_{ij} = \frac{1}{\sigma_{ij}} \sum_{k(ij)} \varphi_{ijk}
\end{aligned} \tag{104b}$$

The scalar Laplacian is

$$(-\Delta\varphi)_i = (\mathrm{d}^*\mathrm{d}\varphi)_i = \frac{1}{\sigma_i} \sum_{j(i)} \frac{\sigma_{ij}}{\ell_{ij}} (\varphi_i - \varphi_j) \qquad (105)$$

It enables one, in the absence of boundaries, to rewrite equations (98) and (102) as

$$S = \tfrac{1}{2} \sum_i \varphi_i (-\Delta\varphi)_i \qquad (106)$$

On a compact surface, the action is positive as soon as φ is not a constant, as readily follows from (98), thus the only harmonic functions are constants and $-\Delta$ is a non-negative operator. For pseudoscalar fields, the action, analogous, but distinct, from (102), is

$$\tilde{S} = \tfrac{1}{2} \sum_{(ijk)} \ell_{ijk} \varphi_{ijk} (\mathrm{d}\mathrm{d}^*\varphi)_{ijk} = \tfrac{1}{2} \sum_{(ij)} \frac{\ell_{ij}}{\sigma_{ij}} (\varphi_{ijk_1} - \varphi_{ijk_2})^2 \qquad (107)$$

where (ijk_1) and (ijk_2) are the two triangles which share the link (ij). At least if all σ_{ij}'s are positive and since the surface is orientable, the unique harmonic pseudoscalar is a multiple of η_{ijk}, the orientation tensor with values ± 1 on each (oriented) triangle.

For vector fields φ_{ij}, let us show that harmonicity is equivalent to having both a vanishing curl and divergence

$$[(\mathrm{d}\mathrm{d}^* + \mathrm{d}^*\mathrm{d})\varphi]_{ij} = 0 \Leftrightarrow \{(\mathrm{d}\varphi)_{ijk} = 0, (\mathrm{d}^*\varphi)_i = 0\} \qquad (108)$$

Indeed if φ_{ij} is harmonic, $\psi_i = (\mathrm{d}^*\varphi)_i$ satisfies

$$\mathrm{d}^*\mathrm{d}\psi = \mathrm{d}^*\mathrm{d}\mathrm{d}^*\varphi = -\mathrm{d}^{*2}\mathrm{d}\varphi = 0$$

so that ψ_i is a constant ψ given by

$$\sum_i \sigma_i \psi_i = \psi A = \sum_i \sigma_i (\mathrm{d}^*\varphi)_i = \sum_{(ij)} \sigma_{ij} \varphi_{ij} = 0$$

from $\varphi_{ij} = -\varphi_{ji}$. Hence $\psi = 0$, and we have the second part in equation (108). Therefore $\mathrm{d}^*\mathrm{d}\varphi = 0$, meaning that $\mathrm{d}\varphi$ is harmonic, hence a multiple $\tilde{\psi}$ of η_{ijk}. Again

$$\tilde{\psi} = \sum_{(ijk)} \eta_{ijk} \ell_{ijk} (\mathrm{d}\varphi)_{ijk} = \sum_{(ijk)} \eta_{ijk} (\ell_{ij}\varphi_{ij} + \ell_{jk}\varphi_{jk} + \ell_{ki}\varphi_{ki}) = 0$$

$$(109)$$

where the last equality follows by adding the two contributions for each factor φ_{ij}. Thus $\tilde{\psi} = 0$.

The counting of linearly independent harmonic vector fields yields a classical relation with topology. There are N_1 independent components φ_{ij}. Equation (108) implies $(N_0 - 1) + (N_2 - 1)$

conditions since $\sum_{(ijk)} \ell_{ijk} \eta_{ijk} (\mathrm{d}\varphi)_{ijk} = 0$ and $\sum_i \sigma_i (\mathrm{d}^*\varphi)_i = 0$. Therefore there are $N_1 - N_0 - N_2 + 2 = 2 - \chi = 2g$ linearly independent harmonic vector fields. We recall that the same counting holds in the continuum and, when an analytic structure is prescribed, this amounts to the existence of g holomorphic vector fields or equivalently g abelian differentials of the first kind (i.e. holomorphic differentials).

11.2.2   The conformal anomaly and the Liouville action

Given a compact, connected, oriented surface, without boundaries, with its metric structure, our aim is to promote classical massless field theory to its quantum version, namely to use in the discrete case the exponential of the scalar action (102) as a Boltzmann weight in a path integral over the scalar field φ_i. For the time being, the geometry is frozen and we assume that the surface is regular enough for the various σ's to be positive.

We proceed by setting up the Gaussian path integral in such a way that it yields a result proportional to $[\det'(-\Delta)]^{-1/2}$ where the prime indicates the omission of the zero mode, or equivalently the determinant of $-\Delta$ in the space orthogonal to a constant φ with respect to the square norm $\sum_i \sigma_i \varphi_i^2$, it being understood that φ_i's are real. For a finite triangulated surface, the operator Δ is a finite $N \times N$ matrix, with $N \equiv N_0$ the number of vertices.

Let $E_1, \ldots, E_{N-1}$ be the positive eigenvalues of $-\Delta$ and $E_0 = 0$ corresponds to the unique zero mode. The associated orthonormal set of eigenfunctions will be denoted $\varphi^{(0)}, \varphi^{(1)}, \ldots, \varphi^{(N-1)}$. Expanding the field as

$$\varphi_i = \sum_{0 \leq n \leq N-1} C_n \varphi_i^{(n)} \tag{110}$$

it is seen that

$$S = \tfrac{1}{2} \sum_{1 \leq n \leq N-1} E_n C_n^2 \tag{111}$$

We define the partition function Z as

$$Z = \frac{1}{[\det'(-\Delta)]^{\frac{1}{2}}} = \frac{1}{\left(\Pi_1^{N-1} E_n \right)^{\frac{1}{2}}}$$

$$= \int \prod_0^{N-1} \frac{\mathrm{d}C_n}{\sqrt{2\pi}} \delta(C_0) \exp{-S} \tag{112}$$

The real eigenfunctions are normalized according to

$$\sum_i \sigma_i \varphi_i^{(n_1)} \varphi_i^{(n_2)} = \delta_{n_1,n_2} \tag{113}$$

As a result, the σ_i's being positive,

$$\prod_0^{N-1} dC_n = \left(\prod_0^{N-1} \sigma_i\right)^{\frac{1}{2}} \prod_i d\varphi_i \tag{114}$$

and

$$\varphi^{(0)} = \frac{1}{\sqrt{A}} \tag{115}$$

Thus we rewrite the path integral in (112) as

$$Z = \sqrt{2\pi A} \int \prod_i \frac{d\varphi_i}{\sqrt{2\pi}} \left(\prod_i \sigma_i\right)^{\frac{1}{2}} \delta\left(\sum_i \sigma_i \varphi_i\right) e^{-S(\varphi)} \tag{116}$$

This could readily be generalized to include a source term either in the form $\sum_i J_i \varphi_i$ or $\sum_i \sigma_i j_i \varphi_i$.

This is as far as one can go in general terms in the piecewise linear case. We can however proceed much further if we make the bold assumption that, at the price of some renormalization, the discrete and well-defined path integral is a valid approximation to a continuous theory on a smooth surface with the action written as in equation (94). We then observe that the absence of a mass term implies that it is conformally invariant, in the sense that the combination $\sqrt{g}g^{ab}$ is unaffected by a point dependent rescaling of the metric $g_{ab}(\alpha) \to \rho^2(\alpha)g_{ab}(\alpha)$, which amounts to rescaling lengths locally by the factor ρ. While this holds at the classical level, it fails at the quantum level, leading to the conformal anomaly amply discussed in chapter 9.

In two dimensions, it is always possible in a coordinate patch homeomorphic to the interior of a circle to find appropriate coordinates in which an arbitrary metric takes the form

$$g_{ab}(\alpha) = \rho^2(\alpha)\delta_{ab} \tag{117}$$

For instance, on the upper hemisphere of a sphere of radius r, a stereographic projection yields for the square length element

$$ds^2 = \frac{(d\alpha^1)^2 + (d\alpha^2)^2}{\left(1 + (\alpha_1^2 + \alpha_2^2)/4r^2\right)^2} \tag{118}$$

In such a neighbourhood, the coordinates themselves are harmonic functions (these are the so-called thermal coordinates), as they are in flat space. The curvature R (r^{-2} for the sphere) is given by

$$R = -\Delta \ln \rho = -\frac{1}{\rho^2} \Delta_0 \ln \rho \qquad (119)$$

where Δ stands for the Laplacian in curved space and Δ_0 in flat space relative to the metric $g_{ab}^{(0)} \equiv \delta_{ab}$. If in each coordinate patch one uses thermal coordinates, the conformal factor ρ drops out in the expression for the action, but not of course in the Laplace operator.

In the continuous case and for a compact surface, the set of eigenvalues is discrete, and an expansion similar to (110), but extending over an infinite number of modes is still valid. The partition function, $(\det' - \Delta)^{-\frac{1}{2}}$, requires subtractions. If we first ignore them, a variation of the free energy will take the form

$$\delta \ln Z = \delta \ln \left\{ \int \prod{}' \left(\frac{dC_n}{\sqrt{2\pi}} \right) \exp \left(-\tfrac{1}{2} \sum E_n C_n^2 \right) \right\} \qquad (120)$$

The zero mode is given by equation (115).

We concentrate on a variation with respect to an infinitesimal conformal change of the form $g_{ab} \to (1 + 2\delta\varepsilon)g_{ab}$, where $\delta\varepsilon$ varies from point to point. This induces a change in the eigenvalues and eigenfunctions, and, for a fixed field $\varphi(\alpha)$, in the components C_n, but the combination $\sum E_n C_n^2$ is invariant, this being precisely the consequence of classical conformal invariance. Thus in equation (120) the variation arises solely from the Jacobian of the transformation from C_n to $C_n + \delta C_n$. Since the zero mode had to be subtracted, we find

$$\delta \ln Z = \left\langle \left(\sum_0^\infty \frac{\partial \delta C_n}{\partial C_n} - \frac{\partial \delta C_0}{\partial C_0} \right) \right\rangle \qquad (121)$$

From the invariance of the field φ

$$\delta\varphi = \sum_0^\infty \left(\delta C_n \varphi^{(n)} + C_n \delta\varphi^{(n)} \right) = 0$$

Since

$$\int d^2\alpha \sqrt{g} \varphi^{(n_1)} \varphi^{(n_2)} = \delta_{n_1, n_2}$$

we have

$$\frac{\partial \delta C_n}{\partial C_n} = -\int \mathrm{d}^2\alpha\sqrt{g}\varphi^{(n)}\delta\varphi^{(n)} = \int \mathrm{d}^2\alpha\sqrt{g}\left(\frac{1}{4}\frac{\delta g}{g}\right)\left[\varphi^{(n)}\right]^2$$

and

$$0 = \int \mathrm{d}^2\alpha\sqrt{g}\varphi^{(n)}\delta\varphi^{(n)} + \int \mathrm{d}^2\alpha\sqrt{g}\left(\frac{1}{4}\frac{\delta g}{g}\right)\left[\varphi^{(n)}\right]^2$$

Hence formally

$$\delta \ln Z = \int \mathrm{d}^2\alpha\sqrt{g}\frac{1}{4}\frac{\delta g}{g}\sum_0^\infty\left[\varphi^{(n)}\right]^2 - \frac{1}{2}\frac{\delta A}{A} \tag{122}$$

The Gaussian average implied in equation (121) has dropped, but the quantity $\sum_0^\infty \left[\varphi^{(n)}\right]^2$ is infinite, as a consequence of the infinite number of modes, itself due to the continuum theory. This is an ultraviolet divergence and, instead of curing it by a ζ-function regularization, we shall employ an equivalent device, namely replace $\sum_0^\infty \left[\varphi^{(n)}(\alpha)\right]^2$ by the heat kernel at coinciding points $U(s;\alpha,\alpha)$

$$\sum_0^\infty\left[\varphi^{(n)}(\alpha)\right]^2 \to U(s;\alpha,\alpha) = \sum_0^\infty\left[\varphi^{(n)}(\alpha)\right]^2\exp-sE_n \tag{123}$$

This is the trick used by Fujikawa in various similar circumstances. In general the symmetric heat kernel is given by

$$U(s;\alpha,\beta) = U(s;\beta,\alpha) = \sum_0^\infty \varphi^{(n)}(\alpha)\varphi^{(n)}(\beta)\mathrm{e}^{-sE_n}$$

$$\left(\frac{\partial}{\partial s} - \Delta_\alpha\right)U(s;\alpha,\beta) = 0 \qquad\qquad s > 0 \tag{124}$$

$$\lim_{s\to+0}U(s;\alpha,\beta) = \delta(\alpha,\beta)$$

The invariant δ-function satisfies

$$\varphi(\alpha) = \int \mathrm{d}^2\beta\sqrt{g(\beta)}\delta(\alpha,\beta)\varphi(\beta) \tag{125}$$

Thus we have

$$\delta \ln Z_{\mathrm{reg}}(s) = \int \mathrm{d}^2\alpha\sqrt{g}\frac{1}{4}\frac{\delta g}{g}U(s;\alpha,\alpha) - \frac{1}{2}\frac{\delta A}{A} \tag{126}$$

and we ask for the small s behaviour. We expect to find terms of order s^{-1} plus finite ones in this two-dimensional case, in the

absence of boundaries (which would generate terms of order $s^{-\frac{1}{2}}$).
In flat space

$$U_{\text{flat}}(s;\alpha,\beta) = \frac{\exp\left(-(\alpha-\beta)^2/4s\right)}{4\pi s} \tag{127}$$

Therefore $U_{\text{flat}}(s;\alpha,\alpha) = 1/(4\pi s)$ independently of α, which is recognized as a consequence of translational invariance. In curved space, if $R(\alpha)$ stands for the curvature, we expect on purely dimensional grounds

$$U_{\text{curved}}(s;\alpha,\alpha) = \frac{1}{4\pi s}\left[1 + \lambda s R(\alpha) + \cdots\right]$$

since s has dimension [length]^2 and R [length]$^{-2}$, while λ must be a universal constant. It is therefore sufficient to know the result for any specific curved manifold, a sphere of radius r for instance, to deduce the value of λ.

For the metric (118) on the sphere in the vicinity of the point with coordinates $\alpha^1 = \alpha^2 = 0$, one readily derives that the heat kernel is of the form

$$U_{\text{sphere}}(s;\alpha,0) = \frac{e^{-\alpha^2/4s}}{4\pi s}\left[1 + \tfrac{1}{3}Rs\left(1 + \frac{\alpha^2}{4s} + \frac{(\alpha^2)^2}{8s^2}\right) + \cdots\right] \tag{128}$$

Thus

$$U_{\text{curved}}(s;\alpha,\alpha) = \frac{1}{4\pi s}\left[1 + \tfrac{1}{3}R(\alpha)s + \cdots\right] \tag{129}$$

With a metric locally of the form (117), it follows from equation (119) that

$$\begin{aligned} U_{\text{curved}}(s;\alpha,\alpha) &= \frac{1}{4\pi s} - \frac{1}{12\pi}\Delta\ln\rho + \cdots \\ &= \frac{1}{4\pi s} - \frac{1}{12\pi}\frac{1}{\rho^2}\Delta_0\ln\rho \end{aligned} \tag{130}$$

With this choice of metric, we have $g = \rho^4$, hence $\tfrac{1}{4}\delta g/g = \delta\ln\rho$. Upon inserting these expressions into the regularized form (126), we obtain, neglecting terms which vanish with s,

$$\delta\ln Z_{\text{reg}}(s) = \delta\left[\frac{A}{8\pi s} - \tfrac{1}{2}\ln\frac{A}{A_0} + \frac{1}{24\pi}\int d^2\alpha\rho^2\ln\rho(-\Delta\ln\rho)\right] \tag{131}$$

The short time expansion (129) for the heat kernel at coinciding points can be slightly generalized to near coincidence as follows

$$U_{\text{curved}}(s;\alpha,\beta) = \frac{\exp(-d_{\alpha\beta}^2/4s)}{4\pi s} \left[1 + \tfrac{1}{3} R_\alpha \left(s + \tfrac{1}{4} d_{\alpha\beta}^2 \right) + \cdots \right]$$

(132)

with $d_{\alpha\beta}$ is the geodesic distance from α to β. The factor $(4\pi s)^{-1}$ in the flat case can be interpreted as the inverse of the area of a circle of radius $r_{\text{eff}} = 2\sqrt{s}$. This is in agreement with U being thought as a probability density for a Brownian motion. On a curved surface (curvature R), a circle with the same small radius has an area $\pi R^{-1} 2(1 - \cos\theta)$ where $\theta = R^{\frac{1}{2}} r_{\text{eff}} = 2R^{\frac{1}{2}}\sqrt{s}$, i.e. an area $4\pi s \left(1 - \tfrac{1}{3} Rs + \cdots \right)$. This provides an interpretation for the leading correction in equation (129). Can the reader generalize the argument (i) to include the second correction in equation (132) (ii) to a curved d-dimensional manifold.

The infinite term proportional to s^{-1} in (131) has to be renormalized to a finite one with an arbitrary coefficient of dimension [length]$^{-2}$. The safest way to make the derivation sensible is to assume that the scale variation was confined to a compact support within a coordinate patch in thermal coordinates.

The reader should be warned of the pitfalls in using the expression found above or rather, that it should be handled with due respect to boundary conditions. For instance, on a sphere with a point deleted, consider the system of coordinates with elementary arc length square given in equation (118). While it is true that the integral

$$\int_{\mathcal{R}^2} d^2\alpha \rho^2 \ln\rho(-\Delta \ln\rho) \equiv \int_{\mathcal{R}^2} d^2\alpha \ln\rho(-\Delta_0 \ln\rho)$$

is finite, the one obtained through an integration by parts

$$\int_{\mathcal{R}^2} d^2\alpha \left[\left(\frac{\partial}{\partial\alpha^1} \ln\rho \right)^2 + \left(\frac{\partial}{\partial\alpha^2} \ln\rho \right)^2 \right]$$

is not. In general, since $-\Delta \ln\rho = R$, it is clear that if the integral of R does not vanish, i.e. if the Euler characteristics in nonzero, from

$$\chi = \frac{1}{2\pi} \int d^2\alpha \rho^2 R$$

(133)

it follows that the quantity $\ln\rho$ cannot be defined everywhere as a nonsingular function. To avoid this difficulty, one could assume that the compact surface has the topology of a torus ($\chi = 0$) in which case it admits a flat metric.

The conclusion is that, while the classical massless field theory exhibits a local scale invariance, this is not true of the quantum theory where the renormalized free energy has a dependence on the local scale which assumes for a compact surface the following form

$$\ln \frac{Z}{Z_0} = -\tfrac{1}{2} \ln \frac{A}{A_0} + \frac{1}{24\pi} \int d^2\alpha \rho^2 \left[\ln \rho(-\Delta \ln \rho) + \mu^2 \right] \quad (134)$$

In this expression, A_0 is an arbitrary area to make $\ln Z$ dimensionless, and μ is an arbitrary mass scale since $\int d^2\alpha \rho^2 = A$. As was emphasized, equation (134) makes better sense when comparing two choices of local scales the ratio being of compact support within a thermal coordinate patch. The second term in (134) is called the Liouville action, and the above result is due to Polyakov (1981). The scale independent term might also require a (logarithmic) renormalization, as a more careful analysis will show in the sequel.

Equation (134) implies a lack of conformal invariance of the renormalized free energy of the form

$$\delta \ln \left[\left(\frac{A}{A_0} \right)^{\frac{1}{2}} Z \right] = \frac{1}{12\pi} \int dA \, R \, \delta\psi \quad (135)$$

where $dA = d^2\alpha \sqrt{g}$ is the invariant infinitesimal element of area, and the metric is rescaled as $g \to e^{2\delta\psi} g$.

An alternative form of the Liouville action is as follows. Let $G(\alpha, \beta)$ be the subtracted Green function on the surface

$$G(\alpha, \beta) = \sum_1^\infty \frac{1}{E_n} \varphi^{(n)}(\alpha) \varphi^{(n)}(\beta)$$

$$-\Delta G(\alpha, \beta) = \delta(\alpha, \beta) - \frac{1}{A} \quad (136)$$

Then

$$-\Delta \int dA_\beta G(\alpha, \beta) R_\beta = R_\alpha - \frac{2\pi}{A} \chi \quad (137)$$

Thus, if we assume for simplicity that the Euler characteristic χ vanishes,

$$\ln \left[\left(\frac{A}{A_0} \right)^{\frac{1}{2}} \frac{Z}{Z_0} \right] = \frac{1}{24\pi} \int \int dA_\alpha R_\alpha G(\alpha, \beta) \, dA_\beta R_\beta + \frac{1}{24\pi} \int dA \mu^2 \quad (138)$$

This admits of a simple discretized form (again for $\chi = 0$). The second term is proportional to the area. Set

$$dAR \to q_i = \left(1 - \frac{\theta_i}{2\pi}\right) \tag{139}$$

such that $\sum_i q_i = 0$. Define a (massive) discrete symmetric Green's function such that

$$\sum_{k(i)} \frac{\sigma_{ik}}{\ell_{ik}} \left[G(i,j;m^2) - G(k,j;m^2)\right] + m^2\sigma_i G(i,j;m^2) = \delta_{ij} \tag{140}$$

Then the first term on the r.h.s. in (138) assumes the form

$$S_{\text{Liouville}} = \frac{1}{24\pi} \lim_{m^2 \to 0} \sum_{i,j} q_i G(i,j;m^2) q_j \tag{141}$$

The zero mode appearing in the limit $m^2 \to 0$ is harmless since $\sum_i q_i = 0$. This nonlocal action is reminiscent of a Coulomb gas with the charges being replaced by the local curvature.

11.2.3   Sums over smooth surfaces

We present now Polyakov's continuous model of random surfaces which serves as a foundation for string field theory. The applications to a tentative unified theory of forces (including gravity) in particle physics will not be covered here, but motivates the formalism.

The surprising outcome of the analysis will be that dimension 26 is an upper bound in which the expressions make sense. In this dimension the model enjoys an additional quantum reparametrization invariance. This holds for the bosonic version discussed below, but turns into $d = 10$ for a generalized version including fermionic degrees of freedom. This extension might also be of interest in statistical physics, since there is a hope to interpret some three-dimensional systems, such as the Ising model, in terms of random fermionic surfaces using their dual gauge version.

The amount of analytical sophistication required to cope with the continuous theory is unfortunately rather heavy. The unfamiliar but interested reader is urged to refer to some of the reviews quoted in the notes, for instance those of Friedan and Alvarez which expanded on Polyakov's original work and provide the mathematical background.

In the continuum limit, rather than using the arc length, it was found that a natural action for a random path is proportional to

$$S_{path} = \int d\tau \left(\frac{d\mathbf{x}\,(\tau)}{d\tau} \right)^2 \tag{142}$$

with τ a parameter along the curve, the proper time in relativistic quantum mechanics, or length according to the metric induced from ambiant space. Although this is not invariant under reparametrization, one could generalize this expression by introducing a monotonic function $\tau = f(s)$ and formulate the path integral in such a way that, after gauge fixing, the action recovers the form (142).

For smooth surfaces embedded in Euclidean space $\mathcal{R}^d$, one wishes similarly to replace the Nambu–Goto action (proportional to the area computed using the induced metric), by a suitable form analogous to (142). Rather than choosing a specific metric, Polyakov suggested, by analogy with quantum gravity, to average over all possible metrics defining the (square) infinitesimal arc length

$$ds^2 = g_{ab}(\alpha)\,d\alpha^a\,d\alpha^b \tag{143}$$

Thus the path integral takes the form

$$Z = \sum_{\text{topologies}} \lambda_0^\chi \int_{\text{metrics}} \mathcal{D}g \int_{\text{embeddings}} \mathcal{D}(\mathbf{X})$$

$$\exp - \left[S(g, \mathbf{X}) + \mu_0^2 \int d^2\alpha \sqrt{g} \right] \tag{144}$$

Here λ_0 is a bare activity conjugate to the characteristic χ given by equation (133). The term proportional to the area $A = \int d^2\alpha\sqrt{g}$, computed with the metric g, leaves us the option of absorbing an infinite similar counterterm in a suitable renormalization.

The vacuum functional in (144) involves only compact orientable surfaces (the latter for simplicity) and we shall also assume a restriction to connected surfaces. Otherwise one should presumably include combinatorial factors. The d fields $X^\mu(\alpha)$ map the surface in $\mathcal{R}^d$, and the main part of the action reads

$$S(g, \mathbf{X}) = \tfrac{1}{2} \int d^2\alpha \sqrt{g}\, g^{ab}(\alpha) \sum_\mu \frac{\partial}{\partial\alpha^a} X^\mu \frac{\partial}{\partial\alpha^b} X^\mu \tag{145}$$

The target space could be replaced by a d-dimensional manifold with metric $G_{\mu\nu}(\mathbf{X})$ as indicated in equation (95). We refrain from doing so.

The quadratic action (145) is similar to the one-dimensional one with the appropriate geometric factor, $g = \det g_{ab}$, and g^{ab} the inverse of g_{ab}. It enjoys important invariance properties. These are of three types.

i) Reparametrization invariance, accompanied of course by the corresponding changes in the metric.

ii) Rescaling of the metric $g_{ab} \rightarrow \rho^2 g_{ab}$, under which the combination $\sqrt{g}\, g^{ab}$ is invariant. This is also called Weyl invariance and affects lengths and areas but not angles. Metrics which differ by a scale factor are therefore said to belong to the same conformal class. It should be stressed that if we only look at the action on metrics, the effect of reparametrization can partly be identified with a rescaling and this will play a role in the sequel.

iii) The action is also invariant when $\mathbf{X}$ is transformed by an element of the Euclidean group. Only the noncompact translation part is a dangerous one and will be avoided by separating the center of mass motion or constant mode.

Constrained by the requirement of involving only $\mathbf{X}$ through its gradient, the action (145) supplemented by the terms conjugate to $\ln \lambda_0$ and μ_0^2 contains all possible renormalizable combinations compatible with these symmetries, given that a physical length scale has been absorbed in $\mathbf{X}$.

Our task is to make sense of the formal expression (144), more precisely of each piece individually. Therefore in the following, we deal with a fixed topology and wish to extract a factor corresponding to the infinite reparametrization group (more pedantically the diffeomorphism group). In other words, we want to fix the gauge, keeping the benefits of gauge invariance. It will be seen that at the quantum level with regard to scale invariance (assuming the renormalized μ^2 to vanish), this can only be achieved in dimension 26 as announced.

When μ_0^2 vanishes, the classical equations of motion obtained by varying in turn g_{ab} and X^μ are

$$T_{ab} = -\sum_\mu \partial_a X^\mu \partial_b X^\mu + \frac{1}{2} g_{ab} \sum_\mu \partial^c X^\mu \partial_c X^\mu = 0 \quad (146)$$

$$\Delta X^{\mu} \equiv \frac{1}{\sqrt{g}} \partial_a \sqrt{g} \partial^a X^{\mu} = 0 \qquad (147)$$

We recognize in (146) the vanishing of the symmetric traceless energy momentum tensor which requires g_{ab} to be proportional to the metric $\sum_{\mu} \partial_a X^{\mu} \partial_b X^{\mu}$ induced from the ambiant Euclidean space on the surface. The field coordinates X^{μ} have then to be harmonic for this metric. This is the condition for minimal surfaces (i.e. surfaces which minimize the area locally). Reinserted in the action, these expressions yield the induced total area. Clearly this would make more sense for surfaces with boundaries.

To begin with, we recall a property stated in equation (117), namely that any metric can be taken locally in a diagonal form by an appropriate choice of coordinates. One can even further insure that at some point the conformal factor is unity and that its gradient vanishes. An additional stronger property is that each conformal class of metrics contains a metric g_0 of constant curvature. This says that given g, there exists a choice of ψ such that $g = e^{2\psi} g_0$ with R_0 constant, i.e. from equation (119) this requires solving for ψ the classical Liouville equation

$$-\Delta \psi = R - e^{-2\psi} R_0 \qquad (148)$$

The reader will recall that the uniformization of Riemann surfaces implies that for $\chi = 0$ a torus may be obtained by taking the quotient of the complex plane by a lattice of translations, while for $\chi < 0$ it may be obtained as the quotient of the upper half complex plane by an appropriate discrete crystallographic subgroup of $SL(2R)$. A reference metric is $dz\,d\bar{z}$, or $dzd\bar{z}/(\text{Im}z)^2$ respectively, in appropriate coordinates. The case of the sphere has been exemplified above. To see what is involved, use complex coordinates $z = \alpha^1 + i\alpha^2$, $\bar{z} = \alpha^1 - i\alpha^2$. A general metric takes the form $ds^2 = g_{zz}^2 dz^2 + g_{\bar{z}\bar{z}} d\bar{z}^2 + 2g_{z\bar{z}} dzd\bar{z}$, $(g_{\bar{z}\bar{z}} = \overline{g_{zz}},\ g_{z\bar{z}} = \bar{g}_{z\bar{z}})$ with the positivity condition being $g_{z\bar{z}}^2 > g_{zz}g_{\bar{z}\bar{z}}$, $g_{z\bar{z}} > 0$. Set

$$\frac{\mu}{\bar{\mu}} = \frac{g_{\bar{z}\bar{z}}}{g_{zz}} \qquad \frac{1}{2}(1 + \mu\bar{\mu}) = \left|\frac{g_{z\bar{z}}}{g_{zz}}\right|^2 \left(1 - \sqrt{1 - \left|\frac{g_{zz}}{g_{z\bar{z}}}\right|^2}\right)$$

$$\lambda = \frac{2g_{z\bar{z}}}{1 + \mu\bar{\mu}}$$

$$(149a)$$

Positivity means $\lambda > 0$ and $|\mu| < 1$, while the metric reads

$$ds^2 = \lambda\,|dz + \mu d\bar{z}|^2 \qquad (149b)$$

The above property amounts to saying (with due attention paid to the crystallographic invariance for $\chi < 0$) that it is possible to find a smooth change of variable $u = u(z, \bar{z})$, $\bar{u} = \overline{u(z, \bar{z})}$, such that in the variables u, $\bar{u}$, the metric assumes the (anti-) diagonal form $\mathrm{d}s^2 \propto \mathrm{d}u\mathrm{d}\bar{u}$, i.e. to solve the Beltrami equation $\partial u/\partial\bar{z} = \mu\,\partial u/\partial z$.

We return to a specific term in the sum (144) with the aim to "fix the gauge" by taking g conformally equivalent to a given metric g_0 of constant curvature (i.e. $g = \mathrm{e}^{2\psi}g_0$ with ψ varying). To do so, we study variations δg around g. One defines a square norm on deformations (with an obvious generalization to tensor fields)

$$\|\delta g\|^2 = \int \mathrm{d}^2\alpha\sqrt{g}g^{ac}g^{bd}\delta g_{ab}\delta g_{cd} \tag{150}$$

This enables one to obtain a volume element $\mathcal{D}g$ by analogy with Riemann manifolds in a standard fashion. This definition implies the metric g on the surface. The variation δg can be split into two orthogonal parts according to (150), a symmetric traceless part δh and a piece proportional to g itself (the trace part) as

$$\delta g = 2\delta\varphi\, g + \delta h \tag{151}$$

Since these pieces are orthogonal, $\mathcal{D}g$ can be factored into

$$\mathcal{D}g = \mathcal{D}\varphi\mathcal{D}h \tag{152}$$

both factors being reparametrization invariant. One has in mind that this corresponds to a factorization into an integration on the conformal factor and reparametrization both acting on the reference metric g_0. To see this, we examine how an infinitesimal reparametrization $\alpha \to \alpha + \delta\alpha$ affects infinitesimally the metric

$$(g_{ab} + \delta g_{ab})\,\mathrm{d}\alpha^a\mathrm{d}\alpha^b = g(\alpha + \delta\alpha)_{ab}\,\mathrm{d}\left(\alpha^a + \delta\alpha^a\right)\mathrm{d}\left(\alpha^b + \delta\alpha^b\right)$$

i.e.

$$\delta_v g_{ab} = \delta\alpha^c\frac{\partial}{\partial\alpha^c}g_{ab} + g_{ac}\frac{\partial\delta\alpha^c}{\partial\alpha^b} + g_{cb}\frac{\partial\delta\alpha^c}{\partial\alpha^a} = D_b v_a + D_a v_b \tag{153}$$

where we have set

$$v_a = g_{ab}\delta\alpha^b \tag{154}$$

for the infinitesimal (covariant) vector field associated to the reparametrization, and D is the covariant derivative involving Γ

the Riemann–Christoffel symbol

$$D_a v_b = \partial_a v_b - \Gamma^c_{ab} v_c$$

$$\Gamma^c_{ab} = \tfrac{1}{2} g^{cd} \left\{ \frac{\partial g_{ad}}{\partial \alpha^b} + \frac{\partial g_{bd}}{\partial \alpha^a} - \frac{\partial g_{ab}}{\partial \alpha^d} \right\} \tag{155}$$

This abuse of tensor indices will readily be remedied by the judicious choice of reference metric. We also have under rescaling

$$\delta_\psi g_{ab} = 2\delta\psi g_{ab} \tag{156}$$

The sum $(\delta_v + \delta_\psi)g$ can now be compared to the general orthogonal decomposition (151), observing that we have to subtract from $\delta_v g$ its trace. Thus

$$\left(\delta_v + \delta_\psi \right) g = 2\delta\varphi g + \delta h \tag{157}$$

$$\delta\varphi = \delta\psi + \tfrac{1}{2} g^{ab} D_a v_b \qquad \delta h = D_a v_b + D_b v_a - g_{ab} \left(g^{cd} D_c v_d \right)$$

The standard notation is

$$\delta h = P_1 v \tag{158}$$

where P_1 is a first order differential operator which maps vectors (hence the index 1) onto symmetric traceless 2-tensors. We now trade the variables φ and h for ψ (associated to a conformal factor) and v which represents reparametrization, taking care of a Jacobian functional determinant (the Faddeev–Popov determinant)

$$\mathcal{D}\varphi \mathcal{D} h = \mathcal{D}\psi \mathcal{D} v \left| \frac{D(\varphi, h)}{D(\psi, v)} \right| \tag{159a}$$

with a symbolic presentation

$$\frac{D(\varphi, h)}{D(\psi, v)} = \det \left| \begin{matrix} I & X \\ 0 & P_1 \end{matrix} \right| = \det P_1$$

As we are only interested in the absolute value, we insert in (159) instead of $\det P_1$ the positive quantity $(\det P_1^\dagger P_1)^{\frac{1}{2}}$ where the operator $P_1^\dagger P_1$ has the advantage that it maps vectors fields on vector fields. Thus

$$\mathcal{D}\varphi \mathcal{D} h = \mathcal{D}\psi \mathcal{D} v \left(\det P_1^\dagger P_1 \right)^{\frac{1}{2}} \tag{159b}$$

One could represent the determinant as a Grassmannian integral over "ghost" fields. This familiar device is useful for further developments in string field theory, especially to give a compact form to the expansion (144) interpreted as a perturbative series of a would-be second quantized interacting theory. The analogy is with the Feynman expansion expressed in terms of one-dimensional graphs. The Grassmannian integral, similar to the one over the **X**-fields, assumes a simple form using complex coordinates introduced below, emphazing its conformal invariance. We leave as an exercise for the reader to establish its expression.

To define the measure $\mathcal{D}v$, we use a norm on vector fields, which has the merit of being independent from the conformal factor ψ, setting by analogy with equation (150)

$$\|v\|^2 = \int \mathrm{d}^2\alpha \sqrt{g} g^{ab} v_a v_b = \int \mathrm{d}^2\alpha \sqrt{g_0} g_0^{ab} v_a v_b$$

The (infinite and formal) integral $\int \mathcal{D}v$ may be interpreted as the volume of the reparametrization group.

We have to pay attention also to the fact that the operator P_1 may have zero modes i.e. may annihilate certain vector fields (called conformal Killing fields) corresponding to reparametrizations which just rescale the metric. Those are already included in the integral over ψ. To avoid double counting, and to define a positive Jacobian $(\det' P_1^\dagger P_1)^{\frac{1}{2}}$, we must integrate not over the full space of v's but on the orthogonal complement of the (finite-dimensional) space of zero modes. This has the drawback that the corresponding integral yields only part of the "volume" of the reparametrization group. Finally, it was implicitly assumed that the most general δg can be obtained in the form $\left(\delta_\psi + \delta_v\right) g$. This need not always be the case and must be investigated as a separate issue. The trace part does not present a problem. The question is whether every traceless δh belongs to the image of P_1. This is in general not the case and leaves (for $\chi \leq 0$) a finite-dimensional space, with a measure denoted $\mathcal{D}h_M$, which is called the Teichmuller or moduli space, with an intricate global structure. To be precise, the Teichmuller space is a covering of the space of moduli. So we have more precisely

$$\mathcal{D}g = \mathcal{D}h_M \mathcal{D}v^\perp \mathcal{D}\psi \left(\det' P_1^\dagger P_1\right)^{\frac{1}{2}} \tag{159c}$$

where the prime refers to the determinant over nonzero modes.

Let the series in (144) be written as a sum of contributions with different topologies

$$Z = \sum_\chi Z_\chi \tag{160}$$

An individual term can be written

$$Z_\chi = \lambda_0^\chi \int \mathcal{D}v^\perp \int \mathcal{D}h_M \mathcal{D}\psi \int \prod_\mu \mathrm{d}C_0^\mu$$

$$\mathrm{e}^{-\mu_0^2 \int \mathrm{d}^2\alpha\sqrt{g}} \left(\det' P_1^\dagger P_1\right)^{\frac{1}{2}} \left(\det'[-\Delta]\right)^{-\frac{1}{2}d} \tag{161}$$

where we have performed the integral over the **X**-fields, separating the integral over constant normalized zero modes C_0^μ (hence the prime over the determinant of the Laplacian, computed of course with the metric g). Since all the quantities involved are reparametrization invariant, the integral $\int \mathcal{D}v^\perp$ factors out and will simply be omitted in a renormalized version of Z_χ which is still a formal one, since the determinants are of course ill-defined and also require renormalization. We follow the heuristic physicist's tradition of proceeding from nonsensical infinite quantities towards renormalized ones, rather than an axiomatic presentation of the final product which would hide the motivation. The reader might also worry (and we worry with him) that different topologies have different integrals over the reparametrization group, so that at this stage the way to combine the various Z_χ's into a unique sum remains problematic.

Our main concern is with the definition of the determinants and with the integral over moduli. We expect from section 2.2 to see the Liouville action emerging again from the ψ dependence. This leads us into a technical detour of intrinsic interest, which elaborates on the schematic discussion of the massless field given above.

To investigate the determinants occurring in equation (161), we use complex coordinates $z = \alpha^1 + \mathrm{i}\alpha^2$, $\bar{z} = \alpha^1 - \mathrm{i}\alpha^2$. The metric reads $g = \mathrm{e}^{2\psi}g_0$, with a reference metric g_0 such that locally $\mathrm{d}s^2 = \mathrm{e}^{2\psi_0}\,\mathrm{d}z\mathrm{d}\bar{z}$, corresponding to a constant curvature as agreed above. Thus only $g_{z\bar{z}}$ is nonzero and positive while $g^{z\bar{z}} = g_{z\bar{z}}^{-1}$.

We study analytic transformations acting on tensors with contravariant and covariant indices (the keyword is analytic). These transformations leave the metric structure invariant. The metric g allows us to raise or lower indices (its transformation law is of course

$g_{z\bar{z}} \, \mathrm{d}z\mathrm{d}\bar{z} = g'_{z\bar{z}} \, \mathrm{d}z'\mathrm{d}\bar{z}'$). Thus we can restrict ourselves to covariant indices say, in which case (as in chapter 9) the transformation law is specified by requiring the combination $t_{h,\bar{h}} \, \mathrm{d}z^h \mathrm{d}\bar{z}^{\bar{h}}$ to be invariant. The suffixes h and $\bar{h}$ (both integers here) are the numbers of z and $\bar{z}$ lower indices. The simplification brought about by analytic transformations, to which we limit ourselves henceforth, is that a component $t_{h,\bar{h}}$ is simply multiplied by a factor. It is slightly more convenient in the present context to trade the $\bar{h}$ lower $\bar{z}$ indices for $\bar{h}$ upper z indices. We denote the tensor $t^{\bar{h}}{}_h = (g^{z\bar{z}})^{\bar{h}} \, t_{h,\bar{h}}$, so that the transformation law reads

$$\left(t^{\bar{h}}_h\right)' = \left(\frac{\mathrm{d}z'}{\mathrm{d}z}\right)^{\bar{h}-h} t^{\bar{h}}_h \tag{162}$$

To conform to usage, we shall adopt the opposite convention from chapter 9 by using n (and not s) for the difference $\bar{h} - h$ ($s = -n$). For fixed n (positive or negative) the set of (one-component) tensorial quantities which transform as indicated will be called $\mathcal{T}^n$.

The operators we consider such as $-\Delta$ or P_1 map symmetric traceless tensors onto symmetric traceless tensors. The statement is empty for the scalar Laplacian and for P_1, it maps vectors on symmetric traceless 2-tensors. Such real tensors have only two independent components in two dimensions.

To see this in Euclidean geometry, associate to such a tensor the homogeneous polynomial $P_n(x,y) = \sum_{k+k'=n} T_{11...,22...} x^k y^{k'}$ with *a priori* $n + 1$ coefficients. The tensor is traceless if the ordinary Laplacian $\partial^2/\partial x^2 + \partial^2/\partial y^2$ acting on P_n gives zero. This provides us with a homogeneous polynomial of degree $n - 2$, hence with $n - 1$ conditions (or zero if $n = 1$), thus leaving two independent coefficients such that $P_n(x,y) = t_+(x - iy)^n + t_-(x + iy)^n$. If the tensor is real the components are complex conjugate. The argument goes through in curved space. In complex coordinates, one component belongs to $\mathcal{T}^n$, call it t, while the second component $\bar{t}$ would have n $\bar{z}$ upper indices. To conform to the above convention, it is multiplied by $(g_{z\bar{z}})^n$ to obtain an element of $\mathcal{T}^{-n}$. By complexification, the set of (real) symmetric traceless tensors of rank n can therefore be considered as belonging to the direct sum $\mathcal{T}^n \oplus \mathcal{T}^{-n}$, with an association

$$T \equiv \{t, (g_{z\bar{z}})^n \, \bar{t}\} \tag{163}$$

where it is understood that t is in $\mathcal{T}^n$. The corresponding space of associated real symmetric traceless tensors is denoted $\mathcal{S}^n$. The simpler case $n = 0$ with only one component is set aside. Using the special form of the metric, the covariant derivative has two components D^z and D_z (with an appended suffix n if necessary to recall on which space it operates, as D_n^z and D_z^n)

$$T^n \xrightarrow{D_z^z} T^{n+1} \qquad t \to g^{z\bar{z}} \bar{\partial} t$$

$$T^n \xrightarrow{D_z^z} T^{n-1} \qquad t \to \left(g^{z\bar{z}}\right)^n \partial_z \left(g_{\bar{z}z}\right)^n t \qquad (164)$$

and an invariant scalar product is derived from the norm

$$\|t\|^2 = \int \mathrm{d}^2\alpha \sqrt{g} \left(g_{z\bar{z}}\right)^n \bar{t}t \qquad (165a)$$

One can easily verify these formulas and check the transformation laws.

In essence D^z is a decoration of the $\bar{\partial}$ operator and the factors in D_z compensate for the passage from $D^{\bar{z}}$ to D_z when acting on the appropriate space. A similar remark applies to the norm which we extend (with a reality condition) to S_n by requiring for T written as in equation (163) that

$$\langle T_1, T_2 \rangle = \int \mathrm{d}^2\alpha \sqrt{g} \left(g_{\bar{z}z}\right)^n \left(\bar{t}_1 t_2 + t_1 \bar{t}_2\right) \qquad (165b)$$

Using these notations, the adjoint of D^z is acting from T^{n+1} to T^n and

$$\left(D_n^z\right)^\dagger = -D_z^{n+1} \qquad (166)$$

There are now two possible definitions of the Laplacian acting on T^n. Either we can climb one rung on the ladder then retreat, or do the opposite. In formulas

$$T_n \to T_n \qquad \begin{aligned} -\delta_n^{(+)} &= -2D_z^{n+1} D_n^z \equiv 2\left(D_n^z\right)^\dagger D_n^z \\ -\delta_n^{(-)} &= -2D_{n-1}^z D_z^n \equiv 2D_{n-1}^z \left(D_{n-1}^z\right)^\dagger \end{aligned} \qquad (167)$$

The factor 2 is included to agree with the flat space definition (surely had Laplace lived long enough, he would have included a minus sign into the definition of his operator to make it positive!). These operators are semidefinite positive.

On the space S^n, operators are acting on the two conjugate components

$$S^n \overset{P_n}{\underset{P_n^\dagger}{\rightleftharpoons}} S^{n+1} \qquad \begin{aligned} P_n &= \left(D_n^z, D_z^{-n}\right) \\ P_n^\dagger &= \left(-D_z^{n+1}, -D_{-(n+1)}^z\right) \end{aligned} \qquad (168)$$

and P_1 and $P_1^\dagger$ are those occurring in formula (161). The associated Laplacians on S are

$$S_n \to S_n \qquad -\Delta_n^{(+)} = 2P_n^\dagger P_n = \left(-\delta_n^{(+)}, -\delta_{-n}^{(-)}\right)$$

$$S_{n+1} \rightarrow S_{n+1} \qquad -\Delta_{n+1}^{(-)} = 2P_n P_n^\dagger = \left(-\delta_{n+1}^{(-)}, -\delta_{-(n+1)}^{(+)}\right) \qquad (169)$$

The relation between the spectra of $P_n^\dagger P_n$ and $P_n P_n^\dagger$ (in particular for $n = 1$) is crucial for the following. Both operators are non-negative but may have zero modes which span their kernel. For an operator O, the kernel of $O^\dagger$ is orthogonal to the image of O, and the same relation holds with O and $O^\dagger$ interchanged. Thus in terms of orthogonal complements

$$\left(\ker - \Delta_n^{(+)}\right)^\perp \leftrightarrow \left[\ker - \Delta_{n+1}^{(-)}\right]^\perp \qquad (170)$$

The left-hand side is also the complement of $\ker P_n$ the right-hand side of $\ker P_n^\dagger$. To make this correspondence precise, let $\{T_r\}$ be an orthonormal set of nonzero eigenmodes of $-\Delta_n^{(+)}$ with corresponding eigenvalues λ_r (the scalar product being understood in the sense of equations (165)), then

$$2P_n^\dagger P_n T_r = \lambda_r T_r \qquad \lambda_r > 0 \qquad (171)$$

Thus $2P_n P_n^\dagger (P_n T_r) = \lambda_r (P_n T_r)$ and $P_n T_r$ is an eigenvector of $-\Delta_{n+1}^{(-)}$ with the same eigenvalue λ_r. These modes are orthogonal. To normalize them, we note that

$$\langle P_n T_r, P_n T_{r'} \rangle = \langle T_r, P_n^\dagger P_n T_{r'} \rangle = \tfrac{1}{2}\lambda_r \delta_{r,r'}$$

Therefore an orthonormal basis for $\left[\ker\left(-\Delta_{n+1}^{(-)}\right)\right]^\perp$ is

$$\left\{\tilde{T}_r = \sqrt{\frac{2}{\lambda_r}} P_n T_r\right\}$$

justifying equation (170).

In the particular case $n = 1$, the zero modes of P_1 are the conformal Killing vectors, while the harmonic traceless 2-tensors, zero modes of $P_1^\dagger$, span those deformations of the metric which can not be described by rescaling or changes of coordinates and are associated to the moduli, which left a finite-dimensional measure in equations (159c) and (161). In general the zero modes are harmonic tensors in their respective spaces. We turn to the determinants themselves and use as before the heat kernel method with

$$\ln \det' \left(-\Delta_n^{(+)}\right) = -\int_\varepsilon^\infty \frac{dt}{t} \, \text{Tr}' \exp t\Delta_n^{(+)} \qquad (172)$$

since

$$-\int_\varepsilon^\infty \frac{dt}{t} e^{-t\lambda} = \ln \varepsilon\lambda + \gamma + o(\varepsilon) \qquad (173)$$

This is a (brute force) regularization for $\sum_r' \ln \lambda_r$, the sum running over positive eigenvalues, and ε is an ultraviolet cutoff, the square of an elementary length scale. One should recall that $-\Delta_n^{(+)}$ operates in the direct sum of two spaces (for $n \neq 0$). The method to extract information from (172) is, as in section 2.2, to study the short time behaviour of the appropriate kernel.

The calculation parallels the one already presented. We shall nevertheless give some details pertaining to the modified Laplacians occurring here, using perturbation theory. Let α denote a point at which we want to estimate the diagonal heat kernel $G(\alpha, \alpha; t) = (\alpha | e^{t\Delta_n^{(+)}} | \alpha)$. This is really a two-by-two diagonal matrix. We recall that G behaves as t^{-1} for small t, and that length scales as $t^{1/2}$. We look for terms up to order t^0. Thus

$$(\alpha | e^{t\Delta_n^{(+)}} | \alpha) = \frac{G_{-1}}{t} + G_0 + \cdots \qquad (174a)$$

where $G_{-1}, G_0, \ldots$ are two-by-two diagonal matrices which depend of course on the point under consideration. Inserting this expansion in (173), we have

$$\ln \det' \left(-\Delta_n^{(+)} \right) = -\frac{1}{\varepsilon} \int d^2\alpha \sqrt{g} \, \mathrm{Tr} \, G_{-1}$$

$$+ \ln \varepsilon \left[\int d^2\alpha \sqrt{g} \, \mathrm{Tr} \, G_0 - \dim \mathrm{Ker} \left(-\Delta_n^{(+)} \right) \right] + \text{finite terms}$$

$$(174b)$$

The second term in $\ln \varepsilon$ arises from the original prescription of summing over nonzero modes.

Suppose that we vary the scale factor ψ in $g = e^{2\psi} g_o$. Suppressing all indices, for a nonzero eigenmode in S^n

$$\lambda \langle T, T \rangle = 2 \langle PT | PT \rangle$$

reads explicitly

$$\lambda \int d^2\alpha \sqrt{g} \, (g_{z\bar{z}})^n \, t\bar{t} = 2 \int d^2z \sqrt{g} \, (g_{z\bar{z}})^{n+1} \left(g^{z\bar{z}} \partial t \right) \left(g^{z\bar{z}} \partial \bar{t} \right)$$

Hence from the normalization, we have

$$\delta\lambda + \lambda 2(n+1) \langle T, \delta\psi T \rangle = 4n \langle PT, \delta\psi PT \rangle$$

Recalling that $\sqrt{\frac{2}{\lambda}} T = \tilde{T}$ is the corresponding normalized eigenmode in S^{n+1}

$$\delta\lambda = 2\lambda \left\{ n \left\langle \tilde{T}, \delta\psi\tilde{T} \right\rangle - (n+1) \langle T, \delta\psi T \rangle \right\}$$

and

$$\delta \ln \det{}' \left(-\Delta_n^{(+)} \right) = \int_\varepsilon^\infty dt \sum_r{}' \delta\lambda_r e^{-t\lambda_r}$$

$$= 2 \int_\varepsilon^\infty dt \frac{d}{dt} \sum_r{}' \left[(n+1) \langle T_r, \delta\psi T_r \rangle - n \left\langle \tilde{T}_r, \delta\psi \tilde{T}_r \right\rangle \right] e^{-t\lambda_r}$$

$$= 2n \operatorname{Tr}' \delta\psi e^{\varepsilon\Delta_{n+1}^{(-)}} - 2(n+1) \operatorname{Tr}' \delta\psi e^{\varepsilon\Delta_n^{(+)}} \qquad (175)$$

To evaluate this quantity we need the analog of equation (174a) for the operator $\Delta_{n+1}^{(-)}$. Explicitly, writing in a column the two diagonal elements of the operator

$$-\Delta_n^{(+)} = 2(g_{z\bar{z}})^{-1} \left[\begin{matrix} [\partial\bar{\partial} + n(\partial \ln g_{z\bar{z}})\bar{\partial}] \\ [\partial\bar{\partial} - n(\partial \ln g_{z\bar{z}})\bar{\partial} - n(\partial\bar{\partial}\ln g_{z\bar{z}})] \end{matrix} \right]$$

An alternative derivation of a result similar to equation (130) uses a perturbative method by setting $g_{z\bar{z}} = \frac{1}{2}e^{2\psi}$ with a coordinate system adapted to the point α at which we evaluate the diagonal element, this point being chosen at the origin with $\psi(0) = 0$, $\partial\psi(0) = \bar{\partial}\psi(0) = 0$. With reference to the traditional (Euclidean) notation of quantum mechanics, we write $\mathbf{P}^2$ for (minus) the flat Laplacian $\mathbf{P}^2 = -4\partial\bar{\partial}$

$$-\Delta_n^{(+)} = \left(\begin{matrix} \mathbf{P}^2 + (e^{-2\psi} - 1)\mathbf{P}^2 - 8n\partial\psi e^{-2\psi}\bar{\partial} \\ \mathbf{P}^2 + (e^{-2\psi} - 1)\mathbf{P}^2 + 8n\partial\psi e^{-2\psi}\bar{\partial} + 8ne^{-2\psi}(\partial\bar{\partial}\psi) \end{matrix} \right)$$

$$\psi(z,\bar{z}) = \tfrac{1}{2}z^2\partial^2\psi(0) + z\bar{z}\partial\bar{\partial}\psi(0) + \tfrac{1}{2}\bar{z}^2\bar{\partial}^2\psi(0) + \cdots$$

Since $|z|$ is of order $\sqrt{t}$, we need only to keep terms up to second order and perturbation theory up to first order will suffice in the form

$$e^{-\mathbf{P}^2 t - tV} = e^{-\mathbf{P}^2 t} - \int_0^t dt' e^{-\mathbf{P}^2(t-t')} V e^{-t'\mathbf{P}^2} + \cdots$$

For the upper and lower components respectively, we have

$$V_{up} = V_0 - V_1$$
$$V_{down} = V_0 + V_1 + V_2$$
$$V_0 = -\left[z^2\partial^2\psi(0) + 2z\bar{z}\partial\bar{\partial}\psi(0) + \bar{z}^2\bar{\partial}^2\psi(0) \right]$$
$$V_2 = 8n\partial\bar{\partial}\psi(0)$$

We are only interested in $\operatorname{Tr}\left(0 \left| e^{t\Delta_n^{(+)}} \right| 0 \right)$, where the trace refers to the sum of contributions of upper and lower components, so the contribution of V_1 cancels. The contribution of V_0 has already been obtained in equation (130). The reader might enjoy doing the computation in the present formalism. It gets doubled here due to the trace, as does the zeroth order term. Finally $V_2 = 2n\Delta\psi(0)$ contributes to the trace with a factor

$$-\int_0^t dt' \int \frac{d^2p}{(2\pi)^2} e^{-p^2t} = -\frac{1}{4\pi}$$

Putting everything together

$$\text{Tr}\left(\alpha \left| e^{t\Delta_n^{(+)}} \right| \alpha\right) = \frac{1}{2\pi t} - \frac{1}{6\pi}\Delta\psi(\alpha) - \frac{n}{2\pi}\Delta\psi(\alpha) + o(t)$$

where we have successively distinguished the various contributions. Thus with f diagonal in configuration space, and recognizing that the curvature is given by equation (119), $R = -\Delta\psi$,

$$\text{Tr}\, f e^{\varepsilon\Delta_n^{(+)}} = \int d^2\alpha\sqrt{g} f \left\{ \frac{1}{2\pi\varepsilon} + \frac{(3n+1)}{6\pi}R \right\} + o(\varepsilon) \qquad (176a)$$

In a similar vein

$$\text{Tr}\, f e^{\varepsilon\Delta_{n+1}^{(-)}} = \int d\alpha\sqrt{g} f \left\{ \frac{1}{2\pi\varepsilon} - \frac{(3n+2)}{6\pi}R \right\} + o(\varepsilon) \qquad (176b)$$

Finally for $n = 0$, if Δ stands for the scalar Laplacian, we found before that

$$\text{Tr}\, f e^{\varepsilon\Delta} = \int d\alpha\sqrt{g} f \left\{ \frac{1}{4\pi\varepsilon} + \frac{1}{12\pi}R \right\} + o(\varepsilon) \qquad (176c)$$

An immediate consequence of these computations is the following beautiful observation leading to a special case of the (Atiyah–Singer) index theorem. The index is defined as

$$I_n = \dim\left[\ker P_n\right] - \dim\left[\ker P_n^\dagger\right] \qquad (177)$$

Since $-\Delta_n^{(+)} = 2P_n^\dagger P_n$ and $-\Delta_{n+1}^{(-)} = 2P_n P_n^\dagger$ have an identical nonzero spectrum, for any t,

$$I_n = \text{Tr}\, e^{t\Delta_n^{(+)}} - \text{Tr}\, e^{t\Delta_{n+1}^{(-)}}$$

Being t-independent, this quantity can be computed in the limit $t \to 0$ with the help of equations (176). Setting $f \equiv 1$, the t^{-1} terms cancel, and from the finite terms emerges the relation with topology through the Euler characteristic

$$I_n = \frac{(2n+1)}{2\pi} \int d^2\alpha\sqrt{g} R = (2n+1)\chi \qquad (178)$$

giving the difference between the number of linearly independent real harmonic symmetric tensors of rank n and $n+1$ and proving that I_n is a topological invariant.

For $n = 1$, the index is $I_1 = 3\chi$. The dimensions involved in (177) are on real numbers. The number of linearly independent conformal Killing vectors $(\dim[\mathrm{Ker}P_1])$ is the (real) dimension of the group of analytic diffeomorphism called the global conformal group in chapter 9. Thus $\dim[\mathrm{Ker}P_1] = 6, 2, 0$ if $\chi = 2, 0$ or negative, corresponding to the groups $SL(2, \mathcal{C})$ for the sphere or complex translations in the case of the torus. Compact Riemann surfaces of genus larger than one admit at most a finite group of automorphisms. Thus $\dim\left[\mathrm{Ker}P_1^\dagger\right] = \dim(\mathrm{Ker}P_1) - 3\chi$ is zero for genus zero, two for genus one and $6(g-1)$ for genus larger than one, all these on real numbers. But this is precisely the dimension of the space of those traceless tensors h_{ab} which are not obtained by rescaling or reparametrizations, in other words the (real) dimension of the moduli space. Of course this result is familiar for $g = 0$ and 1.

We have now all the tools to extract from the determinants all divergent terms as well as the finite ones that depend on the conformal scale ψ. From equations (174) and (176a)

$$\ln\det'\left(-\Delta_n^{(+)}\right) \tag{179}$$

$$= -\frac{1}{2\pi\varepsilon} \int \mathrm{d}^2\alpha\sqrt{g} + \ln\varepsilon\left[\left(n + \tfrac{1}{3}\right)\chi - \dim\left[\mathrm{Ker}P_n\right]\right] + \text{finite}$$

This is dimensionally correct since ε has dimension $[\text{length}]^2$ and could be written in terms of an ultraviolet cutoff $\varepsilon = \Lambda^{-2}$. The coefficient of the $\ln\varepsilon$ term is of course a pure number. This is still insufficient, since we are also interested in the finite part which contains the anomaly. Up to the logarithmic term multiplied by $\dim(\mathrm{Ker}P_n)$ (which shall soon be disposed of) the remaining divergences can be taken care by a renormalization of the potential λ_0 and the bare "tension" μ_0^2 in equation (161).

The variational formula (175) involves primed traces. If we use instead full traces as computed in (176ab), we have to subtract the contributions from the zero mode subspaces. Let $\left\{T_s^{(0)}\right\}$ be a basis of the kernel of P_n. Consider the finite symmetric matrix

$$(H_n)_{ss'} = \left\langle T_s^{(0)}, T_{s'}^{(0)} \right\rangle = \int \mathrm{d}^2\alpha\sqrt{g}\,(g_{z\bar{z}})^n \left\{t_s^{(0)}\overline{t_{s'}^{(0)}} + c.c.\right\} \tag{180}$$

Recall that $t_s^{(0)}$ are analytic in z and can be choosen as ψ-independent functions. We have therefore

$$\delta H_{ss'} = 2(n+1)\left\langle T_s^{(0)}\delta\psi T_{s'}^{(0)}\right\rangle$$

The projector on $\ker P_n$ can be written in an obvious notation

$$\mathcal{P}_n = \sum_{ss'} \left| T_s^{(0)} \right\rangle \left(H_n^{-1} \right)_{ss'} \left\langle T_{s'}^{(0)} \right|$$

and

$$2(n+1)\,\mathrm{Tr}\,\delta\psi\mathcal{P}_n = 2(n+1)\sum_{s,s'} \left\langle T_s^{(0)}\delta\psi T_{s'}^{(0)} \right\rangle \left(H^{-1} \right)_{s's}$$

$$= \mathrm{Tr}\,\delta H_n H_n^{-1} = \delta\,\mathrm{Tr}\ln H_n = \delta\ln\det H_n \tag{181}$$

For the projector $\tilde{\mathcal{P}}_{n+1}$ on $\ker P_n^\dagger$, the corresponding functions $\tilde{t}_u^{(0)}$ are such that $(g_{z\bar{z}})^{n+1}\,\tilde{t}_u^{(0)} = \tilde{t}_u^{(0)}$ are analytic functions of $\bar{z}$. Choose these $\tilde{t}$'s independent of ψ. The same reasoning applied to the matrix

$$\left(\tilde{H}_n \right)_{uu'} = \left\langle \tilde{T}_u^{(0)}, \tilde{T}_{u'}^{(0)} \right\rangle \tag{182}$$

shows that

$$-2n\,\mathrm{Tr}\,\delta\psi\tilde{\mathcal{P}}_{n+1} = \delta\ln\det\tilde{H}_n \tag{183}$$

Combining equation (175) with (176*ab*) and the above results, we find that under a scale variation and using the reference metric $g = e^{2\psi}g_0$, with $R = e^{-2\psi}\left[R_0 - \Delta_0\psi_0\right]$

$$\delta\ln\det'\left(-\Delta_n^{(+)}\right)$$

$$= \frac{-1}{\pi}\int \mathrm{d}^2\alpha\sqrt{g_0}\delta\psi\left[\frac{e^{2\psi}}{\varepsilon} + \frac{6n(n+1)+1}{3}\left(R_0 - \Delta_0\psi\right)\right]$$

$$+ \delta\ln\det H_n\det\tilde{H}_n \tag{184}$$

$$\ln\det'\left(-\Delta_n^{(+)}\right)$$

$$= -\frac{1}{2\pi\varepsilon}\int \mathrm{d}^2\alpha\sqrt{g_0}e^{2\psi}$$

$$- \frac{6n(n+1)+1}{6\pi}\int \mathrm{d}^2\alpha\sqrt{g_0}\left\{g_0^{ab}\partial_a\psi\partial_b\psi + 2R_0\psi\right\}$$

$$+ \ln\det H_n\det\tilde{H}_n + f_n \tag{185}$$

with f_n a term independent of ψ (but still singular). We obtain as expected Liouville's action as part of the finite contribution and observe that the ψ-dependent singular terms agree with those in equation (179). Comparing both expressions, we conclude that

$$\ln\det'\left(-\Delta_n^{(+)}\right)$$

$$= -\frac{1}{2\pi\varepsilon} \int \mathrm{d}^2\alpha \sqrt{g_0} e^{2\psi} + \ln\varepsilon \left(\tfrac{1}{3}(3n+1)\chi - \mathrm{dimKer}P_n\right)$$

$$-\frac{6n(n+1)+1}{6\pi} \int \mathrm{d}^2\alpha \sqrt{g_0} \left[g_0^{ab}\partial_a\psi\partial_b\psi + 2R_0\psi\right]$$

$$+ \ln\left(\det H_n \det \tilde{H}_n\right) + F_n \tag{186}$$

With F_n a finite, ψ -independent contribution (therefore function only of the moduli, or equivalently of the reference metric). In the final analysis this is the hardest term to obtain, but also the most interesting one. In chapter 9 when discussing free fields on a torus we had an example of such an evaluation. The expression (186) is needed for $n = 1$. That $-\Delta_1^{(+)}$ differs of $P_1^\dagger P_1$ by a factor 2 is of course irrelevant since any constant factor 2, $\sqrt{2\pi}$... can be absorbed in the definition of the integration measure. The analogous result for the scalar Laplacian, with the conventions on regularization of this section and the extra care paid to topology is left as an exercise. It reads

$$\ln\det'(-\Delta) = -\frac{1}{4\pi\varepsilon} \int \mathrm{d}^2\alpha \sqrt{g_0} e^{2\psi} + \left[\tfrac{1}{6}\chi - 1\right]\ln\varepsilon$$

$$-\frac{1}{12\pi} \int \mathrm{d}^2\alpha \sqrt{g_0} [g_0^{ab}\partial_a\psi\partial_b\psi$$

$$+ 2R_0\psi] + \ln \int \mathrm{d}^2\alpha \sqrt{g_0} e^{2\psi} + F_0 \tag{187}$$

where F_0 is finite and only depends on the reference metric. This expression agrees with equation (134) (since $\ln Z = -\tfrac{1}{2}\ln\det'(-\Delta)$) in its dependence on the scale factor.

The final ingredient before putting everything together is the integration measure. The finite-dimensional measure on moduli was written $\mathcal{D}h_M$. We can now use an expansion of h in $\ker P_1^\dagger$ over the zero modes of $P_1^\dagger$ to write

$$\mathrm{d}h = \sum_u \mathrm{d}\tau_u \tilde{T}_u \tag{188}$$

Therefore

$$\mathcal{D}h_M = \prod_u \mathrm{d}\tau_u \left(\det \tilde{H}_1\right)^{\frac{1}{2}} \tag{189}$$

This term is absent for $g = 0$, is a double integral for $g = 1$, and otherwise the index u runs from 1 to $6(g - 1)$.

We now collect the results from equations (186), for $n = 1$ and (187), where we emphasize that F_1 and F_0 are functions of τ's, to write

$$Z_\chi = \left[\lambda_o \varepsilon^{\frac{1}{12}(8-d)}\right]^\chi \left(\int \mathcal{D}v^\perp\right) \int \mathcal{D}\psi \prod_{\mu=1}^{d} \left[\left\{\frac{\varepsilon}{\ln \int d^2\alpha\sqrt{g_0}e^{2\psi}}\right\}^{\frac{1}{2}} dC_0^\mu\right]$$

$$\times \prod_u d\tau_u \det \tilde{H}_1 \det H_1^{\frac{1}{2}} \varepsilon^{-\frac{1}{2}\dim(\ker P_1)}$$

$$\times \exp -\left\{\frac{(26-d)}{24\pi}\int d^2\alpha\sqrt{g_0}\left[g_0^{ab}\partial_a\psi\partial_b\psi + 2R_0\psi\right] + \right.$$

$$\left.\left(\mu_0^2 + \frac{1}{4\pi\varepsilon}\left(1 - \tfrac{1}{2}d\right)\right)\int d^2\alpha\sqrt{g_0}e^{2\psi} + \tfrac{1}{2}\left[dF_0(\tau) - F_1(\tau)\right]\right\}$$

$$(190)$$

We have still been slightly careless in writting the measure. Indeed since we defined dimensionless determinants (cf. equation (173) where the dimensionless combination $\varepsilon\lambda$ occurs) each original Gaussian integral ought to have been integrated with a measure $dx\varepsilon^{-\frac{1}{2}}$. Thus $dC_0^\mu \to dC_0^\mu\varepsilon^{-\frac{1}{2}}$, $(\prod_u d\tau_u) \to (\prod_u d\tau_u)\varepsilon^{-\frac{1}{2}\dim(\ker P_1^\dagger)}$. Furthermore recall that $\int \mathcal{D}v^\perp$ is not the "volume" of the infinite-dimensional group but for $g = 0, 1$ (where the dimension of $\ker P_1$ is nonvanishing) there is an additional piece missing, the infinite "volume" of the global conformal group. Thus one would like to factor out $\int \mathcal{D}v^\perp \times$ "volume of global transformations" and divide out by a renormalized factor (we take it to be one) times $\varepsilon^{-\dim(\ker P_1)}$ (where $\dim(\ker P_1)$ is of course the dimension of this group). That a factor $\frac{1}{2}$ is missing in the exponent is a consequence of the fact that the vanishing eigenvalues of $\left(P_1^\dagger P_1\right)^{\frac{1}{2}}$ would have dimension 1. In any case, these subtleties just imply that the factor $\left(\lambda_0\varepsilon^{\frac{1}{12}(8-d)}\right)^\chi \varepsilon^{-\frac{1}{2}\dim(\ker P_1)}$ goes over into $\left(\lambda_0\varepsilon^{\frac{1}{12}(8-d)}\right)^\chi \varepsilon^{I_1/2}$ with $I_1 = 3\chi$. Altogether this means

$$\left(\lambda_0\varepsilon^{\frac{1}{12}(8-d)}\right)\varepsilon^{-\frac{1}{2}\dim(\ker P_1)} \to \left(\lambda_0\varepsilon^{\frac{1}{12}(26-d)}\right)$$

We are now in position to define the renormalized quantities

$$\lambda = \lambda_0\varepsilon^{\frac{1}{12}(26-d)}$$

$$\mu^2 = \mu_0^2 + \frac{1}{4\pi\varepsilon}\left(1 - \tfrac{1}{2}d\right) \tag{191}$$

The measure on the moduli space is

$$dM_\tau = \prod_u d\tau_u \det \tilde{H}_1(\det H_1)^{\frac{1}{2}} \tag{192}$$

It would seem that the different pieces in this quantity depend on the scale ψ according to the variational equations (181) and (183), and everything (including the metric g_0) depends on the moduli. For $g > 1$, $\det H_1 = 1$. After the cancellation of ε factors, the measure over translational zero modes can then be written

$$\prod_{\mu=1}^{d} \left\{ \frac{\mathrm{d}C_0^{\mu}}{\left(\int \mathrm{d}^2\alpha \sqrt{g_0} e^{2\psi} \right)^{\frac{1}{2}}} \right\} = \prod_{\mu=1}^{d} \mathrm{d}X_0^{\mu} \qquad (193)$$

where X_0^{μ} is the center of mass (or translational) coordinate since the expansion of X^{μ} involved the zero mode contribution $C_0^{\mu} \times \left(\int \mathrm{d}\alpha \sqrt{g_0} e^{2\psi} \right)^{-\frac{1}{2}} + \cdots$. Finally we rescale Z_χ to Z_χ^{Ren} by the "volume" of the full reparametrizations group and write

$$S(\tau) = \tfrac{1}{2} \left[dF_0(\tau) - F_1(\tau) \right] \qquad (194)$$

the unknown, but probably most significant, part of the action.

The end result of this analysis is the expression (Polyakov, Friedan)

$$Z_\chi^{Ren} = \lambda^\chi \int \mathcal{D}\psi \int \prod_1^d \mathrm{d}X_0^\mu \int \mathrm{d}M_\tau \exp -S_{tot}$$

$$S_{tot} = S(\tau) + \frac{26-d}{24\pi} \int \mathrm{d}^2\alpha \sqrt{g_0} \left[g_0^{ab} \partial_a \psi \partial_b \psi + 2R_0 \psi \right]$$

$$+ \mu^2 \int \mathrm{d}^2\alpha \sqrt{g_0} e^{2\psi} \qquad (195)$$

There are admittedly many points which ought to be clarified in the derivation. In particular finding the precise domain of integration in moduli space is a difficult question. Moreover the superposition of terms with different topologies might suffer from serious drawbacks. One point is obvious however. The proposed bosonic continuous theory for random surfaces does not make sense in dimension higher than 26, since the kinetic contribution of the ψ-field becomes negative. In dimension smaller than 26, there appears an essential degree of freedom, namely the ψ-field, due to the quantum scale anomaly. This has been called the Liouville theory. Dimension 26 is marginal. At the price of setting $\mu^2 = 0$, the ψ contribution factors out (possibly by rescaling correctly the measure $\mathrm{d}M_\tau$) and can be absorbed into a further renormalization of Z_χ.

There are two types of observables that can be introduced to study the behaviour of these random surfaces. The nonlocal ones

such as occur in gauge theories are boundary loops. The whole
derivation in the continuum can be carried over in the presence of
prescribed boundaries with due care paid to reparametrization
of the boundary itself. Depending on the type of boundary
conditions, extra terms may occur, proportional to the area as well
as the length of the boundary. In the context of particle physics,
this yields the open string model. Other types of observables are
local in the sense that the surfaces are required to go through
fixed points $\{x^{\mu}\}$ in target space. This introduces in the path
integral factors of the type $\delta^{(d)}(\mathbf{X} - \mathbf{x}) = \int(\mathrm{d}^d p/(2\pi)^d)\exp(-i\mathbf{p} \cdot \mathbf{x})\exp(i\mathbf{p} \cdot \mathbf{X})$, the so-called vertex-operators, which allow to
estimate correlations on the surface.

The continuous theory of surfaces is obviously involved and con-
siderable sophisticated work (still in progress) is needed in order
to extract information. Discretized versions partly amenable to
numerical simulations are therefore of great interest in providing
an alternative probe.

11.2.4   Discretized models

It would seem that a discretized approach avoids the question of
reparametrization invariance. It will turn out that this is a little
too optimistic.

It is worth remarking that a triangulation does not *a priori*
approximate as uniformly a smooth surface as does a piecewise
linear path with respect to a curve. A classical example is to
approximate a vertical cylinder by triangles obtained by joining
the vertices of horizontal regular polygons with n sides, centered
on the axis at points of ordinate kb and inscribed in circles of radius
a, each one rotated by an angle π/n with respect to the previous
one (figure 5). For a layer of K such families of $2n$ triangles, the
area of the cylinder is $2\pi abK$. On the other hand, each triangle
has an area $\frac{1}{2}.2a\sin(\pi/n) \times \sqrt{b^2 + 4a^2\sin^2(\pi/2n)}$, hence the ratio
of triangular area to the continuous one is

$$\frac{n}{\pi}\sin\frac{\pi}{n}\left\{1 + \frac{4a^2}{b^2}\sin^2\frac{\pi}{2n}\right\}^{\frac{1}{2}}$$

Let n go to infinity. If a^2/b^2 is kept fixed, the ratio of areas tends
to unity. If however a^2/b^2 is allowed to vary with n, we can obtain

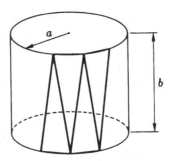

Fig. 5   Approximating a cylinder by a regular triangulation

any result (in this case larger to one) and even infinity, even though
the area of each elementary triangle tends to 0.

Let us describe first the simplest approach. Fix the topology, a
torus ($\chi = 0$) being the easiest one to treat, with a map in $\mathcal{R}^d$ for
each point of a given triangulation. Take an action proportional
to the embedded area, each triangle contributing a term ℓ_{ijk}, and
integrate over all sites but one

$$Z = \int {\prod}' d^d X_i e^{-\beta S}$$
$$S = \sum_{(ijk)} \ell_{ijk} \tag{196}$$

We have set a short distance cutoff (analogous to $\varepsilon^{\frac{1}{2}}$ in the
continuum) equal to unity, or absorbed it in the definition of β.
We are interested in the case where the mesh of the triangulation
becomes small. The most regular one is a regular triangular lattice
in base space on a torus with N sites, $3N$ links and $2N$ triangles.
On the base space, we have then a discrete translational symmetry.
If we scale each $\mathbf{X}$ by λ, the partition function Z transforms
according to

$$Z(\beta) = \lambda^{d(N-1)} Z(\lambda^2 \beta)$$

i.e.

$$\langle \beta S \rangle = 2\beta N \left\langle \ell_{ijk} \right\rangle = \tfrac{1}{2} d(N-1)$$

For N large,

$$\beta \left\langle \ell_{ijk} \right\rangle = \tfrac{1}{4} d \tag{197}$$

Hence a saddle point method is valid for large d. One introduces conjugate parameters for each link, enforcing the condition $\left(\mathbf{X}_i - \mathbf{X}_j\right)^2 = \ell_{ij}^2$. Recall that the area ℓ_{ijk} is given in terms of the ℓ_{ij}'s by equation (99). The partition function reads

$$Z = \int \prod{}' d^d X_i \int \prod_{(jk)} \left(\frac{d\lambda_{jk}}{2i\pi} d\ell_{jk}\right)$$

$$\exp - \left\{ \sum_{(jk)} \lambda_{jk} \left[\left(\mathbf{X}_j - \mathbf{X}_k\right)^2 - \ell_{jk} \right] + \beta S \right\} \quad (198)$$

Assuming a homogeneous mean field ($\lambda_{jk} = \lambda$ and $\ell_{jk} = \ell$ independent of the link), the saddle point condition leads to

$$Z \sim \exp - S_{eff}(\lambda, \ell)$$
$$S_{eff} = N \left[\tfrac{1}{2} d \ln \lambda - 3\lambda\ell^2 + \tfrac{1}{2}\sqrt{3}\beta\ell^2 \right] \quad (199)$$

with stationary conditions in ℓ and λ requiring

$$\lambda = \frac{d}{6\ell^2} \qquad \beta\sqrt{3}\ell^2 = d \quad (200)$$

in agreement with equation (197). Therefore $S_{eff} = \tfrac{1}{2}Nd$ $\ln(\beta/2\sqrt{3})$. With these values for λ and ℓ, the model reduces to d-uncoupled Gaussian massless fields when investigating any observable depending on the coordinates $\mathbf{X}$. For instance, since $\langle (\mathbf{X}_j - \mathbf{X}_k)^2 \rangle = \ell^2$, the radius of gyration defined as

$$R_N^2 = \frac{1}{N} \left\langle \sum_j \mathbf{X}_j^2 \right\rangle \quad (201a)$$

assumes the value

$$R_N^2 = \frac{1}{\pi} \frac{\sqrt{3}}{4} \ell^2 \ln N = \frac{1}{\pi} \langle \text{elementary area} \rangle \ln N \quad (201b)$$

with a logarithmic behaviour arising from the corresponding infrared singularity of the two-dimensional propagator. This is to be contrasted with the linear behaviour in N for Brownian curves. In equation ($201b$) the prefactor $(1/\pi)\langle$elementary area$\rangle$ is specific to the triangulation, and of order d. It is corrected by terms of order $1/d$. In the present case, the leading correction in $(1 + (2/d) + \cdots)$ has been observed in numerical simulations.

We conclude that a model based on a fixed topology and fixed triangulation, with a Boltzmann weight in $\exp(-\text{Area})$, yields as a consequence of equation (201) completely collapsed surfaces (of infinite Hausdorff dimension) and no simple modification of the action, as for instance by including a discretized Liouville term, seems to modify this conclusion. What is required to obtain a more sensible picture, closer to the continuous theory, is to re-examine the discrete analog of reparametrization invariance. This can only be to admit for a fixed genus all possible inequivalent triangulations $\mathcal{T}$ rather than pick a fixed one, which led us to the above catastrophic collapse. These triangulations appear as random lattices on an abstract base space of fixed topology. In order to simplify matters, and in analogy with the continuous case, a reference metric is assigned to such a triangulation with each link of equal length. The triangles are equilateral and therefore to each site is assigned an element of area $\sigma_i = \frac{1}{3}n_i$ in appropriate units (the equivalent of $\sqrt{g}$), where n_i is the number of its neighbours (equivalently the number of incident triangles or links). The total (intrinsic) area is equal to the total number of triangles N_2

$$A = \sum_i \sigma_i = N_2 \tag{202}$$

The curvature at a site in base space is given by the defect angle

$$\sigma_i R_i = 2\pi - \tfrac{1}{3}\pi n_i = \tfrac{1}{3}\pi n_i \left(\frac{6}{n_i} - 1 \right) \tag{203}$$

and is such that

$$\frac{1}{2\pi} \sum_i \sigma_i R_i = \chi \tag{204}$$

A term involving the square of the curvature would read

$$\sum_i \sigma_i R_i^2 = \sum_i \tfrac{1}{3}\pi^2 \frac{(6 - n_i)^2}{n_i} \tag{205}$$

To construct the action as was done in the continuum, one supplements a contribution proportional to the area (202) with a quadratic term in the embedding fields $\mathbf{X}_i$ using the base metric. For a fixed characteristic χ, one sums over all inequivalent triangulations $\mathcal{T}$ (annealed average) with arbitrary many triangles and in addition one integrates over the $\mathbf{X}_i$ with a measure

$$\mathcal{D}\mathbf{X} = \prod_i \left[\left(\frac{\sigma_i}{2\pi} \right)^{\frac{1}{2}d} d^d\mathbf{X}_i \right] (2\pi)^{\frac{1}{2}d}\delta^d \left(\frac{1}{A}\sum \sigma_i \mathbf{X}_i \right) \qquad (206)$$

omitting for simplicity the integral over the center of mass. Thus

$$Z_\chi = \sum_T \frac{1}{k(T)} e^{-\beta A} \int \mathcal{D}\mathbf{X} \exp -\tfrac{1}{2} \sum_{(ij)} \left(\mathbf{X}_i - \mathbf{X}_j \right)^2 \qquad (207)$$

The sum over triangulations replaces the sum over metrics, and $k(T)$ is a combinatorial factor equal to the order of the symmetry group of a given triangulation. An extra weight including R^2–terms such as in equation (205) could be added to test for their irrelevance. The coefficient β is the analog of μ_0^2 in the preceding section or, equivalently, of a bare string tension.

The discrete operator occurring in the quadratic X-piece of the action has diagonal matrix elements $n_i = 3\sigma_i$ and off-diagonal ones zero or minus one if they correspond to a link. This is not quite the discrete scalar (minus) Laplacian obtained by multiplying both sides by the diagonal matrix $\sigma_i^{-\frac{1}{2}}$. Thus

$$(-\Delta)_{ii} = 3$$
$$(-\Delta)_{ij} = \begin{cases} 0 \\ -(\sigma_i\sigma_j)^{-\frac{1}{2}} \end{cases} \qquad (208)$$

With the omission of the zero mode, one finds therefore that

$$Z_\chi = \sum_T \frac{1}{k} e^{-\beta A} \left[\frac{\det'(-\Delta)}{A} \right]^{-\frac{1}{2}d} \qquad (209)$$

where each piece depends on T. That $\det'(-\Delta)$ is divided by the area A is a consequence of subtracting the center of mass integral. The dimension d appears as a parameter. Similarly one could consider local Green's functions tying the surfaces to a set of points $\{\mathbf{X}_a\}$

$$G(\{\mathbf{X}_a\}) = \sum_T \frac{1}{k(\tau)} e^{-\beta A} \int \mathcal{D}(\mathbf{X}) \exp \left(-\tfrac{1}{2} \sum_{(ij)} \left(\mathbf{X}_i - \mathbf{X}_j \right)^2 \right)$$
$$\times \prod_a \left[\sum_j \sigma_j \delta \left(\mathbf{X}_j - \mathbf{X}_a \right) \right] \qquad (210)$$

appropriately normalized, assuming of course the lattice to contain more points than the set $\{\mathbf{X}_a\}$.

For a fixed genus, the partition function Z_χ is a power series in $e^{-\beta}$ since A is an integer and is expected to have a finite radius of convergence as some examples will soon show. Thus the model makes sense at least for $\beta > \beta_c$. At the critical point β_c, some singularities will occur, and one is interested as usual in the critical behaviour.

The one-point function G_1 is independent of $\mathbf{X}$ by translational invariance in target space and agrees with $-\partial_\beta Z_\chi$. The analogy with statistical physics is that the second derivative of Z_χ is a susceptibility (rather than a specific heat) and is proportional to the mean area. At criticality, one defines the exponent γ through

$$\frac{\partial^2}{\partial\beta^2} Z_\chi \sim \frac{1}{(\beta - \beta_c)^\gamma} \qquad (211)$$

If γ is positive the mean area diverges, but remains bounded if $\gamma < 0$.

For β close to β_c, the two-point function is expected to scale as

$$G_2\left(\mathbf{X}_{12}\right) \sim \frac{1}{|\mathbf{X}_{12}|^{d-2+\eta}} e^{-m(\beta)|\mathbf{X}_{12}|} \qquad (212)$$

$$m\left(\beta\right) \sim (\beta - \beta_c)^\nu \qquad (213)$$

with exponents η and ν. For a fixed triangulation one can define a radius of gyration

$$\langle \mathbf{X}^2 \rangle_T = \frac{1}{A} \sum_{i,j} \sigma_i \sigma_j \left\langle \left(\mathbf{X}_i - \mathbf{X}_j\right)^2 \right\rangle$$

where the bracket means an average with respect to the $\mathbf{X}$ part of the measure. A generating function with respect to all triangulations reads

$$Z_{\langle \mathbf{X}^2 \rangle} = \sum_T \frac{1}{k\left(T\right)} e^{-\beta A} \langle \mathbf{X}^2 \rangle_T$$

If one performs the expansion

$$Z_\chi = \sum e^{-n\beta} a_n \qquad Z_{\langle \mathbf{X}^2 \rangle} = \sum e^{-n\beta} b_n \qquad (214)$$

the ratio b_n/a_n is an average radius of gyration for fixed (intrinsic) area A (equal to n)

$$\langle \mathbf{X}^2 \rangle_A = b_n/a_n \qquad (215)$$

A possible definition of the Hausdorff dimension d_H is obtained in the limit $n \to \infty$ through

$$\langle \mathbf{X}^2 \rangle_A \underset{A \to \infty}{\sim} A^{2/d_H} \tag{216}$$

Show that, if $m\,(\beta_c)$ vanishes, one expects the following scaling relations

$$\gamma = \nu\,(2 - \eta) \qquad \nu = 1/d_H \tag{217}$$

For definiteness, we assume in the sequel planar topology (i.e. $\chi = 2$ with the plane compactified to a sphere to yield a compact surface). It was noticed by David that the model of randomly triangulated surfaces is equivalent, up to irrelevant terms, to a planar ϕ^3 theory with an exponential propagator, generalizing the discussion given in section 4 of chapter 10. Here ϕ is an $N \times N$ Hermitian matrix, and N tends to infinity. The ϕ-field theory is based on the (unbounded) action in Euclidean target space

$$S_\phi = \int d^d \mathbf{X} \ \mathrm{Tr}\, \phi \ e^\Delta \phi + \frac{g}{\sqrt{N}}\,\mathrm{Tr}\,\phi^3 \tag{218}$$

which is only used as a mean to generate a perturbative series. In the planar ($N \to \infty$) limit, the diagrams are dual to finite triangulations $\mathcal{T}$ of the plane (figure 6). One can modify the Lagragian to eliminate tadpoles, and even self-energy insertions by including a second degree polynomial in ϕ. The ordinary combinatorial rules yield the factor $1/k\,(\mathcal{T})$ occuring in equation (207). The remaining contribution arises from the Feynman integral with a Gaussian configuration space propagator $\exp - (\mathbf{X}_1 - \mathbf{X}_2)^2$ between vertices of the planar diagrams.

One has to integrate over all $\mathbf{X}$'s (but one). This is not recognized immediately as the $\mathbf{X}$-part in the integral (207), the reason being that the $\mathbf{X}$'s are indexed by sites of the planar graph rather than by those of the dual triangulation. Up to a factor, the Feynman integral here is of the form $(\det' K)^{-\frac{1}{2}d}$, where $\det' K$ is any minor of the matrix K related to the incidence matrix on the planar graph, $K_{ab} = -1$ if a and b are neighbours, $K_{aa} =$ number of neighbours of vertex a, and otherwise zero. According to Kirchoff's theorem (chapter 7), $\det' K$ is the number of spanning trees on the planar graph, each one in one-to-one correspondence

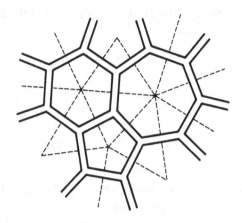

Fig. 6   Duality between planar diagrams (double lines) and triangulations (broken lines).

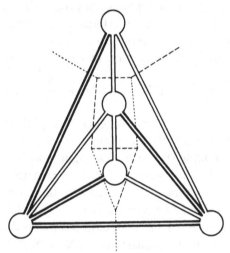

Fig. 7   An example of duality between spanning trees of planar graphs (indicated by heavy double or broken lines).

with a spanning tree on the dual lattice, i.e. the triangulation T. This is a specific property of planar topology which would require modifications to be implemented on surfaces of higher genus. On figure 7, we show an example of such duality between spanning trees.

As a result, $\left(\det{}' K\right)^{-\frac{1}{2}d} = \left(\det{}' K_T\right)^{-\frac{1}{2}d}$ differs from the expected result $[(\det -\Delta)/A]^{-\frac{1}{2}d}$ only by factors arising in the last case from the integration measure (206) i.e. $\prod_i \sigma_i^{\frac{1}{2}d} =$ $\exp \frac{1}{2} d \sum_i \ln \sigma_i$. Since $\sigma_i = \frac{1}{3} n_i = 2 \left[1 - \sigma_i R_i / \pi \right]$,

$$\prod_i \sigma_i^{\frac{1}{2}d} = \exp \left\{ \frac{1}{4} dA \ln 2 - (2 - \ln 2) \frac{1}{2} d\chi + \cdots \right\} \qquad (219)$$

using the properties of a planar triangulation $(2N_1 = 3N_2,$

$$N_0 - N_1 + N_2 \equiv \chi \equiv \frac{1}{2\pi} \sum_i \sigma_i R_i$$

incidentally, this shows that N_2 is even). The higher order terms involve higher powers of the curvature and are presumed irrelevant (or could be taken care of by suitably modifying the bare action in equation (207)). The first term on the r.h.s. in (219) is a shift in β and the second involves the expected topological invariant.

Thus from the point of view of renormalization, the planar ϕ^3 model defined through equation (218) is equivalent to the discretized Polyakov model. A renormalization of the propagator would be required to avoid ϕ^3 self-mass insertions (and triangulations with vertices having only two neighbours which could be omitted) and an added linear term in ϕ to eliminate tadpoles, where in the triangulations some vertices would be identified.

Consider the vacuum connected functional in the ϕ^3 version. The series in $e^{-\beta}$ is equivalent to an ordinary perturbative series in the coupling constant g for the planar theory where we have some experience in the convergence properties as well as singularities. A smooth extrapolation in the dimension d is afforded by equation (209). In particular, when $d = 0$, we can use the results of chapter 10 and in particular equation (10.290) which shows a finite radius of convergence in $g^2 = e^{-2\beta}$, with a behaviour

$$d = 0 \qquad Z_\chi = \frac{1}{2} \sum_1^\infty \frac{\left(72 e^{-2\beta} \right)}{(k+2)!} \frac{\Gamma \left(\frac{3}{2} k \right)}{\Gamma \left(\frac{1}{2} k + 1 \right)}$$

$$e^{-2\beta_c} = \frac{1}{108\sqrt{3}} \qquad (220)$$

The general term in the series behaves as $e^{-2(\beta - \beta_c)k} k^{-\frac{7}{2}}$. As a result the singular part in Z_χ close to β_c is in $(\beta - \beta_c)^{\frac{5}{2}}$ and the exponent γ defined through the second derivative in β is negative

$$d = 0 \qquad \gamma = -\frac{1}{2} \qquad (221)$$

meaning that the average area at criticality remains finite. Both conclusions are not affected by suppressing tadpole or self-energy graphs. Of course, this is only an irrealistic toy model.

It turns out that similar tricks can be applied to the case $d = -2$ (Kazakov, Kostov, Migdal) where one gets a sum weighted by the number of spanning trees, equivalent to a (complex) fermionic theory. The corresponding series has also a finite radius of convergence with

$$d = -2 \qquad \gamma = -1 \qquad (222)$$

On the other hand (Zamolodchikov), the continous theory can be used perturbatively to lowest order at large unphysical negative d to produce a value

$$d \to -\infty \qquad \gamma = \tfrac{1}{6}(d - 7), \qquad \nu = 0 \qquad (223)$$

In the references, much more information is extracted from the soluble cases $d = 0$ and $d = -2$, where the correlation length $m\,(\beta_c)^{-1}$ seems to diverge at the transition with a very small exponent, meaning a large Hausdorff dimension. This seems also to hold in larger dimensions.

A mean field (large d) theory can also be studied on a lattice (chapter 6). It assumes a very rough spiky form of a branched polymer of cubes, with

$$d \to +\infty \qquad \gamma = \tfrac{1}{2} \qquad \nu = \frac{1}{d_H} = \tfrac{1}{4} \qquad (224)$$

A consistent picture for the exponent γ seems therefore that it increases up to $\tfrac{1}{2}$ for $d \to \infty$, while the Hausdorff dimension would be always large if not infinite.

Random surface theory is a rich and complex subject which cannot said to have reached a stable status. For instance, we did not discuss the sum over topologies, nor did we include detailed information on the ambient space. In physical applications, one may want to consider situations where the string tension is vanishing or small and where terms involving the mean curvature (the sum of inverse principal radii of curvature) plays a role. Additional intrinsic degrees of freedom of a random membrane may also be involved. Thus one has a large spectrum of problems to tackle. Random geometry looks a very promising field.

Notes

Statistical geometry is described by L.A. Santalo in *Integral Geometry and Geometrical Probability*, Addison-Wesley, Reading (1976). For the role of these concepts in condensed matter physics, see the contribution by R. Collins in *Phase Stability of Metals and Alloys*, P.S. Rudman, J. Stringer and R.I. Jaffee eds, McGraw-Hill, New York (1967), by R. Zallen in *Fluctuation Phenomena*, E.W. Montroll and J.L. Lebowitz eds, North-Holland, Amsterdam (1979) and the book by J.M. Ziman *Models of Disorder*, Cambridge University Press (1979).

The Poissonian random lattices were studied by N.H. Christ, R. Friedberg and T.D. Lee, *Nucl. Phys.* **B202**, 89 (1982) **B210** (**FS6**) 310, 337 (1982). The work of J.L. Meijering appeared in *Philips Research Report* **8**, 270 (1953). For a related work, see H.G. Hanson, *J. Stat. Phys.* **30**, 591 (1983). The contributions of the authors in collaboration with M. Bander appear in *New Perspectives in Quantum Field Theories*, J. Abad, M. Asorey and A. Cruz eds., World Scientific, Singapore (1986). For a study of statistical models on random two-dimensional lattices, see D. Espriu, M. Gross, P.E.L. Rakow and J.F. Wheater, *Nucl.Phys.* **B265** [**FS15**], 92 (1986). Dirac–Kähler fermions are discussed in P. Becher and H. Joos, *Z. Phys.* **C15**, 343 (1982).

Discretized gravity was formulated in the influential paper by T. Regge, *Nuovo Cimento* **19**, 558 (1961). Curvature on piecewise flat manifolds and its continuous limit is investigated in depth by J. Cheeger, W. Müller and R. Schrader, *Comm. Math. Phys.* **92**, 405 (1984). For related work on quantum gravity, see G. Feinberg, R. Friedberg, T.D. Lee and H.C. Ren, *Nucl. Phys.* **B245**, 343 (1984).

In the context of particle physics, the dual model for strong interactions gave birth to a surface theory after the pionneering work of Y. Nambu reported in *Symmetries and Quark Models*, R. Chaud ed., Gordon and Breach, New York (1970), T. Goto, *Prog. Theor. Phys.* **46**, 1560 (1971) and several others. The study of random surfaces as a generalization of Brownian motion was given a new impetus by A.M. Polyakov, *Phys. Lett.* **103B**, 207 (1981). For a review including references, motivations and exact results, see J. Fröhlich in *Applications of Field Theory to Statistical Mechanics*, L. Garrido ed., Lecture Notes in Physics **216**, Springer, Berlin (1985). Our presentation is based on the

810	11	Random Geometry

analysis by O. Alvarez, *Nucl. Phys.* **B216**, 125 (1983) and the 1982 lectures of D. Friedan at Les Houches published in *Recent Advances in Field Theory and Statistical Physics*, J.-B. Zuber and R. Stora eds, North-Holland, Amsterdam (1984). The inclusion of extrinsinc curvature terms in the action can be traced to the work of W. Helfrich, *Z. Naturforsch.* **C28**, 693 (1973) and *J. Physique* **46**, 1263 (1985). See also A.M. Polyakov, *Nucl. Phys.* **B268**, 406 (1986).

For effective means to compute anomalies in the path integral approach see K. Fujikawa, *Phys. Rev.* **D21**, 2848 (1980), **D23**, 2262 (1981).

The study of interfaces between coexisting phases has a long history. Some recent representative works are D.J. Wallace and R.K.P. Zia, *Phys. Rev. Lett.* **43**, 808 (1979), M.J. Lowe and D.J. Wallace, *Phys. Lett.* **93B**, 433 (1980), *J. Phys.* **A13**, L381 (1980), F. David, *Phys. Lett.* **102B**, 193 (1981).

Discrete models of surfaces were studied in the context of gauge theories by J.-M. Drouffe, G. Parisi and N. Sourlas in a paper quoted in chapter 6, by D. Weingarten, *Phys. Lett.* **90B**, 285 (1980), T. Eguchi and H. Kawai, *Phys. Lett.* **110B**, 143 (1982), **114B**, 247 (1982), by B. Durhuus, J. Fröhlich and T. Jonsson, *Nucl. Phys.* **B225** [FS9], 185 (1983), *Phys. Lett.* **137B**, 93 (1984). D.J. Gross, *Phys. Lett.* **138B**, 185 (1984), A. Billoire, D.J. Gross and E. Marinari, *Phys. Lett.* **139B**, 75 (1984) and B. Duplantier, *Phys. Lett.* **141B**, 239 (1984) considered models involving a fixed triangulation, while a discretized version of Polyakov's model and its connection to the planar limit were investigated by F. David, *Nucl. Phys.* **B257** (**FS14**), 543 (1985), V.A. Kazakov, I.K. Kostov and A.A. Migdal, *Phys. Lett.* **157B**, 295 (1985) and J. Ambjørn, B. Durhuus and J. Fröhlich, *Nucl. Phys.* **B257** (**FS14**), 433 (1985). Related computations were carried out in the continuous version by A.B. Zamolodchikov, *Phys. Lett.* **117B**, 87 (1982) and J. Jurkiewicz and A. Krzywicki, *Phys. Lett.* **148B**, 148 (1984). This list is by no means exhaustive.

As regards the literature on string (and superstring) field theory it is by now enormous. Many early papers are collected in reprint volumes such as *Dual Theory*, M. Jacob ed., North Holland, Amsterdam (1974 and 1984) and *Superstrings*, J.H. Schwarz ed., World Scientific, Singapore (1985). M.B. Green, J.H. Schwarz and E. Witten present a comprehensive view in the two volumes of *Superstring Theory*, Cambridge University Press (1987).

INDEX

A

A–D–E classification, 579, 754
Abe, 157
Abnormal dimension, 176
Adjoint representation, 625
Airy's equation, 653
Alvarez, 780
Amplitude ratios, 314
Amputation, 239
Anderson, 646
Andrews, Baxter, Forrester, 588
Annealed average, 802
Anomalies, 268
Anomalous dimension, 241, 274
Anomaly, 516
Anticommuting variables, 48
Area law, 345
Articulated diagrams, 250
Ashkin–Teller model, 554, 594
Asymptotic freedom, 329, 437, 727
Atiyah–Singer index theorem, 793
Avron, Seiler, 729
Axial gauge, 333

B

Background gauge, 340
Baker, Nickel, Green, Meiron, 317
Baker–Campbell–Haussdorff formula, 334

Bare parameters, 237
Baxter, 558
Baxter, Andrews, Forrester, 579
Becchi–Rouet–Stora (BRS), 647, 657, 717
Belavin, Polyakov, Zamolodchikov, 501, 578
Beltrami equation, 784
Bender, Wu, 447
Berezin, 48
Berezinskii, 200
Berlin–Kac, 140
Bernal, 739
Berry's phase, 734
Bethe approximation, 121
Bethe lattice, 419
Block spin method, 168
Bogoliubov, Parasiuk, Hepp, Zimmermann renormalization scheme, 264
Born amplitude, 282
Bosonization, 97, 722
Brownian motion, 1, 649, 778, 801
Brézin, Le Guillou, Zinn-Justin, 311

C

Caianiello, 95

Callan–Symanzik equation, 233, 273, 285, 309
Canonical degree, 237
Canonical dimension, 23, 241
Cardy, 546, 601
Cartan matrix, 622
Cartan subalgebra, 619
Cartan–Killing, 579, 619
Casimir effect, 519, 564
Casimir invariants, 583, 625, 635
Casimir operator, 332
Cauchy determinant, 84
Cayley tree, 419
Central charge, 515
Character expansion, 374
Cheeger, Müller, Schrader, 768
Cheng–Wu, 88
Chevalley, Serre, 622
Christ, Friedberg, Lee, 738
Classical Coulomb gas, 193
Classical Heisenberg model, 26, 107
Clausius–Mossotti formula, 109
Clifford algebra, 52, 63
Co-roots, 623
Coleman, 197
Configuration number, 439
Confinement, 345
Conformal anomaly, 773
Conformal invariance, 502
Conformal weights, 511
Connected correlation functions, 239
Contragredient form, 527
Coordination number, 1
Corrections to scaling laws, 301
Correlation length, 40
Coulomb gas, 203, 691
Coulombic partition function, 593
Coxeter numbers, 585, 624, 626
Coxeter–Dynkin diagrams, 622
Creutz, 473
Critical amplitudes, 189

Critical exponent, 6
Critical exponents, 125, 280
Critical point, 40, 274
Critical slowing down, 460, 471
Cross ratio, 518
Crossover exponent, 720
Cumulant expansion, 77, 239, 412
Curie transition, 58

D

Darboux–Chritoffel kernel, 699
David, 805
Decimation, 176
Deconfining transition, 354
Dedekind's function, 528, 549, 611
Descendant, 526
Di Francesco, Saleur, Zuber, 577
Diagram, 405
Diffusion equation, 3
Dimensional analysis, 236
Dimensional regularization, 250
Dimensional transmutation, 716
Dirac equation, 93
Dirac field, 721
Dirac matrices, 394
Dirichlet–Voronoi construction, 739
Discrete Laplacian, 3
Dominant weight, 635
Dotsenko, 561, 563
Dotsenko, Dotsenko, 716, 720, 728
Dual complex, 348
Dual Coxeter number, 624
Dual group, 347
Duality transformation, 180, 204
Duality, 59, 345
Dynamical dimension, 242
Dyson, 310, 647, 665, 690

E

Effective potential, 156, 240, 249
Eguchi, Ooguri, 578
Elitzur's theorem, 341
Elliptic functions, 609
Energy–momentum tensor, 507
Enveloping algebra, 583
ε-expansion, 235, 290, 311
Equation of state, 166
Equations of motion, 145
Euclidean fields, 21
Euler characteristic, 740, 765, 793
Euler's formula, 705
Euler's pentagonal identity, 549, 610
Euler's relation, 409

F

Faddeev–Popov determinant, 785
Faddeev–Popov ghost fields, 337, 436
Feigin–Fuchs, 533
Ferdinand, Fisher, 574
Fermi levels, 714
Feynman integrals, 246
Feynman rules, 154, 244, 337
Finite size effects, 478
Fisher circles, 139
Fisher, 233, 235, 290, 480
Fisher–Gaunt, 157
Fock–Bargmann space, 50, 676
Fokker–Planck equation, 487, 649
Free fields, 22
Freudenthal formula, 626
Friedan, 780, 798
Friedan, Qiu, Shenker, 543, 592, 618, 638
Frustrated partition function, 72, 86
Frustration, 352, 588
Fujikawa, 776

G

Gaudin, 699
Gauge fields, 329
Gauge fixing, 435
Gauge invariance, 329
Gauss series, 561
Gauss sum, 612
Gauss–Bonnet formula, 766
Gaussian discrete model, 205
Gaussian integrals, 22
Gaussian model, 33, 550, 592
Gell-Mann–Low formula, 522
Generating functionals, 236
Genus, 705
Gepner, Qiu, 582
Gepner, Witten, 581
Glueball mass spectrum, 390
Goddard, Kent, Olive, 636, 642
Goldstone modes, 107, 118, 120, 127
Goldstone's theorem, 298, 300, 435, 495
Grading, 524
Graph, 405
Grassmannian integrals, 48
Green function, 8
Green functions, 238
Gross–Neveu model, 490, 720
Group characters, 331
Group theoretical factors, 259

H

Haar measure, 427
Hadron spectrum, 494
Haffnian, 95
Hall conductance, 729
Halperin, 647
Hard hexagon model, 558
Hardy–Ramanujan formula, 548
Harris criterion, 716, 720
Hausdorff dimension, 6, 802, 808
Heat bath algorithm, 460

Heat kernel, 332, 776, 790
Hierarchical model, 192
Higgs fields, 350
High temperature expansion, 33, 421
Highest weight vector, 526
Hilbert–Einstein action, 768
Hioe, Montroll, 715
Hohenberg, 197, 219
Homology and cohomology groups, 349
Hopping parameter, 492
Hopping probability, 759
Huse, 588
Hypergeometric equation, 561
Hyperscaling hypothesis, 165

I

Importance sampling, 458
Infrared fixed point, 171
Interacting fields, 25
Interface energy, 480
Intersection probability, 16
Irrelevant operators, 174, 307
Ising model, 33, 58, 573, 605

J

Jordan–Wigner representation, 63
Julia set, 193

K

Kac character formula, 640
Kac determinant, 528, 532
Kac table, 555, 559, 589
Kac–Moody algebra, 526, 581, 619, 638
Kadanoff, 168
Kadanoff–Ceva, 87
Kaufmann, 574
Kazakov, Kostov, Migdal, 808

Killing field, 786
Killing vector, 790, 794
Kirchoff's theorem, 410, 805
Kirkwood–Yvon theory, 110
Kosterlitz–Thouless transition, 193, 200
Kramers–Wannier duality, 61
Kronecker's formula, 567
Kubo's formula, 732

L

Lagrangian, 233
Landau levels, 675
Landau–Ginzburg criterion, 158, 166
Langevin equation, 486
Laplace transform, 153
Laplace–Beltrami operator, 332
Lattice fermions, 393
Lattice gauge fields, 328
Lattices, 43
Laughlin, 688, 729
Lee–Yang singularity, 125, 129, 131, 547, 566, 588
Legendre transformation, 155, 159, 239, 415
Level spacings, 696
Level, 581
Liapunov exponent, 666
Lie algebra, 330, 579, 619
Lifshitz tail, 654
Liouville action, 773, 779, 795
Liouville theory, 798
Lipatov, 310, 447
Localization length, 670
Loop expansion, 246
Loop group, 626
Low temperature expansion, 430
Lower critical dimension, 143

M

Ma, 456
Majorana field, 93, 554, 721
Majority rule, 183
Marginal operators, 174, 218, 307
Markov process, 5, 458
McBryan–Spencer, 222
McCoy–Wu, 78
Mean field approximation, 108, 352
Meijering, 743
Mermin–Wagner theorem, 143, 193, 197, 219
Metropolis algorithm, 462
Microcanonical simulations, 465
Migdal–Kadanoff approximation, 178
Minimal coupling, 330
Minimal models, 546
Minimal subtraction scheme, 337
Modular group, 567
Modular invariance, 564
moduli, 786
Monodromy, 562
Monte Carlo renormalization group, 456, 481
Monte Carlo sweep, 460
Multicritical points, 317
Möbius transformation, 510

N

n-vector model, 25, 118, 259, 427, 436, 595
Nambu–Goto action, 781
Neveu–Schwarz and Ramond boundary conditions, 577, 616
Nickel, 311
Nielsen–Ninomiya theorem, 396
Niemeijer–Van Leeuwen, 183
Noether current, 508
Noether's theorem, 97

nonlinear σ-model, 26, 107, 436, 769
Number of loops, 247

O

One particle irreducible Green functions, 239
One-loop correction, 155
Onsager, 48, 58, 68, 75
Orbifold, 554, 594
Order–disorder variables, 87
Oscillator wavefunctions, 698

P

Padé approximants, 448
Painlevé equations, 98
Parisi, Sourlas, 686, 715
Parisi, Wu Yong Shi, 486
Pasquier, 579, 588, 595
Path integral, 12
Peierls' argument, 432
Perimeter law, 345
Perturbation theory, 242
Perturbative renormalization, 264
Peter–Weyl theorem, 427
Pfaffian, 57
Phenomenological renormalization, 224
Pippard–Ginsberg, 165
Planar approximation, 703
Planar limit, 805
Plaquette, 60, 331
Poisson distribution, 665
Poisson formula, 204, 580
Polyakov loops, 469
Polyakov, 738, 779, 780, 798
Potts model, 176, 557, 595
Power counting, 261
Primary divergent diagrams, 262
Primary field, 512
Pseudofermion method, 494

Q

Quadratic differential, 518
Quantum Chromodynamics, 329, 490
Quantum Hall effect, 675
Quasiprimary field, 512
Quenched approximation, 491

R

Radial quantization, 521
Radius of gyration, 801, 804
Random geometry, 738
Random lattice, 664, 739
Random matrices, 690
Random number generator, 463
Random potential, 647, 679
Random surfaces, 764
Random walk, 1
Real space renormalization, 162, 168
Regge, 768
Relevant operators, 174
Renormalization flow, 213, 270
Renormalization group, 270
Renormalized Green functions, 265
Replica trick, 658
Reproducing kernel, 684
Resolvent, 680
Ricatti equation, 649
Riemann's ζ function, 569
Riemann–Christoffel symbol, 785
Roots, 620
Roughening transition, 91, 205, 385

S

Scaling dimension, 505
Scaling fields, 174
Scaling hypothesis, 165
Scaling laws, 162

Scaling regime, 89
Schwarzian derivative, 517, 551
Schwinger functions, 238
Schwinger term, 631
Second order phase transition, 70
Self-avoiding walk, 31
Semicircle law, 690, 708
Shankar, 721, 729
Simply laced Lie algebra, 579
Sine-Gordon model, 206
Six vertex model, 554
Slater determinant, 698
Solid on solid (SOS) model, 82, 205, 386
Spherical model, 140
Spin wave approximation, 197
Spontaneous magnetization, 61, 72
Spontaneous symmetry breaking, 112
Staggered Kogut–Susskind fermions, 399
Star–triangle transformation, 183
Stochastic quantization, 486
String tension, 344, 385
Strong coupling expansion, 371
Strong coupling series, 33
Sugawara, 634
Sum over paths, 9, 27
Sum over surfaces, 780
Super-renormalizability, 262
Super-Schwarzian derivative, 615
Superconformal algebra, 613
Superficial convergence, 262
Superfield, 683, 718
Supersymmetry, 682
Surface tension, 85
Swendsen, 456, 482
Szegö's lemma, 74, 76

T

't Hooft loop, 352
't Hooft, 703
Tadpole graph, 244
Teichmuller space, 786
Thermalization time, 460
Thermodynamic limit, 23
θ-series, 609
Toeplitz matrix, 74
Topological complex, 345
Transfer matrix, 37, 62, 136
Tree diagrams, 154, 248
Triangulated manifold, 765
Tricritical Ising model, 618

U

Universality, 33
Upper critical dimension, 144

V

Vacuum diagrams, 244
Vandermonde determinant, 77
Verma module, 526
Vertex functions, 239
Vertex operator, 554
Villain action, 204, 332
Virasoro algebra, 521, 523
Virasoro characters, 547
Virial theorem, 663
von Neumann–Wigner theorem, 734
Vortices, 197, 201, 553

W

Ward identity, 145, 508, 513
Ward–Slavnov identity, 337
Watson's quintuple product, 611
Wavefunction renormalization, 264
Wegner, 679
Weierstrass function, 612
Weights, 624
Weinberg's theorem, 274
Wess–Zumino–Witten model, 626
Weyl character formula, 625
Weyl group, 622
Weyl invariance, 782
Wick ordering, 247, 536
Wick's theorem, 24, 57, 94, 99, 243, 434
Wigner, 647, 690
Wilson fermions, 396
Wilson loop, 344, 765
Wilson, 233, 235, 290, 311, 328, 456, 482
Wilson's action, 331
Wulff construction, 89

X

XY-model, 193, 594

Y

Yang, 75
Yang, Mills, 328, 335

Z

Z_n clock model, 474
Zamolodchikov, 436, 587, 808
ζ-regularization, 568

Printed in the United States
By Bookmasters